AF771862

Rearing Animal and Plant Pathogen Vectors

Rearing Animal and Plant Pathogen Vectors

Editors
Karl Maramorosch
Professor Emeritus, Department of Entomology
Cook College, Rutgers–The State University of New Jersey
New Brunswick, NJ, USA

and

Farida Mahmood
Department of the Army
US Army Public Health Command Region–South
JBSA Fort Sam Houston, TX, USA

CRC Press
Taylor & Francis Group
Boca Raton London New York

CRC Press is an imprint of the
Taylor & Francis Group, an **informa** business
A SCIENCE PUBLISHERS BOOK

CRC Press
Taylor & Francis Group
6000 Broken Sound Parkway NW, Suite 300
Boca Raton, FL 33487-2742

© 2014 Copyright reserved
CRC Press is an imprint of Taylor & Francis Group, an Informa business

No claim to original U.S. Government works

Printed in the United States of America on acid-free paper

International Standard Book Number: 978-1-4665-9821-8 (Hardback)

This book contains information obtained from authentic and highly regarded sources. Reasonable efforts have been made to publish reliable data and information, but the author and publisher cannot assume responsibility for the validity of all materials or the consequences of their use. The authors and publishers have attempted to trace the copyright holders of all material reproduced in this publication and apologize to copyright holders if permission to publish in this form has not been obtained. If any copyright material has not been acknowledged please write and let us know so we may rectify in any future reprint.

Except as permitted under U.S. Copyright Law, no part of this book may be reprinted, reproduced, transmitted, or utilized in any form by any electronic, mechanical, or other means, now known or hereafter invented, including photocopying, microfilming, and recording, or in any information storage or retrieval system, without written permission from the publishers.

For permission to photocopy or use material electronically from this work, please access www.copyright.com (http://www.copyright.com/) or contact the Copyright Clearance Center, Inc. (CCC), 222 Rosewood Drive, Danvers, MA 01923, 978-750-8400. CCC is a not-for-profit organization that provides licenses and registration for a variety of users. For organizations that have been granted a photocopy license by the CCC, a separate system of payment has been arranged.

Trademark Notice: Product or corporate names may be trademarks or registered trademarks, and are used only for identification and explanation without intent to infringe.

Visit the Taylor & Francis Web site at
http://www.taylorandfrancis.com

CRC Press Web site at
http://www.crcpress.com

Science Publishers Web site at
http://www.scipub.net

Preface

This volume is an update of the previously published volume entitled Maintenance of Human, Animal, and Plant Pathogen Vectors. The aim of this volume has been to collate, in one manual, existing information on laboratory handling of vectors that transmit pathogens of human, animal, and plant diseases. The editors felt the need to bring together the standard breeding and manipulation methods as well. No similar handbook has been available until now.

Specialists from America, Asia and Europe were invited to prepare descriptions of methods used for the cultivating, breeding and handling of mosquitoes, ticks, black flies, triatimine bugs, aphids, whiteflies, leafhoppers, piesmids, thrips, nematodes, fungal vectors, sterile techniques and cell culture of vectors. The authors of chapters have prepared a thoughtful treatment of experimental data and pertinent literature, including as yet unpublished data, personal experiences, and conclusions. The chapters deal with widely separated areas of interest and contain descriptions of standard, as well as highly specialized methods used by medical and veterinary entomology and plant pathology experts. The artificial borders between the diverse fields of interest have been bridged to acquaint researchers with the work and methods of investigators in distant, and yet related domains. The familiarity with the different techniques and materials used for a particular group of vectors often can be shared and applied to other vector groups. Pitfalls and difficulties encountered during laboratory maintenance of vectors, safety measures to prevent contamination and escape of insects, ticks, nematodes and fungal vectors have been stressed. The chapters are well documented by tables, illustrations, and up-to-date references.

Personal experiences and conclusions provide a large body of information that will serve actively working researchers and students at universities, research institutes, experiment stations and commercial laboratories. The maintenance of vectors will be of special interest to both established workers and to those who intend such work for the first time. Medical and veterinary researchers, microbiologists, entomologists, virologists, parasitologists, plant pathologists, geneticists, biocontrol specialists, as well as graduate students in related fields of biomedical, veterinary and agricultural research will find this work of interest. The editors hope that this manual will benefit all who are interested in the use of vectors of diverse disease agents.

Karl Maramorosch
Farida Mahmood

Contents

1

Laboratory Maintenance of Mosquitoes

Farida Mahmood

ABSTRACT

I have discussed different methods of mosquito colonization using adults, larvae, and eggs collected from natural, as well as artificial container breeding habitats. Important biosafety measures applied for constructing biologically safe and secure insectaries are also presented. I have also discussed general larval rearing techniques that work for most mosquitoes as well as fine-tuning of such techniques for rearing hard to colonize species of *Anopheles, Aedes, Culex* and *Culiseta* mosquitoes. Also proven methods of artificially inseminating mosquitoes that do not mate under caged conditions are discussed. Blood feeding techniques using live animals as well as *in vitro* feeding techniques are shown for successful colony maintenance as well as determination of their vector competence to arboviruses and other parasites.

Key Words: Field Collection methods, Mosquito colonization, artificial insemination, *in vitro* blood-feeding

---

Department of the Army, US Army Public Health Command Region-South, 2899 Schofield Rd. Fort Sam Houston, TX 78234, USA.

Views expressed in this chapter are solely of the author and does not reflect the views of US Army Public Health Command Region-South or US Department of Defense. The specific materials described for mosquito colony maintenance were chosen based on availability, effectiveness and cost efficiency. The resulting usage is not to promote the individual companies mentioned; users can substitute similar equipment or substances manufactured by other companies. Mentioning of a proprietary product does not constitute an endorsement or recommendation by the author or US Army Public Health Command Region-South or the US Department of Defense for its use.

Introduction

Anyone interested in mosquito colonization is already familiar that mosquitoes had caused men loosing battles in history and causing major disease outbreaks yearly by vectoring deleterious diseases like Malaria, Dengue, West Nile virus, Filariasis, and other arboviruses in the world. Mosquitoes undergo complete metamorphosis and 3 stages of their life cycle, eggs, larvae (4 instars), and pupae are aquatic forms that breed in various permanent and transitory bodies of water in nature. Terrestrial winged adults that are independent of the aquatic environment emerge from the pupae and both sexes sugar feed on nectar from flowers and sometimes sugars secreted by aphids to sustain flight activity as well as a source of nutrition. Females take a blood meal on various mammals, birds, snakes, turtles, and at times on frogs or alligators to obtain blood proteins that are converted into yolk proteins for developing eggs. Mosquitoes pick up many pathogens while blood feeding on their various reservoir hosts (Beaty and Marquardt 1996). The pathogens are amplified in susceptible vector mosquito species and later transmitted by them to healthy hosts during blood feeding thus resulting in disease outbreaks in human, animal and avian populations.

Certain arboviruses are transmitted by infected female mosquitoes to their progeny via their infected eggs and such mosquitoes serve as a reservoir for the renewed activity of the virus in nature. Arboviruses like LaCross virus can be transmitted from infected males *Ochlerotatus triseriatus* to uninfected females during copulation. Thus an in-depth knowledge of the physiology, biology, blood feeding habits, and the disease carrying reservoir stages of mosquitoes is essential for their safe colonization and later maintenance of colonized species for experimentation in the insectaries. Colonized mosquitoes are also used for testing efficacy of new pesticides and repellents because of the continuous rapid development of insecticide resistance in wild mosquito species.

As mentioned above, mosquito species differ in mating, blood feeding, breeding as well as resting habitats in nature. Field collection techniques and later colonization procedures that work for one species may not work well for another species. In this chapter, I have presented various prerequisites for successfully maintaining mosquito colonies and addressed basic limitations for working simultaneously with vectors and their pathogens (SALS 1980).

Field Collection Methods

Collection of Eggs

Depending on available resources the easiest and least expensive method for colonizing urban pest species like *Culex pipiens, Cx. quinquefasciatus, Cx. restuans,* and the most common *Aedine* pests species like *Ae. albopictus* and *Ae. aegypti* is from their eggs. Eggs of urban species can be collected by manipulating their oviposition behavior and queues used by them for selecting an oviposition site. Physiological queues used by container breeding *Culex* and *Aedes* species had been successfully manipulated to create an artificial aquatic oviposition medium in the laboratory (Reiter 1983, 1987) for collecting their eggs. This oviposition attractant is added to a tub, bucket, or oviposition cup that is placed in areas where adult biting activity and larval breeding is detected (Figs. 1–4). The egg rafts of *Cx. pipiens, Cx. quinquefasciatus* and *Cx. restuans* can be collected from ponds and ditches with naturally polluted water as well as by using a plastic tub that is ¼ filled with fermented oviposition medium or attractant (Fig. 1) and is left overnight in a suitable habitat frequented by mosquitoes.

Figs. 1–4. Field collection methods of mosquito eggs from their breeding habitats. Fig. 1. A plastic tub containing oviposition attractant for collecting egg rafts of urban *Culex* species. **Fig. 2.** Preparation of oviposition medium with water and dried hay or Bermuda grass clippings in a one gallon bottle. **Fig. 3.** *Culex tarsalis* egg rafts. **Fig. 4.** An oviposition cup lined with filter papers showing groups of individually deposited *Ae. albopictus* eggs.

The artificial oviposition medium is prepared by adding ¼ to one cup of rabbit food pellets or different amounts of hay, wheat straw, sod, or dried grass clipping to one to 8 gallon plastic bottles respectively and adding de-chlorinated or aged tap water. This mixture is allowed to ferment for 3 to 5 days in a closed bottle (Fig. 2).

Mosquitoes are easy to find in areas around cattle sheds, horse stalls, near polluted standing water ponds, and by an area with large number of containers filled with rain water and rotting leaves. The plastic tub or other containers filled with the above infusion can be placed in such habitats. The oviposition tub or bucket should be checked daily and *Culex* egg rafts (Fig. 3) can be collected either using a small plastic spoon or a wooden applicator. Egg rafts of *Culex* and *Culiseta* species can be reared individually or in groups of 5 depending on the stage of mosquitoes used for identifying the species (larvae or emerging adults) for final colonization. Always use chlorine free aged water to hatch the egg rafts. *Aedes albopictus* and *Ae. aegypti* eggs can be collected using black plastic oviposition cups lined with filter paper (Fig. 4), a wooden ice cream paddle, or a paddle made from hard card board, and adding enough aged water to cover ½ of the filter paper or paddle. These cups should be placed near tire dumps or areas with detectable adult *Ae. albopictus* and *Ae. aegypti* activity and near houses (Figs. 5–6). Information about date and site of collection can be recorded on the oviposition paper or paddle using a soft lead pencil. The filter paper or wooden paddle is collected from the oviposition cup few days later or once a week depending on environmental temperature to obtain large number of oviposited eggs. *Aedes,* and *Ochlerotatus,* lay desiccation resistant eggs and oviposition paper or wooden paddle can be stored dry in a large plastic zip-lock bag containing a small wet sponge to avoid complete desiccation of the eggs. The desiccation resistant eggs allow their storage over many weeks and later for hatching a large number of them as a group to obtain many adults for facilitating mating during the colonization of first generation in a caged environment. If some tire dump is identified where larval breeding is detected than the organic debris from such tires can be brought to the laboratory and the larvae can be obtained by adding aged tap water.

Salt marsh *Aedes* and *Ochlerotatus* species as well as *Psorophora* species deposit their desiccation resistant eggs individually on moist soil where eggs survive for many months away from water (Clements 2000) and methods to obtain them are tedious. If all methods of larval and adult collection fail than sod or soil from low areas in tidal marshes, ditches, and natural ponds can be brought in the insectary and samples placed in large pans flooded with water to collect larvae from hatched eggs. If the above method fails for collection of larvae than other egg separating methods can be used (Service 1976). *Anopheles* eggs usually hatch within 2 days of oviposition at room temperature and are not easily collected from their breeding habitats because of being deposited individually on the water surface by the females and obtaining eggs from field collected blood fed mated females is a better technique for their colonization.

Larval Collections

Mosquitoes breed in variety of habitats and different methods of larval collection are used depending on species to be colonized (Laird 1988) (Figs. 5–8). Their larvae can be collected using a 1 pint (450 ml) larval dipper (Bioquip®) (Fig. 7) attached to different sizes of wooden or extendable aluminum handles from most common larval breeding habitats like ponds, ditches, streams, salt and fresh water marshes, and temporary rain water pools. Larvae from tree holes, axils of bromeliads, and tires can be collected using a small disposable plastic pipette, a turkey baster (Fig. 8) or by a small soup ladle.

Figs. 5–8. Different field larval collection methods. Fig. 5. An outdoor pile of discarded tires for collecting larvae of different urban pest *Aedes* species. **Fig. 6.** An urban larval breeding site for collection of *Culex* larvae breeding in discarded containers. **Fig. 7.** A one pint (450 ml) plastic dipper showing *Psorophora* larvae collected from a road side ditch. **Fig. 8.** A turkey baster used for collecting larvae of tree hole breeding mosquito species.

In case of species that breed in tree holes the initial field surveillance for finding tree holes with breeding larvae is very time consuming. Once, such tree holes are detected many species (*Toxiorhynchites* species, *Ae. aegypti*, *Ae. albopictus*, and *Och. triseriatus*) are collected with a turkey baster or large syringe (Figs. 8–9) and an extension tube can be attached to the syringe to reach deep tree holes. Larvae can be concentrated in small buckets or large plastic bottles using various types of sieves or strainers and brought to laboratory for further rearing.

Culiseta melanura larvae are collected from underground crypts formed by the stumps of rotted trees in cedar swamps (Fig. 11) or under roots of large fallen trees in fresh water marshes (Fig. 12). A dipper and manually operated bilge bump is mostly used for larval collections (Fig. 13). Larval collection is not a preferred method for *Cs. melanura* colonization due to few larvae per crypt and time consuming search efforts. Bilge pump is also used to collect *Coquilletidia perturbans* larvae that are attached to the underwater roots of Cattails in polluted ponds. The water thus collected can be passed through a series of decreasing mesh sieves (Fig. 10) to separate large debris from the field collected larval water and finally larvae can be separated in a white enameled pan for laboratory rearing. Another method used for collecting *Coq. perturbans* larvae involves pulling out the plants with their attached roots from the breeding sites and vigorously shaking the roots in

Figs. 9–10. Different field larval collection methods. Fig. 9. A 50 ml syringe with water from a tree hole. **Fig. 10.** Different sizes of sieves for removing organic debris such as rotted leaves and sticks.

buckets of water. This method is not preferred because of the destruction of natural habitats. Wild caught larvae should be reared in water from their original breeding habitat for first few days and later aged tap water added gradually overtime to avoid mortality due to sudden changes in their osmotic regulation. The larval collection of salt marsh *Aedine* species is an efficient method of their colonization.

Field collected larvae are naturally infected with various types of potentially deleterious pathogen to mosquitoes and should be handled in a separate wet laboratory to avoid contamination of already colonized laboratory stocks from contaminated equipment. All equipment used in rearing field collected larvae should be decontaminated using 6% sodium hypochlorite solution and later thoroughly rinsed with tap water. Enameled or steel pans can be washed with tap water and later dried in an oven at high temperature whereas small plastic pans and sieves can be autoclaved for later use.

Figs. 11–13. Different field larval collection methods. Fig. 11. A crypt breeding habitat of *Cs. melanura* in fresh water Cedar swamp. **Fig. 12.** Water under the roots of a fallen tree where *Cs. melanura* larvae breed in fresh water cedar swamp habitats. **Fig. 13.** A bilge pump adapted for the collection of *Cs. melanura* and *Coq. perturbance* larvae from their natural breeding habitats.

Adult Collections

Adult mosquito collection methods depend on the mosquito species to be collected (Service 1976). Different types of equipment such as CDC light trap (Figs. 14 and 17), Reiter's gravid trap (Fig. 15) (Reiter 1983, 1987), BG trap, and one cubic foot square resting boxes (Fig. 16) may be used to collect exophilic species. Similarly, battery powered backpack aspirator, and hand held mechanical aspirators can be used to collect both outdoor and indoor resting mosquitoes. Large number of *Aedes* and *Ochlerotatus* species can be collected using CDC light trap baited with dry ice to provide a steady stream of CO_2 for attracting hungry females. Adult *Cx. pipiens* and *Cx. quiquefasciatus* can be collected by placing CDC light traps in storm drains (Fig. 17) and basements, beside outdoor collections. The best method for collecting endophilic *Anopheles* and *Culex* species is by aspiration using, a battery powered mechanical aspirator, a large backpack aspirator or a mouth aspirator with a filter, from the walls of underground shelters, animal sheds (horse barns and buffalo sheds). Collected mosquitoes are transferred to

Figs. 14–17. Different methods of adult mosquito collection. Fig. 14. A CDC light trap. **Fig. 15.** Reiter's gravid trap. **Fig. 16.** A resting box mosquito collection method for *Cs. melanura*. **Fig. 17.** A CDC light trap hung inside a sewer for collecting *Culex* mosquitoes.

small 1 pint cartons (450 ml) with netting on top and a cotton pad soaked in water or 10% sugar solution are placed on the nettings. The cartons with field collected specimens are stored in a large foam box containing a bag of wet ice during transportation in a vehicle to avoid mosquito death by desiccation and/or overheating.

Wetland species like *Cs. melanura* can be collected using artificial one cubic foot (0.76 cubic meters) resting boxes as well as using CDC light traps. Whereas, gravid *Cx. pipiens*, *Cx. restuans*, and *Cx. quinquefasciatus* females are easy to collect using Reiter's gravid trap (Reiter 1983, 1987). Later, gravid females can be released in a one gallon (4.55 liters) ice cream carton cage or 1 cubic foot (0.76 cubic meters) square cage and an oviposition medium (wheat straw infusion for *Culex* or aged water for *Culiseta)* in glass or plastic cups is placed inside the cage to collect eggs. For experimental purposes individual egg rafts can be obtained by isolating single gravid females in tubes containing ovipostion medium and closed with a piece of cotton or a cap with a screen. A few drops of 10% sugar solution are added to the cotton plug or on top of the screened netting of the tube's cap. Starting a colony from field collected adults is more efficient due to faster adult identification and gravid females are already mated and blood-fed. Eggs rafts laid by each female provide 100–200 larvae. Some field collected gravid females take a second blood meal if an appropriate host (quail, chicken, hamster, and white mice, etc.) is provided overnight in the cage and thus more than one

Fig. 18. A biosafety chamber for handling field collected potentially infected insects and for separating experimentally infected blood fed mosquitoes from unfed mosquitoes.

egg raft per female is collected to improve colonization efforts. Accidental release of potentially infected wild caught adults in the insectary can be avoided either by placing a smaller cage containing mosquitoes in a larger cage or by using a biosafety chamber (Fig. 18).

Laboratory Maintenance

Insectary

An insectary should meet the biological needs of mosquito species to be reared and include: 1) Proper temperature controls such as minimum and maximum temperature required for eggs, larvae, and adult stages. 2) Humidity controls including water vapor barriers, ventilation, air conditioning, humidification, and dehumidification equipment and controls that would not affect environmental temperature of the insectary. 3) Photoperiod controls to provide correct lighting at correct times (dusk, dawn and day light) to initiate swarming and mating of mosquitoes and to avoid conditions that would initiate hibernation in adults or immature stages of the species to be studied.

Second main concern is related to the experimental use of the insectary by the researchers such as: 1) Maintenance of exotic species (not indigenous to the area) in the insectary. Rearing protocols or standard of procedures for their control must be available to individuals working in the insectary for control in case of their accidental escape. Only trained staff should be allowed in an insectary where exotic mosquito species are maintained. 2) The use of human or animal pathogens to study vector competence of local or imported vector also poses a threat. An accidental exposure of staff and public to accidentally released infected mosquitoes can be avoided by training staff and strictly implementing security measures and control protocols. Security of an insectary can be achieved by controlled release of insectary keys or key cards to trained maintenance staff and researchers and by providing them with an option to get vaccinated against the pathogens of concern if a vaccine is available. An initial serum samples should be taken from each individual and antibodies to the pathogen in use must be tested. Part of this sample should be frozen and placed in a cryobank or a proper freezer for later use. Blood of the maintenance staff and researcher working in such a facility should be tested yearly.

The security of the insectary can also be instituted by restricting entry to the premises by posting list of individuals entering the room or by restricting keys to only certain individuals. The list of pathogens used should be posted on the main entry door to the insectary. Federal Bureau of Investigation requires background investigations of all individuals working with biosafety level II or above human and animal pathogens.

Insectary Construction

You can convert any room in a building to an insectary or mosquito rearing facility provided that it is located in an area with minimum temperature fluctuations (Gerberg et al. 1994). The main entrance door to the colony room should not open directly to the hallways and one or more anteroom fitted with negative air pressure curtain or screened doors inside provide containment of the escaped arthropods. More options are available when constructing a new mosquito rearing facility than during conversion of an already existing room into an insectary. While constructing an insectary in an educational facility, an area should be selected that is away from the main hallways, offices and classrooms. The main insectary design should provide for the containment of both vectors and pathogens to avoid accidental human infections. The design should take into consideration factors that affect the safety, health, and comfort of the personnel working inside the insectary. Initial time spent to study the basic engineering design pays off by saving money in the long run (Goodnough and Parnell 1985). In addition to the main insectary or rearing room exit into one or more small anterooms, presence of double curtains on the exit doors of the insectary to the anterooms or a negative air pressure curtain by a fan adds an additional barrier against accidentally released adults. The double curtains can be nailed in or Velcro can also be used to attach the curtains to the frame of the door. Any mosquito, sitting on a worker will be rubbed off while he passes through the double curtains and it will remain inside the insectary. The doors should be tight fitting, should close automatically after exiting of a person, and preferably of metal. Although a metal exit door is ideal for an insectary, a wooden door that is devoid of holes and crevices and painted with a waterproof paint is adequate. All doors of an insectary should open inward and the main insectary door should close automatically after exiting individuals. Brushes or rubber sheet should be fixed to the bottom edges of the doors to restrict escapees. No two doors be opened simultaneously in an insectary and there should be an emergency exit door available if the main entrance closes due to fire or an emergency situation.

Sometimes it is not possible to have a room lacking any windows for conversion to an insectary and window should be converted so outside light is prevented from entering the colony room and affecting the duration of day length or the photoperiod. Changes in photoperiod can affect certain biological mechanisms of arthropods and initiate diapause in eggs, larvae or adults. The insectary windows should be nailed shut and window glass should be painted black to prevent entry of sunlight. A window must be fitted with a double screen having 25 x 25 mesh (per 2.54 square cm) wire screen. A clear sign warning, "Do not open" should be permanently installed next to it or the window opening can be shut permanently using a board from the outside and all crevices or seams plastered.

Interior walls of the insectary should be painted light in color (preferably white), with a mildew resistant, gloss finish paint to provide for easy wiping of surfaces and detection of escaped adults. All crevices and cracks in the walls including the electric inlets should be repaired before painting the room. The ceiling of the room should not be very high and above the reach of a person standing on the floor with an aspirator to catch any escaped mosquitoes. Similarly, all the daylight fixtures should be very simple, recessed and flushed to the ceilings to avoid escaped adult from hiding inside the fixtures and if possible a screen should be installed over the light fixtures as well as the air inlets.

Only essential equipment should be placed in the insectary, thus reducing hiding areas for escaped adults. Cross contamination of different mosquito species can be avoided by constructing screened partitioned small walk in rooms for each species. Each room is provided with shelving for pans and mosquito cages. All supplies and food used for each species and strain are kept separate in these cubicle size walk in screened rooms and are color coded to avoid mixing.

Fluorescent lights installed to provide a uniform daylight source in the main insectary should be attached to a timer for controlling duration of day length. Certain species like *An. culicifacies* and *Cx. tritaeniorhynchus* require dusk and dawn periods provided by gradual dimming of incandescent bulbs that are attached to sunset-sunrise simulation systems or dimmer switches. Other species like *Culiseta melanura* (Mahmood and Crans 1994) and *Cx. salinarius,* mate even if there is a 1-step change of daylight to dusk provided by 15 watts incandescent bulb attached to a simple timer. Such timers are available at many hardware or home appliance stores. Other species like *Cx. pipiens pipiens*, *An. albimanus, Och. atropalpus,* and *Och. triseriatus* mate in the absence of a dusk or dawn and one step change to night or day is sufficient. Electrical outlets should be covered with tape to prevent the entry of escaped mosquitoes inside the holes.

All air entering the insectary should pass through air filters and biosafety level III insectaries should have HEPA filters. For emergency purposes the insectary should be provided with 2 separate heating and cooling units. Presence of a second unit would serve as backup in case the first unit breaks down. The use of main cooling and heating unit for the main building should not be used as a backup system because of fluctuating temperatures resulting due to individual preferences set by individuals in offices, etc., and creates bad working environment. The insectary also requires a humidifier and a 24 hr electronic recorder to note temperature and humidity fluctuations. Such environmental temperature controls are now available that can be controlled from a computer using special software and temperature alerts can be transmitted to a phone alerting the insectary manager. The researchers should keep a log of all environmental data recordings. Experiments requiring special control of certain environmental parameters can be performed in an environmental chamber kept inside the insectary with programmable temperature, humidity, and photoperiod controls (Fig. 19).

An ideal insectary should have a light colored concrete floor that has a drainage hole in the center of the room or the floor can be slightly slanting to one side of the room to a drainage hole for cleaning spills or washing floors with a hose. Such a floor provides for easy cleaning. All the furniture should be on casters to provide for easy rearrangement of shelves and tables according to changing needs of new research protocols. Light colored furniture is ideal because escaped insects can be easily detected. Tables and chairs should be of a very simple design with minimum hiding areas for escaped insects. Tables can be adjustable to various heights to accommodate variable heights of workers.

A large sink that can accommodate various sized rearing pans and adult cages and have a separate hot and cold water supply would suffice most of the insectary needs. Few additional water faucets will be handy when multiple researchers work simultaneously in the insectary. A large drainage

Fig. 19. An environmental chamber with adjustable environmental control for mosquito rearing.

area next to the sink provides for easy drying of rearing pans and cages, etc. A separate small room dedicated as the washing facility and fitted with a good sink, a dish washer that accommodates different sizes of pans and an oven for drying and decontaminating rearing pans in case of parasitic outbreaks in colonies is very handy.

In Bio-safety level II and III insectaries, where infected mosquitoes are handled, a negative air pressure is required and specially designed bio-safety insect separating hood or chamber should be used for handling small infected insects (Fig. 18). Such an insectary should also have triple screens. NIH has special guidelines that should be strictly followed while working with infected mosquitoes. In the USA all universities have special committees or departments that execute such guidelines and all protocols using infected mosquitoes require their approval and specific recommended precautions must be followed.

In an insectary where infected mosquitoes are handled, (BL II, III) a pass through autoclave that can be taken directly inside the room is required for decontamination of contaminated equipment, water, and other equipment that is used for handling infected insects or animals. If a mobile autoclave is not available than all contaminated equipment and infected material should be placed in a covered leak proof container and inserted in a larger bag that is sealed or tightly closed before transferring it to the autoclaving room. Such a container should be directly transferred inside the autoclave without taking out the infected equipment or materials. Certain institutions have special staff or contractors to pick up biohazard trash and the instructions provided by them should be strictly followed. All pathogens (viruses) that are kept frozen in –70°C freezers should be listed in the log book or posted on the freezer. The freezer should be always kept locked and the access to the keys should be through the insectary director or manager. This precaution is necessary to avoid theft of pathogens and also to provide safety for emergency responders in case of a fire.

The name of the insectary director, with telephone numbers to reach her/him in case of emergency should be posted at the entrance door. The warning sign should be posted on the door that no insecticide or pesticide should be brought or used in the insectary. In a biosafety level II and III insectary instructions regarding biosafety level precautions are posted clearly and visibly on the entrance door (Fig. 20). All insectary staff should be well trained, and access should be limited to scientists and maintenance staff directly involved in experiments and laboratory maintenance. Any escaped insects should be considered dangerous even if they are uninfected and should be searched and killed. The technical staff should be instructed not to leave a biosafety insectary, if they see an escaped mosquito without killing it. A separate small room should be constructed in biosafety level II and III insectaries for handling live infected mosquitoes. The entrance door to this room should also have all instructions posted clearly on it (Fig. 21). This room should also have a chill table for manipulation and identification of infected mosquitoes (Fig. 22) and a closed biosafety hood for working with live insects (Fig. 18).

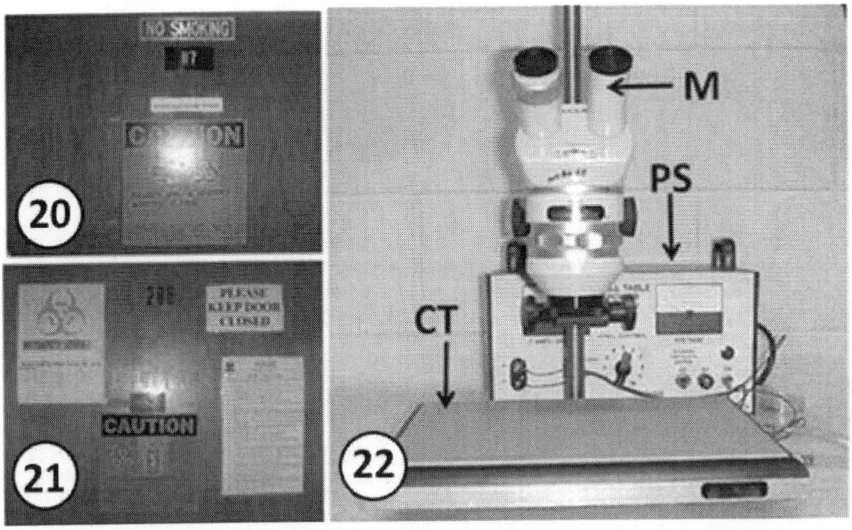

Figs. 20–22. Fig. 20. Posting at the door of a biosafety level II insectary. **Fig. 21.** Instructions posted at the door of biosafety level II laboratory where infected mosquitoes are manipulated. **Fig. 22.** A chill table for the manipulation of live or dead infected mosquitoes. Abbreviations: CT: Chill table; PS: Power supply; M: Microscope.

Basic Techniques for Maintaining Mosquito Colonies

Adult Handling

Most used supplies for handling adult insects are shown in Figs. 23 and 24. Adults can be transferred from one holding container or cage to another by a battery operated commercially available aspirator (Fig. 24) or a simple suction device or manual aspirator can be constructed by using a glass or plastic tube with 12.5 mm outer diameter. To one end of such a tube a piece of photographic film covered with a screen is inserted. This end is inserted inside rubber surgical tubing of any desired length. Depending on the choice of the researcher the other end of the glass tube can be bent to a 45° angle or left straight. Inhaling mosquito scales, feces or dust particles is hazardous to human health and a blowing aspirator designed by "BioQuip Products" is a good alternative. The manual mouth aspirator should never be employed for transferring infected mosquitoes. Mosquitoes

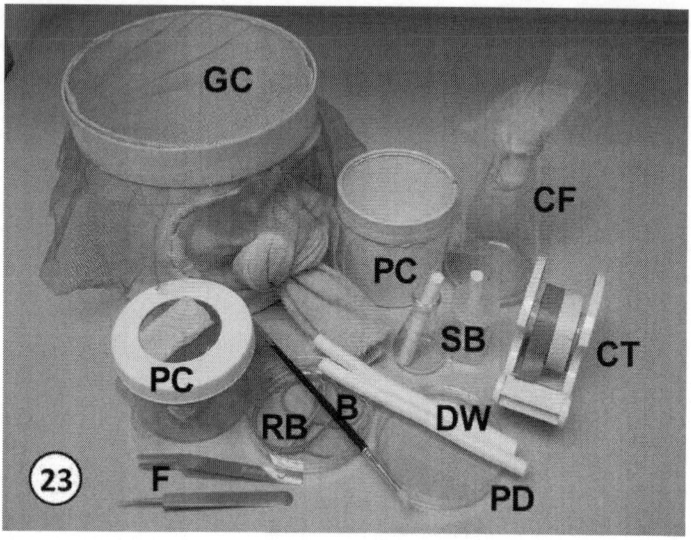

Fig. 23. Different supplies for handling adults. B, Brush; **CF,** Conical glass flask; **CT,** Colored tapes; **F,** Forceps. DW, Dental cotton wicks for inserting in sugar bottles; **GC,** Gallon cage; **PC,** Pint cages; **PD,** Petri dish and RB, Rubber bands.

can be separated into males and females and identified to species using a dissecting microscope. For immobilizing adults, usually CO_2 gas is considered the safest. Many scientists use anesthetic such as chloroform or ether (not recommended) and others have used cold temperatures (2–4°C) for immobilizing mosquitoes.

A small square box covered with cloth at the top and 3/4 of which is covered with a glass plate can be used for CO2 anesthesia. Rubber tubing that is connected to a CO_2 tank is inserted through a side hole in the box.

The CO_2 flow can be controlled through pressure valves on the cylinder. Adults should be exposed to the CO_2 gas for a very short time interval, since gas can enter their bodies through their spiracles and consequently blow up their bodies and cause death. In the absence of CO_2 gas tank or dry ice can be substituted to knock down the mosquitoes by placing a small piece in a cage and covering the cage with a plastic bag.

Mosquitoes can be anesthetized by constructing an apparatus using 2 bent glass tubes, one at an angle of 135° and other at 45°. The tubes can be inserted through a stopper or cork into a conical flask that has a piece of cheesecloth at the bottom. A small amount of anesthetic can be poured inside the conical flask (Fig. 23). The researcher can insert the glass tube that is bent at 45° angle into a second conical flask and blow vapors of the anesthetic into it by using the other glass tube (135°) thus allowing a small amount of anesthetic vapors to enter the second conical flask. Adult mosquitoes can be blown into the second flask and anesthetized in this manner. The adults should be exposed to these vapors only briefly (few seconds) and then taken out in a petri dish and identified or further manipulated according to the requirements of an experimental protocol.

It is again emphasized that adults that are collected from an endemic area of a disease or are provided with an infectious blood meal in the laboratory for experimental purposes should never be manipulated by using mouth aspirators (Fig. 24) and a mechanical aspirator should be used. Infected mosquitoes can be manipulated safely in walk-in cold rooms that are kept at a lower temperature or on a chill table (Fig. 22) or inside a biosafety insect separation chamber or hood (Fig. 18).

Small cages for holding adults can be constructed from 1-pint (450 ml) ice cream paper cartons by fitting with screens. A hole made on the side of the carton is covered with 2 pieces of dental dam that have slits at right angles to each other. Mosquitoes can be released in such a cage through this entrance hole by inserting the aspirator tube through the slits of the dental dam. Various plastic 1-gallon buckets can also be converted into cages. Aluminum cages for maintaining adult mosquitoes are commercially

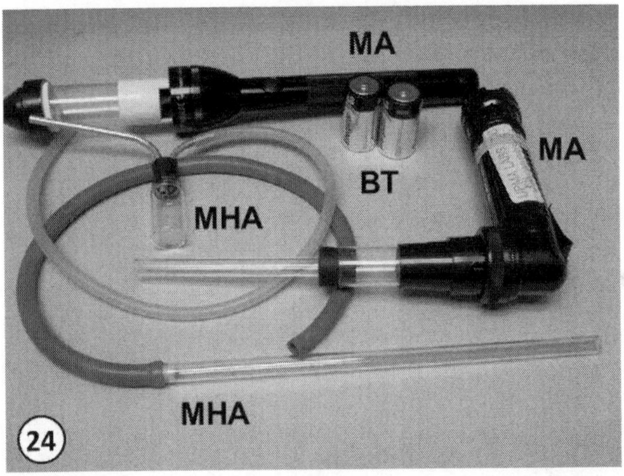

Fig. 24. Aspirators used for handling adult mosquitoes. BT, Batteries; **MA,** Battery powered mechanical aspirators; **MHA,** Mouth aspirators.

available in different sizes. They have an opening, fitted with 25 cm surgical stockinet to allow for placing a sugar source, blood-meal hosts, or releasing adults in the cage (Fig. 25). Infected adults should be double caged and kept in a separate area or room that does not house uninfected mosquito colonies. Infected mosquitoes should always be handled wearing rubber or latex gloves and using feather weight forceps (Fig. 23) inside a biosafety chamber (Fig. 18).

Commercially available 61 X 61 X 61 cm size cages are very convenient for keeping mosquito species that require large areas for mating, such as *An. culicifacies* (Anisely, 1976) or *Cs. melanura* (Mahmood and Crans 1994). These cages (Fig. 25) can be dismantled for easy storage and cleaned easily using tap water and a brush. A smaller version of these cages is available in 30.5 X 30.5 X 30.5 cm size and their screens can be easily replaced if damaged.

A hollow clay tile placed in a pan filled with water inside the cage provides a dark and humid area for the adults to rest (Fig. 26). Some

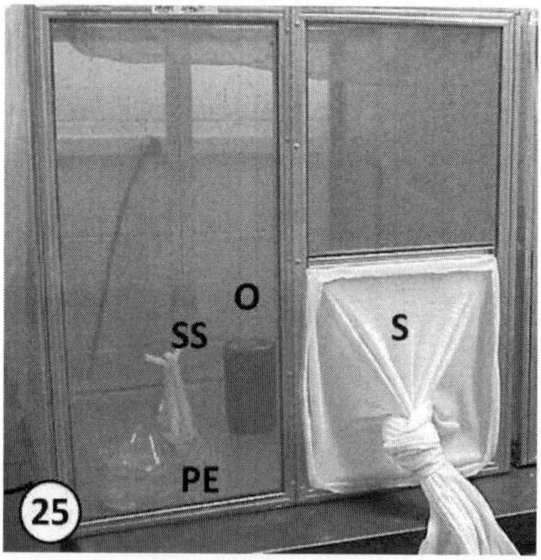

Fig. 25. A 61 cubic cm cage for adult maintenance. **O,** Oviposition cup; **PE,** Pupal emergence cup; **S,** Surgical stockinet fitted at the cage opening; SS, Sugar source.

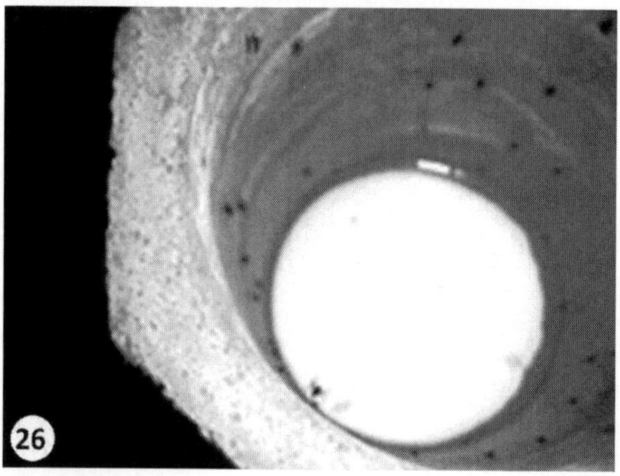

Fig. 26. A hollow clay tile placed in a pan filled with water provides a resting site for *Cs. melanura* inside a cage.

species like *Cs. melanura* prefer to rest in dark humid sites in nature and preferentially select the clay tile to rest under caged condition. A large piece of wet cotton, cheese cloth or a wet towel should be placed on top of the cage and covered finally with a piece of clear plastic. This technique assures high humidity inside the cage and results in increased survival of the adults.

Induced Copulation

A few mosquito species belonging to the genera *Aedes* and *Anopheles*, do not mate readily in the laboratory because of the cage size or the artificial light sources in the insectary. Such mosquitoes can be maintained in the laboratory by induced mating techniques (McDaniel and Horsfall 1957, Baker et al. 1962, Horsfall and Taylor 1967). Equipment required for this technique includes a vacuum pump with a bleeder valve, one glass pipette that is bent at an angle of 45° with a polished tip to hold females, dissection needles for holding males, a dissecting microscope and a chilling source such as a beaker filled with ice or a refrigerator (Figs. 27 and 28). McDaniel and Horsfall (1957) colonized *Ae. stimulans* and *Ae. vexans* by using this valuable technique. They used 72 hr old males that were decapitated and glued to a piece of cardboard by dorsa of their thoraxes leaving their abdomens completely free and movable. They anesthetized the females and cemented

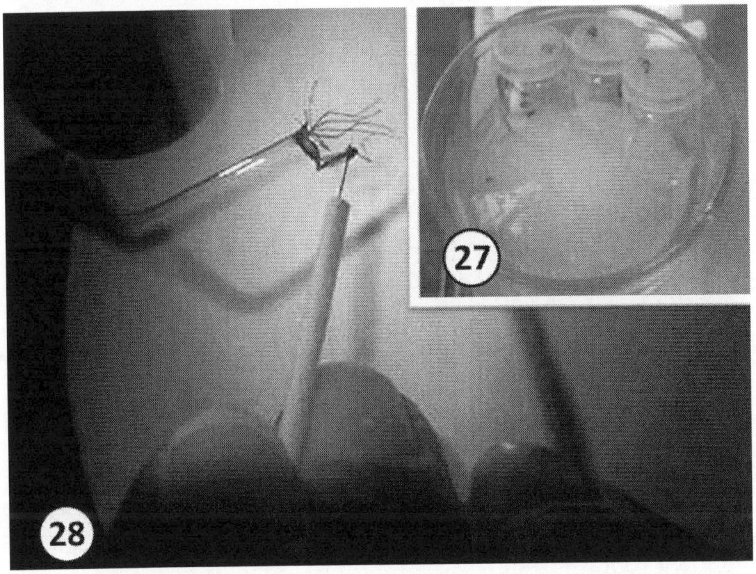

Figs. 27–28. Induced copulation. Fig. 27. A glass dish with frozen water to ice and tubes with mosquitoes. **Fig. 28.** Induced copulation technique showing a chilled female held with a glass pipette attached to a vacuum pump and a chilled decapitated male held with a dissection needle.

dorsa of their thoraxes to separate needles. The apices of abdomens of both sexes were brought near each other with the ventral side of their abdomen in the upper position to initiate mating. This procedure was performed under a 20X stereomicroscope.

McDaniel and Horsfall (1957) were able to inseminate 25 females within an hr. Baker et al. (1962) pretreated *Anopheles* males to 15°C 24 hr before mating. They anesthetized males with CO_2 for 10–15 s before decapitation and anesthetized the females with chloroform for only 5 s. At the same time period, in a different laboratory, Wheeler (1962) cold anesthetized both sexes and glued them to the heads of insect pins. He inserted the pin carrying the glued female to the end of a cork that was glued to a glass slide, thus keeping her in a horizontal position, and brought the male's last abdominal segment containing the sex appendages, to the female's last segment at an angle.

Ow Yang et al. (1963) caught mosquitoes with a fine pipette that was attached to a suction device to catch mosquitoes and then pinned them laterally through the thorax with a minuten needle attached to the end of a 7.5 cm long soft wooden stick. He did not use any anesthesia for immobilizing the adults. Baker (1964) also achieved best results by decapitating males without using any anesthesia. He modified this technique slightly when he caught a male in an aspirator and blew it between layers of cellucotton. He decapitated the male and pinned it in the manner of Ow Yang et al. (1963).

Later modification included the use of blood fed females few to 24 hr after engorgement. This technique proved superior in the sense that in most cases all mated females laid eggs, and blood feeding helped in extrusion of the 8th abdominal segment of the females thus resulting in successful mating (Baker et al. 1962, Baker 1964, Ow Yang et al. 1963, McCustion and White 1976) (Figs. 27 and 28).

Miura (1967, 1969) extruded the 8th abdominal segment of the females by exposing them to double anesthesia with ether. He also prepared 3–4 males simultaneously by pinning them and keeping them in a humid chamber. At the same time he placed 3–4 females with their ventral side up on a block that was covered with methylcellulose. Different types of glues can be used as long as they are fast setting and easily removable. A small drop of hot wax works as well. McCustion and White (1976) held blood fed *Och. sollicitans* females in 1-dram vials and put the tubes on a block of wet ice several hours after blood feeding. They extruded the 8th abdominal segment of the females by using vacuum through a curved glass pipette attached to a vacuum pump and applying a 5 cm mercury vacuum. They cold treated the males too, and as soon as the males became immobile they pinned the males medially through the mesonotum with a minuten insect pin to a 152.4 mm wooden applicator. They brought the adults' terminalia together laterally at an angle of 90–120° (Figs. 27–28).

Age of the males mosquitoes used in induced mating is very critical. Mosquito males rotate their terminalia within the first 24 hr of their life (Provost et al. 1961, Khan and Reisen 1977, Mahmood and Crans 1999). The speed at which male mosquitoes attain sexual maturity is temperature dependent. The age at which male mosquitoes inseminate maximum number of females during one night is 3–6 d after emergence at 28°C (Mahmood and Reisen 1982, 1995). The angle at which males and females are brought together during an induced mating determines the percentage of successful inseminations. An angle of 45° was best for *Anopheles* (Baker et al. 1962, Ow Yang et al. 1963). Maximum mating success for *Aedes* was achieved when their venters were held at an angle of 120° (Horsfall and Taylor 1967).

In species that mate naturally under caged conditions, maximum mating occurs when adults are 3–5 days of age (Mahmood and Reisen 1982, Mahmood et al. 1986, Mahmood and Reisen 1994). Therefore, 4–6 d after emergence is the best time to blood feed many mosquito species. The time of blood feeding varies for different species; some mosquitoes are nocturnal and take a blood meal only at night, whereas others at dusk or dawn or some are day biting species. Some *Anopheles* species expel a lot of blood while blood feeding and a paper towels should be placed in the bottom of the cage to avoid growth of fungi and bacteria. Females of the species that do not mate under caged conditions should be blood fed before initiating an induced mating to increase the chances of obtaining fertilized eggs.

Adult Feeding Requirements

Sugar Feeding

After emergence, both sexes of mosquitoes require a carbohydrate source for energy and survival. Many sources that contain sucrose, such as sugar cubes, a 10% cane sugar solution (100 gm of sucrose/1000 ml de-ionized water), apple slices, and water soaked raisins were used successfully as carbohydrate source by various researchers. Sugar solution is usually added to a 100 ml conical flask or a bottle and braided cotton dental wicks are inserted in it. Sponge wicks and simple wicks made of absorbent cotton can be used in place of braided cotton wicks. Sliced apples are placed inside the cages. Water soaked raisins are placed either inside the cage or mostly on top of the screen netting of the adult holding cage.

Blood Feeding

Colonies of autogenous mosquitoes, such as *Och. atropalpus* can be maintained indefinitely without a blood meal. The majority of anautogenous females require a blood meal for ovarian maturation. Most mosquito species blood feed 4–6 d after emergence at 28°C (Reisen et al. 1979). The blood is required to produce yolk protein during egg development. The rate of ovarian development and, duration of the gonotrophic cycle is directly dependent on the environmental temperature and it differs in different species (Reisen and Mahmood 1979, Mahmood and Crans 1997). In many species, duration of the first gonotrophic cycle is usually longer than subsequent cycles and varies among different species at similar environmental temperature (Mahmood and Reisen 1981).

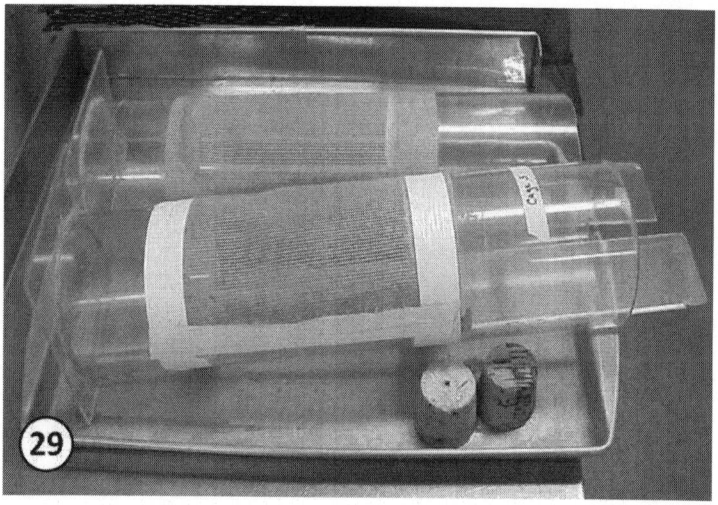

Fig. 29. The guinea pig holder used torestrain guinea pigs while blood feeding mosquitoes in a cage.

Mosquitoes differ in their host preferences for blood feeding. Humans, rabbits, guinea pigs, chickens, white mice, and quail had been successfully used as live hosts. Special types of animal holders are constructed for immobilizing the animals during mosquito blood feeding (Fig. 29). Small animals like albino mice can be restrained in a funnel shaped, wire screens (5 mm wire mesh). The mouse is allowed to crawl inside such a funnel and the end is rolled over to close the opening.

Various types of artificial blood feeding devices had been developed and used to feed hematophagous insects (Collins et al. 1964a,b, Cosgrove et al. 1994, Hastriter et al. 1980, Hastriter and Cavanaugh 1981, Hunt and Mckinnon 1990, Mahmood et al. 2004, 2006, Rutledge et al. 1964, 1976,

Tarishis 1958, Waladde et al. 1991). Blood from cows, sheep, chicken, and rabbit serum has been successfully fed to mosquitoes through an artificial membrane stretched over a blood-feeding device.

Rutledge et al. (1964) (Figs. 30–31), developed the most preferred equipment used now by many laboratories for *in vitro* blood feeding mosquitoes. Hemotek Corporation (Figs. 32–34) manufactures another commonly used apparatus. Many different types of membranes, such as Para film, Baudruche membrane (Trembley 1952), lower belly skin of 14 days old chicken with feathers removed, and salted sausage casings have been used successfully to blood feed mosquitoes. The apparatus developed by Rutledge et al. (1964) consists of a double jacketed glass funnel (Figs. 30–31) with the hollow tube end extending from its vertex. The cylindrical water jacket has a lower inlet and an upper outlet connected to rubber tubing. Warm water (37–40°C) is circulated through the cylindrical jacket of the apparatus by connecting the inlet and outlet tubes to a constant temperature water circulator (Hake Circulator, Series F, Brinkman Instrument Inc.). An artificial membrane and/or a bio-membrane is stretched to the bottom lip of the cone and tied by rubber band, or the membrane can be stretched and taped to the reservoir (Mahmood et al. 2004, 2006). A Baudruche

Figs. 30–31. *In vitro* blood feeding. Fig. 30. A glass water jacket circulator fitted with hog intestinal membrane and attached to a hot water circulator kept at 37°C. **Fig. 31.** A magnified view of the glass water jacket with blood placed on a one gallon ice cream carton cage showing engorged mosquitoes.

membrane can be fixed to the blood feeding reservoir by using hot bee's wax. Heparinized (1 ml of heparin solution per 10 ml of blood) or citrated blood is added from the open end of the apparatus (Fig. 31 and 36).

The funnel is placed on the horizontal screened top of a cage containing hungry mosquitoes (Figs. 31 and 36). In artificial blood feeding, hungry females are initially attracted to the source of heat and imbibe a blood meal. The blood feeding drive of females can be increased substantially by removing the sugar source from the mosquito cage 1 to 2 hrs before blood feeding. Adenosine triphosphate can be added to rabbit or chicken serum to increase feeding success. Adenosine triphosphate acts as a phagostimulant at concentrations below .0050 M, and it acts like a repellent at higher concentrations (Rutledge et al. 1964). The artificial blood feeding apparatus "Hemotek™ Membrane Feeding Systems" (Discovery Workshops, 516 Burnley Road, Accrington, Lancs., BB6JZ England) uses a hollow steel inverted dish as blood containing chamber that is heated directly by an electric source (Figs. 32–34) (Mahmood 1999). A thin collagen membrane is

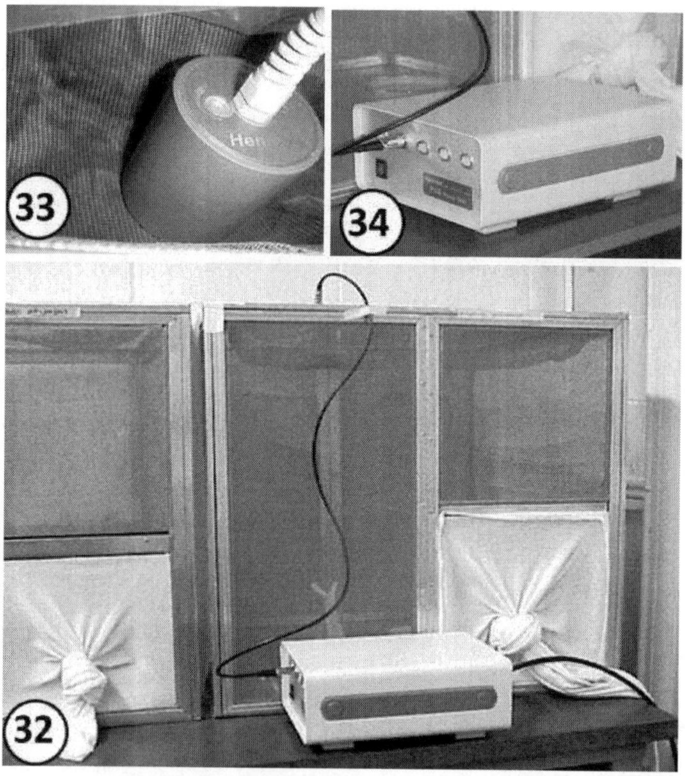

Figs. 32–34. ***In vitro*** **blood feeding using "Hemotek" system. Fig. 32.** blood feeding set up. **Fig. 33.** blood heating unit with attached collagen membrane on top of a cage. **Fig. 34.** Power supply with four outlets for blood feeding.

stretched at the open end of the blood meal reservoir. This membrane is not suitable for artificial blood feeding of *Culex* mosquitoes. Not all mosquito species are attracted to a particular type of membrane and a variety of different membranes must be tried at the start of an artificial membrane feeding attempt for a particular mosquito species. Some *Culex* species are not attracted to any sort of membrane and *Cx. salinarius* females have been successfully blood fed in our laboratory on absorbent cotton pads that were soaked in warm heparinized blood from cows. Besides using cotton pads, blood can be spread on the surface of the cage netting (Russell 1931) (Mahmood et al. 2004, 2006). Mosquitoes can be infected with arboviruses by feeding them through cotton pads or folded pieces of cheese cloth soaked in blood from infected animals (Fig. 37).

Salted sausage casing is an economical source of membranes (Fig. 39). It can be cut in small pieces of appropriate length and split open with a pair of scissors to cover the open end of the glass feeding funnel. The sausage casing membrane should be washed with an ample supply of water to remove all crystallized salt and later stretched on the glass membrane feeder. Similarly, a piece of thoroughly washed sausage casing that is not cut open can be filled with warm blood and tied on both ends to form a sausage (Fig. 39). Such a blood filled sausage is either placed in the mosquito cage

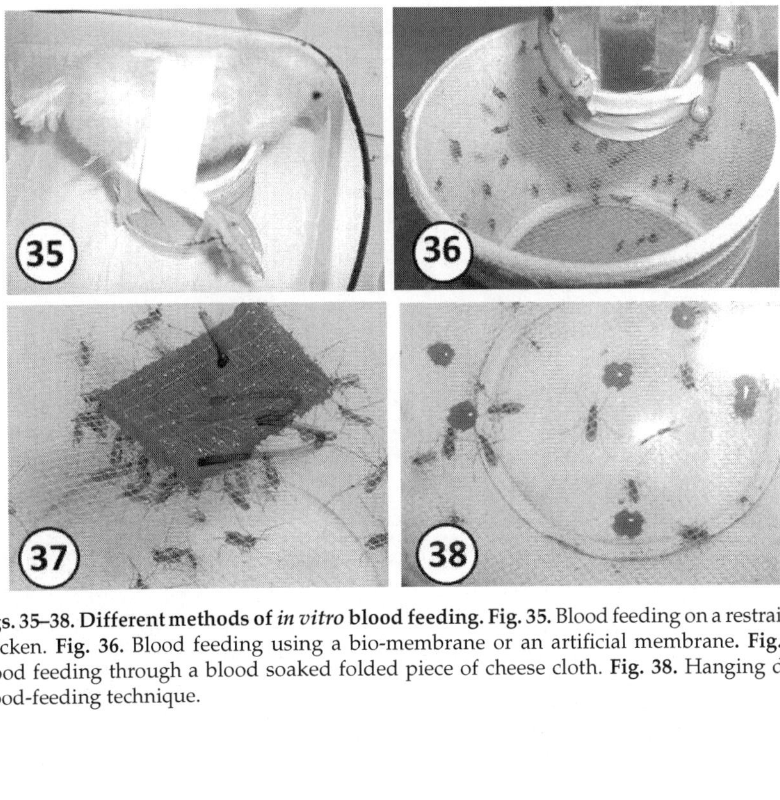

Figs. 35–38. Different methods of *in vitro* blood feeding. Fig. 35. Blood feeding on a restrained chicken. **Fig. 36.** Blood feeding using a bio-membrane or an artificial membrane. **Fig. 37.** Blood feeding through a blood soaked folded piece of cheese cloth. **Fig. 38.** Hanging drop blood-feeding technique.

Fig. 39. ***In vitro*** **blood feeding through a blood sausage. B.** Cow blood; **S.** A blood sausage prepared after cleaning the salted intestine casing of a hog and filling it with heparinized blood from a host animal.

in a petridish. The sausage can be hung from the roof of the cage or placed outside on the screen of the cage.

In genetic crosses single females can be isolated in vials and a nylon screen can be placed on top of the vials using a rubber band. The vial can be placed on a small vertebrate and the female allowed to blood feed. Such females can also be fed either by placing under a small membrane feeder or by placing a drop of warm blood on top of the screen (Fig. 38).

Oviposition

Mosquitoes lay their eggs either individually or in a compact clutch or egg rafts on the surface of water or oviposition substratum. Some *Mansonia* species lay egg rafts on the underside of leaves. Mosquito species that oviposit single eggs either deposit their eggs on the water surface (*Anopheles, Wyeomyia,* and *Toxorhynchites*), or they deposit their eggs on moist substratum, as do most species in the genus *Aedes.* Many different utensils or bowls can be used as oviposition containers (Fig. 40). The number of eggs laid by a single female during one oviposition cycle depends on the blood meal host of the mosquitoes and their genetic background and larval rearing condition.

Fig. 40. Different types of oviposition containers. A. An oviposition container preferred by fresh water *Aedes*. **B.** An oviposition container preferred by *Och. triseriatus*, *Och. atropalpus*, and *Ps. cyanensis*. **C.** An ovipostion container for *Anophelines* is ½ filled with aged tap water and lined with a strip of filter paper. **D.** An oviposition container filled with hay water infusion for egg laying by *Culex* mosquitoes.

Collection of *Aedine* Eggs

Under laboratory confinement, most species of *Aedes* prefer to lay eggs on moist substratum such as moist filter paper, paper toweling, cheese cloth, and absorbent cotton (Beckel 1958, Morlan 1966, Munstermann and Wasmuth 1985, Gerberg et al. 1994). Species like *Ae. aegypti* and *Ae. albopictus* prefer to oviposit on moist filter paper and/or paper toweling that is partially submerged in water (Fig. 40). A 500 ml beaker or an ordinary wide mouth glass jar can be converted into an oviposition container by wrapping black electrical tape on the outside. The jar or beaker is then lined with paper toweling or a piece of 3M filter paper and deionized or aged tap water is added to it so that 1/2 of the strip is above the water surface (Fig. 40). *Ochlerotatus atropalpus* and *Och. triseriatus* as well as *Ps. cyanensis* prefer to lay eggs on wet cotton and or on wet filter paper placed over layers of wet cotton in a black colored tray or a disposable plastic plate (Figs. 40 and 41). *Ochlerotatus taeniorhynchus* prefers to oviposit on moist folded cheesecloth. The cheesecloth is placed in a black colored disposable plastic plate, available in most grocery stores. Some scientists have used 10% seawater for soaking the cheesecloth, while I have successfully obtained eggs from *Och. taeniorhynchus* and *Och. sollicitans* on aged tap water. The appropriate ovipostion medium (Fig. 40) should be placed inside the cage 4–6 days after blood feeding at 28° or 25°C respectively.

The eggs should remain on the oviposition medium, such as wet filter paper for at least 96 hr for embryonation or conditioning (Figs. 40A and B). In case of *Ae. aegypti* and *Ae. albopictus* they can be kept in their oviposition jar and the jar covered with a polyethylene bag (Fig. 43). After embryonation the filter paper should be taken out of the oviposition jar and allowed to dry under a low stream of air for 4–5 hr. The filter papers should be dated and placed in a humidor that contains a saturated solution of Potassium Chlorate at the bottom for maintaining 85% humidity (Fig. 43). Recording of date of oviposition helps in keeping track of the age of the eggs. Considerable

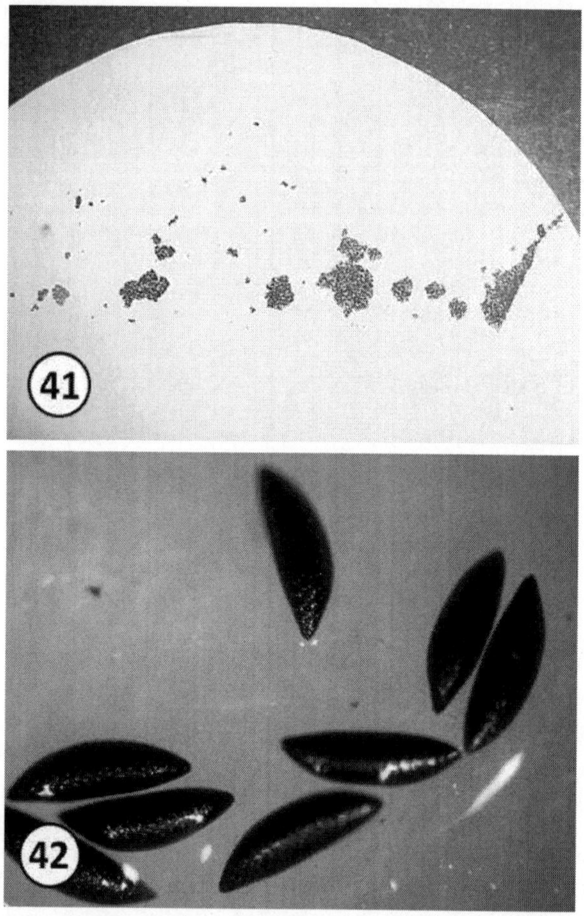

Figs. 41–42. Conditioning of Aedine eggs. Fig. 41. Aedes eggs on a filter paper for storage. **Fig. 42.** *Psorophora cyanensis* eggs on a moist filter paper left for conditioning or embryonation.

reduction in hatch rate occurs in eggs older than 3 months. In case of single family rearing, the females are isolated into oviposition vials that contain the oviposition medium, for example filter papers in water or a small piece of wet cotton in the bottom of the oviposition vial. The open end of the vial can be closed with cotton plug or with a piece of nylon netting fastened by a rubber band. A few drops of sugar solution can be placed on the plug or the netting as a sugar source after isolating a female.

Although eggs of certain *Aedes* and *Ochlerotatus* species can hatch in aged tap water, eggs of most species hatch best in deoxygenated water. Deoxygenated water can be prepared by 2 methods, either by boiling water or by addition of brewer's yeast. Water can be boiled in a big container that in turn contains a vegetable-canning jar. The lid of the canning jar should be tightly closed under water after deoxygenation. Deoxygenated water can also be prepared in a tightly closed container by adding brewer's yeast to the water and letting this suspension age overnight. The oviposition substratum (filter paper, cheesecloth, cotton, or paper towel with eggs) is then placed in the deoxygenated water and eggs are allowed to hatch. Eggs from certain *Aedes* and *Ochlerotatus* species require alternate wetting and drying of eggs for maximum egg hatching.

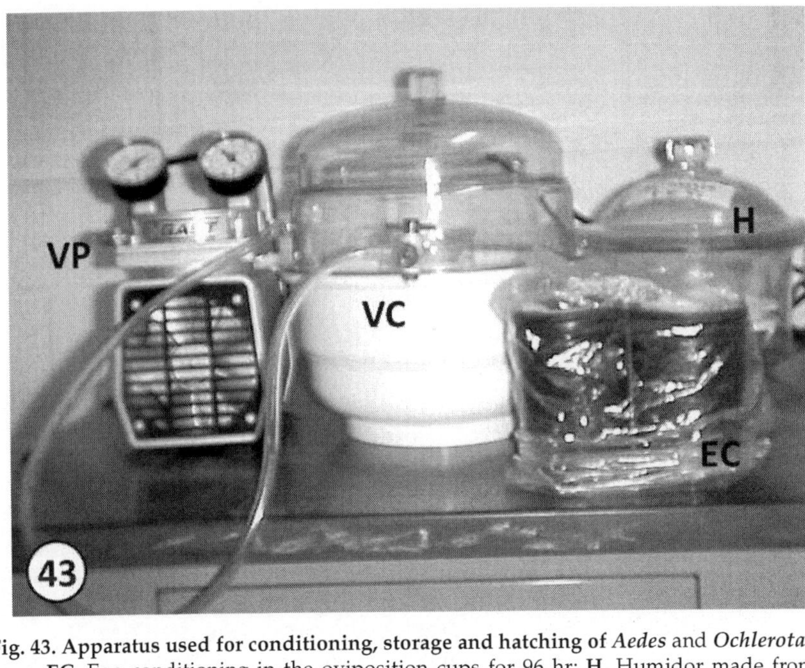

Fig. 43. Apparatus used for conditioning, storage and hatching of *Aedes* and *Ochlerotatus* eggs. EC, Egg conditioning in the oviposition cups for 96 hr; **H,** Humidor made from a desiccator; **VC,**Vacuum apparatus; VP, Vacuum pump.

Eggs from some *Aedes* species hatch in large numbers under vacuum. The eggs are placed in a container with water and then placed in a vacuum apparatus connected to a vacuum pump. The apparatus is closed tightly and vacuum (400 mm of Hg) is applied by the vacuum pump. The eggs are left under vacuum for 1–2 hr or until most of the egg hatch (Fig. 43).

Collection of *Anopheles* eggs

For most *Anopheline* adults, 26–28°C temperature and 80% humidity is ideal for colony maintenance. At this temperature, maximum mating activity of adults is observed at 3–5 days after emergence. After mating females take a blood meal and develop their first batch of eggs. Duration of first gonotrophic cycle after blood feeding varies from 2 to 4 days and is temperature dependent. *Anopheline* lay eggs in deionized as well as aged tap water. For oviposition, a clear cup or glass dish is lined with a filter paper strip and water is added to this container so that the paper strip is half-covered (Fig. 40). If the oviposition container is not lined with a strip of filter paper, the eggs get stuck to the walls of the container and dry out. Filter paper keeps the eggs away from the wall of the oviposition container and prevents desiccation. The eggs should be allowed to embryonate and hatch in the oviposition container. The eggs can be surface sterilized for sterile cultures (Gerberg et al. 1994). This precaution is taken to decontaminate eggs from microsporidian spore infection. After hatching the first instar larvae should be transferred to the larval rearing pans. In the beginning the larvae can be reared at a higher density and actually it helps to avoid overfeeding at the start of the experiment. Later the larval density in the pan can be decreased by splitting the larvae in 2 pans for a more homogeneous culture.

For single family rearing of genetic crosses, a blood fed or gravid female is isolated in an oviposition vial that is lined with a strip of filter paper and then water is added to half-cover the filter paper. The vial is plugged with an absorbent cotton plug soaked with 10% sugar water. After oviposition the female is aspirated out of the vial and placed in a cage for re-feeding or further manipulation according to the needs of an experiment.

Collection of *Culex* Egg Rafts

Culex species lay egg rafts on the surface of water and any large pan with aged or deionized water can be used as an oviposition medium (Fig. 40). Although most species would oviposit on tap water, a few require special oviposition media such as hay infusion for *Cx. nigripalpus* (Nayar 1968), and water containing fresh green algae in case of *Cx. bitaeniorhynchus* (Mohan

1950). Other species prefer polluted water to oviposit. Eggs can be removed from the oviposition container by using wooden applicators and strips of filter paper work as well. Individual egg rafts can be obtained by isolating gravid females individually in oviposition vials. Single family rearing can also be achieved by picking up single egg rafts using wooden applicators and placing them in separate enamel pans for larval rearing.

Larval Rearing

The following larval rearing procedures are applicable to all mosquito species with the exception of *Toxorhynchites* species in which the larvae are predatory on larvae of other mosquito species as well as cannibalistic. Rearing of nonhematophagous species of mosquitoes such as *Tox. ambinensis* are important because they can be intrathoracically inoculated with arboviruses for amplification of the virus without danger of virus transmission by these non-biting mosquitoes (Rosen 1981). The exceptions or variations in rearing procedures of different genera of mosquitoes will be mentioned where necessary. The optimum larval rearing temperature would vary according to the place of origin of the mosquitoes. Larvae for colony maintenance should not be reared at temperatures that are higher

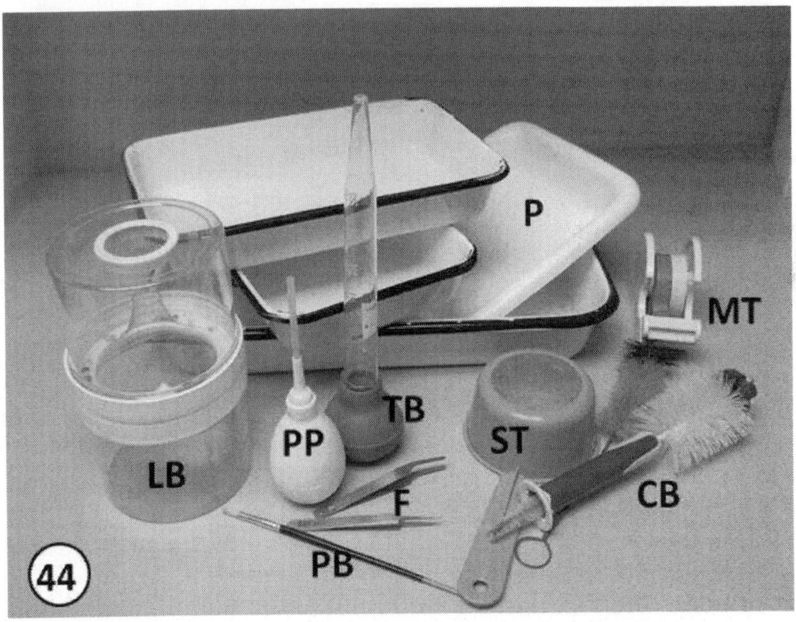

Fig. 44. Supplies for larval rearing. CB, Cleaning brushes; **F,** Forceps; **LB,** Larval breeder (Bioquip); **MT,** Masking tapes of different colors for labeling; **P,** Rearing pans; **PB,** Paint brush for sorting adults; **PP,** Pipette; **TB,** Turkey baster.

than the temperature they are exposed to in nature. Doing so will actually subject them to a selection pressure and only individuals with genetic makeup to withstand higher temperatures will be selected.

Most *Aedes* and *Culex* species can be reared on liver powder, Brewer's yeast pellets of rabbit or hamster food, and pellets or pulverized dog food that is passed through a 40 mesh screen. Special mixtures of food are required for rearing anopheline species. *Anopheles culicifacies* larvae require a mixture of equal parts of wheat germ, Kellogg's Concentrate cereal, and brewer's yeast as food (Ainsely 1976). Slurry of liver powder and baby tetramin fish food is used extensively for *An. albimanus* (Benedict 1997). *Anopheles stephensi* and *An. quadrimaculatus* can be reared without difficulty on liver powder alone.

Duration of immature development is directly dependent upon the rearing temperature, density of larvae per pan and amount of food added to the larval rearing pans (Mahmood et al. 1997, Mahmood and Crans 1998). After colonizing a species' best rearing density should be determined by rearing the larvae at different densities and on different amounts of food to establish base line data. The rearing densities directly affect the fecundity of emerging adults by affecting their size. Similarly, rearing temperature also has a direct effect on the size of emerging adults and consequently on their fecundity and survival under laboratory conditions. Smaller and short-lived adults are produced at high temperatures and under overcrowded conditions. Lower temperatures can slow down larval growth and increased larval mortality might result due to longer duration of the immature development times. Many different types of pans can be used for larval rearing such as plastic shoe boxes, plastic food containers, glass dishes, photographic dishes, enamel pans and porcelain-coated metal trays (Fig. 44).

For larval rearing, besides the pans, sieves, turkey baster, rubber bulbs for pipettes and fish nets for straining first instar larvae are needed (Fig. 44). Under optimal rearing conditions, all *Aedes* and *Culex* larvae should pupate synchronously within 4 days after the start of pupation otherwise sub-optimal rearing conditions are indicated. Overcrowding of larval pans is a most common flaw; the second is overfeeding of larvae. The color of water in rearing pans can be used as an indicator of optimal rearing conditions. A pan with clear light yellow or golden colored water indicates an optimal rearing condition. Cloudiness in the rearing water indicates bacterial overgrowth from overfeeding of the larval pans. The surface of larval rearing pans should be skimmed daily with paper towels to remove scum formed by bacteria. The rearing pans should be covered with a Plexiglas sheet or screen fitted in a plastic frame to avoid larval contamination due to oviposition of escaped females or due to over spilling of pans. The racks with rearing mosquito pans should be covered with polyethylene curtains in the absence of pan covers and one mosquito species or strain should be placed on each rack. Aeration of the larval cultures is necessary for species that require long periods of immature development. Fish tank pumps can be used to aerate a large number of larval rearing pans connected serially (Fig. 45).

Separate pipettes should be assigned for different larval cultures. The pipettes should be cleaned in boiling water after finishing daily chores. This will avoid contamination and carrying over of an infection from one culture to another. Different species of mosquitoes should be kept separate preferably in different rooms. In cases where separate rooms are not available due to space restrictions, larval cultures should be kept separate on different racks. The culture pans can be color coded by allocating different colors of masking tapes or different colors of papers used for preparing labels. Besides physical labeling of the cultures, genetic strains can be kept separated by using some color mutant markers such as larval body color or eye color of adults.

Each larval pan should be labeled clearly with the date of egg hatch, the abbreviation of species name, genetic background of the strain and the geographic site of origin of the species. A logbook should be kept for daily maintenance of the insectary. A record should be kept of the number of egg rafts or the number of hatched larvae that were reared for each species,

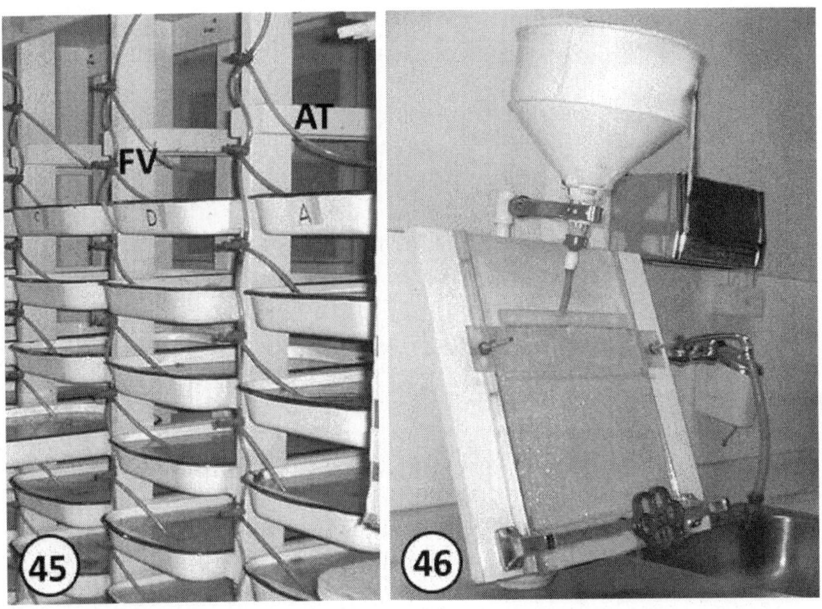

Figs. 45–46. Rearing of immature mosquitoes. Fig. 45. Larval rearing pans; **AT,** Aeration tube. **FV,** Aeration flow control valve. **Fig. 46.** Mechanical pupae separator.

strain, or mutant over time. This information allows the researcher to plan experiments and to have knowledge handy about the size of the mosquito colony. The logbook also helps in planning future experiments. Left over mosquitoes from an experiment should never be added back to the parent colony. Frequent counts should be made to assess the fertility status of the species in culture. This information should be recorded at least every third generation by isolating single females and determining the total number of eggs laid and the percentage of hatched eggs laid by single females.

All infected mosquito species that can transfer a pathogen transovarially to their progeny should be reared in a separate screened room. Species that look morphologically alike and/or strains of a species that differ in their susceptibility to different pathogens or insecticides preferably should be kept in separate small rooms or areas. Such areas can be constructed inside a large insectary by using screens. If possible genetic morphological markers, such as eye color, body color of adults or color of the larvae should be used to keep such strains separate.

Pupae should be passed through a sieve to remove the water from the rearing pans and fresh water added. This precaution is necessary for avoiding bacterial growth at the surface of the water and thus restricting the supply of oxygen to the pupae. Mechanical separators are also designed for separating larvae from pupae (Fig. 46). While mass rearing for genetic release of male pupae only, exploiting the size difference of the two sexes for separation of males from female pupae saves time. Male pupae are smaller than the female pupae in most species when reared at low rearing density. Scientists have used different types of mechanical pupae separators. A sieve with appropriate size of mesh from which only small sized male pupae can pass makes a simple pupae separator. While using size of pupae to separate sexes care should be observed to keep the rearing density of larvae and amount of food added to the pans constant throughout the experiment. Overcrowding can result in small sized females that will pass through the mesh and result in release of females during an experiment. Male and female pupae can be separated using the structure of their genitalia. One can easily learn the structure by isolating large and small pupae individually in small screened container and by comparing the structure of their genitalia from casted skin or exuviae.

Many containers can be used as emergence chambers. Mosquito breeders (BioQuip Products) consisting of 2 chambers are available commercially (Fig. 44). The pupae are kept in the lower chamber. Mosquitoes emerging in the lower chamber can fly through a cone to the upper chamber but cannot fly back to the lower chamber and thus can be collected easily from the upper chamber. This apparatus is useful in studying emergence rhythms of a species during 24 hr (Fig. 44).

While handling large numbers of species or many mutant strains in an insectary, cold anesthesia (using ice-cold water) technique can be used. All contents of a pan are passed through a sieve and larvae and pupae are washed in a bowl filled with ice cold water. The larvae go to the bottom of the bowl and the pupae float at the top. Quick separation of pupae and immediate return of larvae and pupae to normal water temperature in the emergence jar and rearing pans respectively is advised. A longer exposure of larvae to ice-cold water can result in higher mortality of the larvae. Many other devices for mass separation of pupae are described by Gerber et al. (1994).

Newly emerged adults can be released from an emergence container directly in the mosquito colony cage by opening the container inside the cage. During weekends the pupae can be placed directly inside the colony cage in a glass dish. A screen cone can be constructed from a piece of wire screen and placed over the dish (Fig. 47). The emerging adults can fly out of the pupae container but gravid females do not fly inside the wire screen cone to lay eggs in the glass dish containing the pupae. Similar results can be obtained by placing a plastic cone removed from a mosquito breeder over a glass dish containing pupae (Fig. 48).

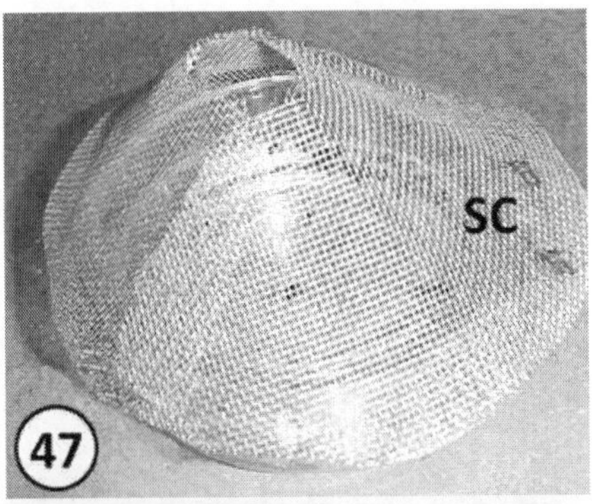

Figs. 47–48. Emergence containers for pupae. Fig. 47. SC, Screen cone placed on a glass dish with pupae. **Fig. 48. PC,** A plastic cone placed over a glass dish; **GD,** containing pupae.

Large scale rearing of mosquitoes for sterile male release programs can be achieved by rearing larvae in large specially constructed pans with spigot at one end for removal of larvae and pupae (Fig. 49). Rearing water with larvae and pupae is collected in buckets and pupae removed using mechanical pupae separator. Larvae for field release experiments should be raised in a room with large glass windows that would acclimatized the emerging adults to natural day length and photo period and the swarming times of males would be similar to those in natural population to be controlled. Thus the laboratory released sterile males would be competitive to the wild males. Adults for large scale sterile male release experiments can be counted by slightly anesthetizing the males and spreading them in clear petri dishes that can be scanned on a simple photocopy machines (Baker et al. 1978). The males can be easily counted from these copies thus reducing injuries to mosquitoes due to handling and to save time for field release procedures. Such a procedure can also be used during capture mark release recapture experiments to determine the size of a wild population of mosquitoes.

Such large pans can also be used for rearing mosquitoes used in experiments where different doses of viruses or parasites are fed to mosquitoes for determining the dose at which mosquitoes become infected in nature. Similar sized mosquitoes raised together would eliminate variation in ingested viral due to large or small sized females affecting the total amount of virus load in the blood meal of females.

Fig. 49. Mass rearing of mosquitoes. B, Bucket; **LP,** Large steel pan with spigot; **S,** Photographer Mr. Tobin Rowland from Vector and Parasite Biology, Entomology Branch, WRAIR.

Time required from egg hatching to the final molt of immature larvae to pupae is genetically controlled in different species (Mahmood et al. 1997, Mahmood and Crans 1998, Clements 2000). Many *Aedes*, *Anopheles* and *Culex* species start pupating 6 to 8 days after egg hatch whereas, species like *Cs. melanura* require at least 18 days at 22–24°C to pupate (Mahmood and Crans 1998). Therefore one must know the genetic background as well as physiological responses and growth rates of species before planning an experiment.

Insectary Administration

If more than 1 person is taking care of colony maintenance then a monthly schedule should be constructed for the maintenance of the colony. A record should be kept of dates of release of newly emerged adults in the cage, the date of sugar change, date of blood feeding, egg hatching and date and number of larval rearing pans set. For most species, the schedule should be such that it would avoid major colony procedures, requiring longer work hours, from falling on weekends. This can be simply done by scheduling dates of oviposition to fall on a Friday or date of egg hatching on a Monday. A checkup sheet should be also present in the insectary so different technicians can check off the assigned duties or chores after finishing them. This allows for a better coordination between people and better colony maintenance practices.

Maintenance of Genetic Variability

For maintaining genetic diversity of a species, the best method is to introduce wild material into the colony at regular intervals. Sometimes collection and edition of new wild caught mosquitoes is not possible such in case of exotic species than the following scheme can be used. During colonization start with isofemale lines and maintain them separately. Also have a mass colony cage and, at regular intervals, introduce individuals from the isofemale lines into the mass colony cage.

The third method is labor intensive and involves a circular mating scheme. For example, you have 4 isofemale lines A, B, C, and D, than mate AxB, BxC, CxD, and DxA and the hybrids are added into the colony cage (R.K. Sakai, personal communication). All these schemes except for the addition of wild material at regular interval are rather labor intensive and may be difficult to maintain unless one has a full time technical staff.

Conclusion

To conclude I hope the above mentioned mosquito maintenance techniques would help many readers to maintain thriving mosquito colonies and help in mosquito control and eradication of various diseases transmitted by them.

Acknowledgements

This chapter is an updated version of the chapter published in 1999 while I was working at Rutgers the State University of New Jersey and some figures used were photographed by the author during her employment (Mahmood 1999). I thank Mr. Tobin Rowland from Vector and Parasite Biology, Entomology Branch, WRAIR for providing Fig. 49. This chapter is dedicated to cherished memories of my parents and my sisters Mrs. Razia Ahmed and Mrs. Zahida Nasim.

References

Ainsely, R.W. 1976. Laboratory colonization of the malaria vector, *Anopheles culicifacies*. Mosquito News 36: 256–258.

Baker, R.H. 1964. Mating problems as related to the establishment and maintenance of laboratory colonies of mosquitoes. Bull. W.H.O. 31: 467–468.

Baker, R.H., W.L. French and J.B. Kitzmiller. 1962. Induced copulation in *Anopheles* mosquitoes. Mosquito News 22: 16–17.

Baker, R.H., W.K. Reisen, R.K. Sakai, C.G. Hays, M.A. Khan, U.T. Saifuddein, F. Mahmood, A. Parveen and S. Javed. 1978. Field assessment of mating competitiveness of male *Culex tritaeniorhynchus* carrying a complex chromosomal aberration. Ann. Ent. Soc. Am. 72: 751–758.

Beaty, J.B. and W.C. Marquardt. 1996. The biology of disease vectors. University Press of Colorado. 632 pp.

Beckel, W.E. 1958. Observations on the rearing of larvae, pupae and adults of some *Aedes* mosquitoes of northern Canada. Canad. J. Zool. 36: 797–898.

Benedict, M.Q. 1997. Care and maintenance of *Anopheline* mosquito colonies. pp. 2–12. *In*: J.M. Crampton, C.B. Beard and C. Louis (eds.). Molecular Biology of Insect Disease Vectors: A Method Manual. Chapman and Hall.

Clements, A.N. 2000. The biology of mosquitoes. 1: CABI Publishing. New York. 511 pp.

Collins, W.E., A.J. Harrison and J.C. Skinner. 1964a. The use of membrane feeding technique to determine infection and transmission thresholds of Semliki Forest virus in *Anopheles quadrimaculatus* and *Anopheles albimanus*. Mosquito News 24: 25–27.

Collins, W.E., G.M. Jeffrey and J.C. Skinner. 1964b. Comparative infectivity of a strain of *Plasmodium falciparum* from Panama to three species of *Anopheles* as studied by membrane feeding. Mosquito News 24: 28–31.

Cosgrove, J.B., R.J. Wood, D. Petric, D.T. Evans and H.R. Abbott. 1994. A convenient mosquito membrane feeding system. J. Am. Mosquito Control. Assoc. 10: 434–436.

Gerberg, E.J., D.R. Barnard and R.A. Ward. 1994. Manual for mosquito rearing and experimental techniques. AMCA Bull. No: 5: 98 pp.

Goodnough, J.L. and C.B. Parnell. 1985. Basic engineering design requirements for ventilation, heating, cooling, and humidification of insect rearing facilities. pp. 137–155. *In*: Pritam Singh and R.F. Moore (eds.). Hand Book of Insect Rearing Vol. 1. Elsevier. New York. 488 pp.

Hastriter, M.W. and D.C. Cavanaugh. 1981. An apparatus for colonizing fleas (Siphonaptera) and collecting pupal cocoons. J. Med. Entomol. 18: 251–252.

Hastriter, M.W., D.M. Robinson and D.C. Cavanaugh. 1980. An improved apparatus for safely feeding fleas (Siphonaptera) in plague studies. J. Med. Entomol. 17: 387–388.

Horsfall, W.R. 1964. Induced copulation of mosquitoes. W.H.O./EBL/28.64: 1–5 pp.

Horsefall, W.R. and M.L. Taylor. 1967. Temperature and age as factors in inducing insemination of mosquitoes (Diptera: Culicidae). Ann. Entomol. Soc. Am. 60: 118–120.

Hunt, G.J. and C.N. McKinnon. 1990. Evaluation of membranes for feeding *Culicoides variipennis* (Diptera: Ceratopogonidae). Springfield, VA: National Technical Information Service. USDA Research Service, ARS-121. 68 pp.

Khan, A.Q. and W.K. Reisen. 1977. Laboratory observations on developmental rhythms in *Culex tritaeniorhynchus*. Mosquito News 37: 637–645.

Laird, M. 1988. The natural history of larval mosquito habitats. Academic Press, New York.

Mahmood, F. 1999. Laboratory Maintenance of mosquitoes. pp. 3–54. *In*: K. Maramorsch and F. Mahmmod [eds.]. Maintenance of Human, Animal and Plant Pathogen Vectors. Science Publishers, Inc. Enfield, USA.

Mahmood, F. and W.J. Crans. 1994. Laboratory colonization of *Culiseta melanura* from southern New Jersey. Proc. N. J. Mosquito Control Assoc. Inc. 81: 94–96.

Mahmood, F. and W.J. Crans. 1997. A thermal heat summation model to predict the duration of the gonotrophic cycle of *Culiseta melanura* in nature. J. Amer. Mosquito Control Assoc. Inc. 13: 92–94.

Mahmood, F. and W.J. Crans. 1998. Effect of temperature on the development of *Culiseta melanura* (Diptera: Culicidae) and its impact on the amplification of eastern equine encephalitis virus in birds. J. Med. Entomol. 35: 1007–1012.

Mahmood, F. and W.J. Crans. 1999. Life history characteristics of *Wyeomyia smithii* from New Jersey. J. Vec. Ecology 24: 1–8.

Mahmood, F., W.J. Crans and N.S. Savur. 1997. Larval competition in *Aedes triseriatus*. J. Vec. Ecol. 22: 90–94.

Mahmood, F. and W.K. Reisen. 1981. Duration of the gonotrophic cycle of *Anopheles culicifacies* Giles and *Anopheles stephensi* Liston, with observations on reproductive activity and survivorship during winter in Punjab Province, Pakistan. Mosquito News 41: 41–50.

Mahmood, F. and W.K. Reisen. 1982. *Anopheles stephensi* Liston (Diptera: Culicidae): Changes in male mating competence and reproductive system morphology associated with aging and mating. J. Med. Entomol. 19: 573–588.

Mahmood, F. and W.K. Reisen. 1994. *Anopheles culicifacies*: effects of age on the male reproductive system and mating ability of virgin adult mosquitoes. Med. Vet. Entomol. 8: 31–37.

Mahmood, F., T. Parveen and W.K. Reisen. 1986. *Culex tritaeniorhynchus* Giles: Changes in male mating competence and reproductive system morphology associated with age and mating experience. Pakistan J. Zool. 18: 273–296.

Mahmood, F., Y. Fang, R.E. Chiles and W.K. Reisen. 2004. Methods for studying the vector competence of *Culex tarsalis* for Western equine encephalomyelitis virus. J. Am. Mosq. Control. Assoc. 20: 277–282.

Mahmood, F., R.E. Chiles, Y. Fang, E.N. Green and W.K. Reisen. 2006. Effect of time after infection, mosquito genotype, and infectious viral dose on the dynamics of *Culex tarsalis* vector competence for western equine encephalomyelitis virus. J. Am. Mosq. Control Assoc. 22: 272–281.

McCuiston, L.J. and D.L. White. 1976. Laboratory colonization of *Aedes sollicitans* (Walker) with a review of the technique of induced copulation. Proc. NJ Mosquito Control Assoc. 63: 164–175.

McDaniel, I.N. and W.R. Horsfall. 1957. Induced copulation of Aedine mosquitoes. Science 125: 745.

Miura, T. 1967. Establishment and maintenance of a laboratory colony of *Aedes nigromaculus* (Ludlow) (Diptera: Culicidae) by induced mating. Calif. Vector Views 14: 57–58.

Miura, T. 1969. Evaluation of techniques used for mass rearing *Aedes nigromaculus* by induced mating. Mosquito News 29: 612–616.

Mohan, B.N. 1950. Certain uncommon habits of *Cx. bitaeniorhynchus*. Indian J. Malariol. 4: 167–173.

Morlan, H.B. 1966. Yellow fever mosquitoes. pp. 585–599. *In*: C.N. Smith (ed.). Insect Colonization and Mass Production. Academic Press, New York. USA.

Munstermann, L.E. and L.M. Wasmuth. 1985. *Aedes triseriatus*. pp. 15–24. *In*: P. Singh and R.F. Moore (eds.). Handbook of Insect Rearing. Vol. II. Elsevier, New York.

Nayar, J.K. 1968. Biology of *Culex nigripalpus* Theobald (Diptera: Culicidae). Part 1. Effects of rearing conditions on growth and the diurnal rhythm of pupation and emergence. J. Med. Entomol. 5: 39–46.

Ow Yang, C.K., Sta F.L. Maria and R.H. Wharton. 1963. Maintenance of a laboratory colony of *Anopheles maculatus* Theobald by artificial mating. Mosquito News 23: 34–35.

Provost, M.W., P.T. Lum and N. Branch. 1961. Rotation of male terminalia in *Aedes taeniorhynchus* (Diptera: Culicidae) as affected by temperature. Ann. Ent. Soc. Amer. 54: 896–900.

Reisen, W.K. and F. Mahmood. 1979. *Anopheles culicifacies* Giles: Some relation among oviposition, refeeding and survivorship. Mosquito News 39: 374–380.

Reisen, W.K., F. Mahmood and T. Parveen. 1979. Laboratory observation on the time of mating of *Anopheles culicifacies* Giles. Mosquito News 39: 328–333.

Reiter, P. 1983. A portable, battery powered trap for collecting gravid *Culex* mosquitoes. Mosquito News 43: 496–498.

Reiter, P. 1987. A revised version of the CDC gravid mosquito trap. J. Am. Mosq. Control Assoc. 2: 219–221.

Rosen, L. 1981. The use of *Toxorhynchites* mosquitoes to detect and propagate dengue and other arboviruses. Am. J. Trop. Hyg. 30: 177–183.

Russell, P.F. 1931. A method for feeding blood meals to mosquitoes, male and female. Preliminary note. Am. J. Trop. Med. 11: 355–358.

Rutledge, L.C., M.A. Moussa and C.J. Belletti. 1976. An *in vitro* blood-feeding system for quantitative testing of mosquito repellents. Mosquito News 36: 283–293.

Rutledge, L.C., R.A. Ward and D.G. Gould. 1964. Studies on the feeding response of mosquitoes to nutritive solution in a new membrane feeder. Mosquito News 24: 407–419.

SALS. 1980. Laboratory safety for arboviruses and certain other viruses of vertebrates. Am. J. Trop. Med. Hyg. 1359–1381.

Service, M.W. 1976. Mosquito Ecology: Field sampling methods. John Wiley and Sons. New York. USA.

Tarishis, B. 1958. Feeding techniques for bloodsucking arthropods. Proc. 10th Internatl. Cong. Entomol. 3: 767–784.

Trembley, H.L. 1952. The distribution of certain liquids in the esophageal diverticula and stomach of mosquitoes. Am. J. Trop. Med. Hyg. 1: 693–710.

Waladde, S.M., S.A. Ochieng and P.M. Gichuhi. 1991. Artificial-membrane feeding of ixodidtick *Rhipicephalus appendiculatus* to repletion. Exper. and Appl. Acarol. 11: 297–306.

Wheeler, R.E. 1962. A simple apparatus for forced copulation of mosquitoes. Mosquito News 22: 402–403.

2

Black Fly Rearing and Use in Laboratory Bioassays#

*Elmer W. Gray** and *Ray Noblet*

ABSTRACT

Black flies (Diptera: Simuliidae) are serious pests of man and animals worldwide. In addition, black flies vector the causative agent of human onchocerciasis, the filarial nematode *Onchocerca volvulus*. Due to their public health significance and worldwide pest status, substantial efforts have been made to develop efficient methods to rear these insects in the laboratory environment. The University of Georgia Black Fly Rearing and Bioassay Laboratory currently has the world's only laboratory colony of black flies. The original field-collected material for this colony, *Simulium vittatum* cytospecies IS-7, was collected as eggs from trailing grasses in Flaxmill Creek, Cambridge, NY. The colony currently produces 200,000–300,000 larvae per week during normal operations. The primary purpose of the colony is to produce uniform larvae for a variety of research activities. Such activities include product development, quality control, and other research associated with the insecticidal proteins of *Bacillus thuringiensis* subsp. *israelensis*. The colony operates in a similar fashion to the Cornell Automated Rearing System incorporating a closed-circulation trough system in an aquatic rearing

Department of Entomology, University of Georgia, Athens, GA, 30602, USA.
*Corresponding author

The specific materials used to operate this colony were chosen based on availability, effectiveness and cost efficiency. The resulting usage is not to promote the individual companies mentioned; users can substitute similar equipment or substances manufactured by other companies. Mentioning of a proprietary product does not constitute an endorsement or recommendation by the University of Georgia for its use.

unit. Water is pumped from a lower reservoir to upper chambers where it flows over a runway creating an ideal larval habitat. Associated systems have been developed to inject food material into the unit, capture emerging adults, induce mating and provide suitable substrate for oviposition. This chapter outlines the construction and use of equipment needed to successfully rear black flies in the laboratory, and their use as experimental subjects for a variety of research purposes.

Key Words: Black flies, *Simulium vittatum*, rearing unit, colony

Introduction

Black flies (Diptera: Simuliidae) have long been recognized as one of the most widespread and troublesome pests of both man and animals. Their ability to vector causal agents of disease and to occur in enormous numbers has resulted in a wealth of research on simuliids. A major concern is their ability to transmit the filarial nematode *Onchocerca volvulus*, the causal agent of human onchocerciasis. This disease affects vast areas of Africa and smaller regions of Central and South America. Research with black flies in both their vectorial and nuisance roles has resulted in significant rearing efforts. The results of these efforts have led to the conclusion that, due to their unique ecology and behavior, black flies are one of the most difficult insects to rear in a laboratory setting (Edman and Simmons 1987). This fact is somewhat ironic for a group of insects whose immatures and adults inhabit such harsh and unpredictable habitats (Edman and Simmons 1985).

A common trait unique to all black fly species is their requirement for flowing water. Consequently, any successful rearing system must provide sufficient current to stimulate larval feeding on suspended food particles and avoid excess accumulation of toxic wastes. While many of the early black fly researchers reared larvae and adults in the laboratory (Puri 1925, Bradley 1935, Dalmat 1955) it appears that Emery (1913) is the earliest published work on rearing of adult black flies from field-collected larvae. Fredeen (1959) was the first to rear a black fly species, *Cnephia dacotensis* (Dyar and Shannon), through a complete generation (Edman and Simmons 1985). However, in order to be considered true colonization, it is necessary to be able to hatch eggs, rear larvae and pupae, and then induce subsequent adults to mate (unless the species is parthenogenic), blood-feed (where necessary), and oviposit (Mokry et al. 1981).

The rearing of black flies has been reviewed extensively starting with Fredeen (1959), Muirhead-Thompson (1966), Raybould and Grunewald (1975), Mokry et al. (1981), Raybould (1981), Edman and Simmons (1985, 1987), Cupp and Ramberg (1997) and Gray and Noblet (1999). The primary methods for creating a current for larval habitat in these papers involve trough systems, magnetic stirrers, mechanical stirrers and compressed air bubbles.

The Cornell Automated Rearing System which incorporated a trough habitat using a closed-system of water circulation was first described by Brenner and Cupp (1980). Modifications of this system have been described by Bernardo et al. (1986), Cupp and Ramberg (1997) and Gray and Noblet (1999). This system is the most reliable, effective and adaptable system developed to date for black fly rearing. The Cornell system has been used to rear several species of black flies including: *Simulium vittatum* Zetterstedt, *S. decorum* Walker, *S. pictipes* Hagen, *S. damnosum* s. l. and *C. dacotensis* (Dyar and Shannon). The *S. vittatum* colony initiated by Dr. Cupp and his associates has been in culture since 1981 and was in its 16th year of continuous colonization and in the 120th generation when reported by Cupp and Ramberg (1997). A sister branch of this colony operated at Clemson University from 1991 to 1999. The Clemson branch of the colony was moved to the University of Georgia in 1999 and continues to produce vigorous larvae and adults that are used in a wide variety of research projects (Gray 2010). Operational protocols of the past twenty years have required the mixing of emerging adults from multiple rearing units and age groups. As a result, an accurate determination of an exact generation is not possible.

Currently the University of Georgia Black Fly Rearing and Bioassay Laboratory operates the only black fly colony in the world. The primary use of this colony is to evaluate larvicidal formulations containing insecticidal proteins of *Bacillus thuringiensis* subsp. *israelensis* (*Bti*). While developing a laboratory system for assaying the potency of *Bti* insecticidal proteins against larval black flies at Clemson University, the need for a monoculture of test specimens was quickly realized. The Clemson University Orbital Shaker Bioassay was initially developed with field-collected larvae from streams near the university campus (Barton et al. 1991). While a repeatable and useful bioassay was developed with the species complexes commonly found in field-collected larvae from the area, the next obvious improvement was to use an individual species to further refine the system.

The original source of material for the University of Georgia black fly colony was eggs collected in September of 1981 from Flaxmill Brook in Cambridge, NY. The colony itself was initiated in December of that year. Specimens from both the original colony and the Georgia colony have been identified as *S. vittatum* cytotype IS-7 (Brockhouse and Adler 2002). No additional introductions of field-collected specimens have been made to the original genetic stock of this colony. The continued viability of the colony has precluded the need for such augmentation. In addition, the risk of introducing a pathogen from outside sources far outweighs any perceived benefit.

The use of *S. vittatum* as a 'model' species has been extremely important in furthering our understanding of black flies in terms of larval control methods and vector capabilities (Barton et al. 1991, Bernardo and Cupp 1986). In addition, specimens from this colony have been instrumental in developing the black fly transcriptome and the black fly genome project (C. Brockhouse, personal communication). Our colony is particularly important for this research as there are no known colonies of natural vector species of black flies (Cupp and Ramberg 1997). As a result of the long-term success of this system and the number of species that have been reared in it, the methods and equipment discussed should work as an effective guide for rearing other black fly species. For our purposes, rearing schedules and protocols have been developed to ensure that large numbers of late instar larvae were available during the middle of the normal work week. The following information will provide detailed descriptions of the aquatic rearing units and all associated systems. The operation and maintenance of all systems is described in addition to troubleshooting measures.

Rearing Unit Structure and Operation

General Description

The University of Georgia rearing system uses nine modified Dayno Aqua-Lab aquatic rearing units. These rearing units have a large lower reservoir (643 liters), a refrigeration unit to cool the water, and a pump to draw water from the bottom of the lower reservoir and move it to the upper chambers where it flows over the runway, creating an ideal larval habitat (Fig. 1). When in operation, water is pumped over a runway that slopes downward at approximately a 10° angle from horizontal (Fig. 2). This arrangement simulates a stream environment for the developing larvae (Bernardo et al. 1986). Since the Clemson branch of the colony was established, various modifications have been made to improve the system. As previously mentioned, the primary use of this system is to produce test specimens for a large-scale larval bioassay laboratory for *Bti* insecticidal protein (IP) formulations. The standard protocol for our bioassay laboratory requires approximately 22,000 late instar larvae for each week's testing. Because of the scale and diversity of our research, our objective has been to maximize and expand the production capabilities of the original Cornell System.

Unfortunately, Dayno Manufacturing apparently no longer manufactures this unit, so transitioning to a new aquatic rearing unit will be likely required. While other sources of similar units may be available (Appendix 1), we agree with Edmans and Simmons (1985) who stated that a comparable unit could be constructed at a reasonable cost. As early as 1963, Hall and Harrod (1963) described a similar system developed to rear *Simulium ornatum*. This

Fig. 1. Aquatic rearing units.

Fig. 2. Runway creating ideal larval habitat.

system, although slightly different in design, is similar in concept to the modified Dayno Aqua-Lab aquatic rearing unit described herein. Systems described by Ham and Bianco (1983) and Tarrant et al. (1987) could also be useful to aid in construction of a smaller and less expensive aquatic rearing unit with similar capabilities. Construction and maintenance of these units would require expertise with wood, fiberglass, refrigeration, electricity and plumbing.

The University of Georgia (UGA) colony currently operates with nine aquatic rearing units. The colony is housed in approximately 900 square feet of laboratory space with no secondary containment being required. *Simulium vittatum* is a highly photo-tactic species. As a result, any flies that escape during handling will immediately move to the brightest light source in the area. During 20 years of colony operation, we have never had a legitimate complaint about escaped black flies.

Rearing Unit Construction and Operation

The aquatic rearing units used in the UGA colony are comprised of two sections. An upper section constructed of 1.27 cm thick Plexiglas® and a lower section comprised of the main reservoir, head wall, and a mechanical deck where the pump and compressor are located. The lower section is constructed of 4 x 10 cm framing and 1.9 cm marine plywood with a waterproof epoxy resin coating on the inner and upper surfaces. The exterior surfaces of the lower portion of the units are finished with Formica®, however, a durable exterior or marine paint would serve equally well. The outside dimensions of the entire rearing unit are 91 cm x 183 cm x 97 cm with an inner reservoir of 87 x 119 x 48 cm which can hold up to 643 L of water. Beside the water reservoir is a head wall with the mechanical deck adjacent to it that houses the refrigeration unit and pump.

The refrigeration unit can maintain water temperatures from 5° to 20° C ± 1°C. This aspect of the rearing unit is the most complex and expensive component to maintain and operate. The refrigeration unit is coupled to chemical resistant, hard plastic coils that are submersed in the reservoir of the unit. Experienced HVAC specialists are required to maintain and service the refrigeration component of the rearing units. Water flow over the runway is maintained by a pump which moves water from the lower reservoir to the upper chambers where it flows onto the runway surface. The pump produces the required flow, ranging from a trickle up to 35 L/min., depending on the stage of larval development and is controlled by the flow-valve setting. Magnetic drive pumps with enclosed motors have proven best suited for this application (Appendix). The circulatory system of the rearing units has two valves to control and direct the water flow: the flow control valve and the drain valve. The drain valve is closed when the unit is in normal operation and the flow control valve is adjusted to the desired level depending on the stage of larval development. When excess water needs to be drained from the lower reservoir, the drain valve can be opened while the flow valve remains unchanged. This stage of operation will split the water flow, with a portion draining from the unit and a portion continuing to flow over the runway. When draining the unit for cleaning, the flow valve is closed and the drain valve is opened and all water flow is pumped out the drain valve.

Runway Description

The runway is a critical component of this rearing system; it provides the surface to which the egg-sheets are attached and on which larval development takes place. All runways are 122 x 40 cm with 9 cm sides. The runway surface is composed of smooth plywood. The sides of the runway consist of 1.5 x 9 cm boards. All surfaces receive a minimum of two thorough coats of polyurethane, taking care that the edges and upper surface of the runway board are well coated. A continuous bead of silicone caulk is applied to the sides of the runway board, after which the sides are screwed onto the plywood runway board, creating a sealed trough that will provide several years of service. In addition, the wooden runway can be dried, sanded, and refinished for continued service. The upper end of the trough has multiple beads of caulk applied from side to side to form a 3 cm wall that prevents water from flowing over the upper end of the runway during operation.

A primary advantage of the plywood runway is that strips of nylon screen (5.0 x 25.0 cm) can be attached to the runway surface with disposable push pins (Fig. 3). This stable surface provides an excellent substrate for larval attachment. Removal of these strips permit collection of large numbers of larvae for research purposes. If larval collection is not a concern, other more durable materials could be used for the runway construction, as larval black flies will readily attach to most any surface (glass, stainless steel, Plexiglas®, plastic) that has a uniform flow over it. Attached to the end of the runway is a 20 x 40 cm piece of plastic sheeting. The sheeting is cut

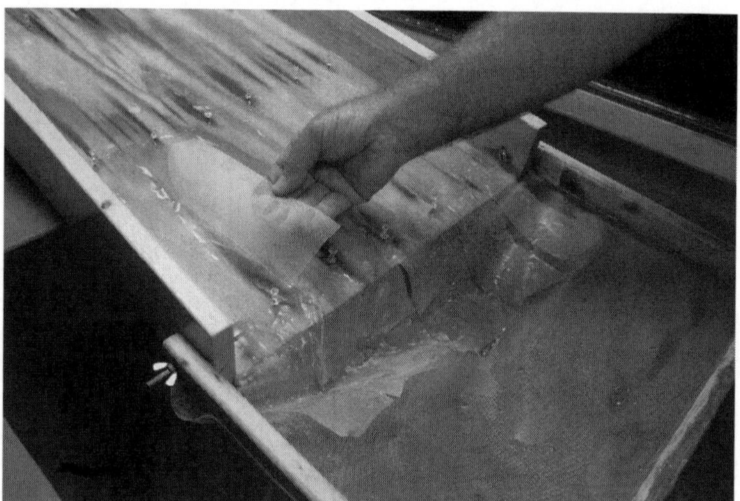

Fig. 3. Runway and screen larval substrate, end of runway "flagging" and catch-screen.

into strips to produce a "flagging" effect as shown in Fig. 3. This material extends the runway slightly and provides additional larval substrate. At the bottom of the runway, a small catch-screen (40 x 25 cm) is attached to 2.5 x 2.5 cm strips of wood that are screwed together to provide a rigid frame (Fig. 3). The catch-screen prevents loose push pins and other debris from being washed into the lower reservoir and drawn into the circulating pump. It is imperative to keep the catch-screen clean throughout the course of a rearing cycle. This practice will allow larvae that wash off the runway the opportunity to move through the screening and drop into the lower reservoir with the normal water flow. If the larvae are drawn to the upper chambers by the pump and water flow, they can be gently brushed back onto the runway with a clean, soft-bristled brush.

Rearing Unit—Upper Plexiglas® Section

With the spillway lip properly sloped and the silicone barrier in place, the upper end of the runway is sealed and all water will flow down the runway creating the larval habitat. The runway is suspended underneath the spillway lip with screw-eyes, chain and S-hooks. It is essential for early larval development that the screw-eyes are installed in the runway at nearly identical locations on each side of the runway and the chains and S-hooks are of identical length and size. Uniformity allows the runway to be suspended in a level manner, thereby allowing even water flow over the entire surface of the runway providing maximum larval habitat. At the down-slope end, the runway empties through a 64 x 71 cm opening cut through the upper Plexiglas® section the unit, into the lower reservoir.

Fig. 4. Water flowing from the spillway lip onto the runway, note caulk of barrier at upper end of runway.

Emergence Hood

Two or three d after observing pupae on the runway, an emergence hood is placed on top of the Plexiglas® portion of the rearing unit (Fig. 5). The emergence hood is constructed of 2.5 x 5 cm wood and # 6, 2.5 cm wood screws. Due to the small size of the framing material, it is best to pre-drill each screw and tighten them by hand. The base dimensions of the emergence hood are the same as the upper surface of the Plexiglas® section of the rearing unit (183 x 91 x 30 cm). In addition, there is a tower type area that rises 81 cm above the upper edge of the Plexiglas® section of the rearing unit. The tower area is centered above the chambers where the water flows onto the runway. It has a 23 cm square, framed structure, at the highest point that serves as an attachment site for the emergence funnel and the black cloth and plastic that covers the emergence hood. Braces run from the square at the apex of the tower to the opposite end of the emergence hood frame creating a tapered, tent-like effect. The outside of the emergence hood is covered with black plastic sheeting to help maintain a high level of humidity inside the hood, which enhances adult survival (Rutschke and Grunewald 1984). Weather stripping (high density foam, 1.3 x 1.9 cm) is applied to the bottom edge of the emergence hood to provide a tight seal to the rearing unit. Doors on each end of the emergence hood are also sealed with weather stripping (0.6 x 1.0 cm) and held closed with turn buttons or hook and eyes. All surfaces of the emergence hood frame are painted flat black prior to covering the frame with the black cloth and plastic.

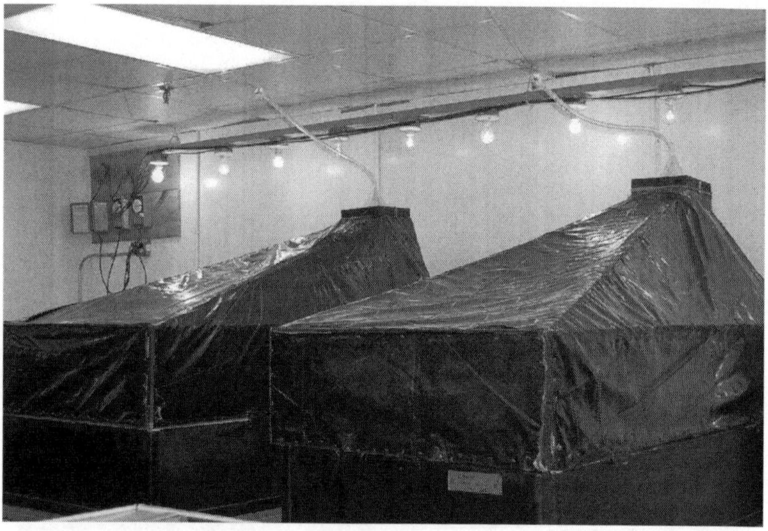

Fig. 5. Emergence hoods with the lighting system in the background.

Emergence Funnel

Because adult black flies are highly phototactic, collection of the emerging adults is relatively simple. The only light allowed to enter the emergence hood is through a clear, glass emergence funnel attached at the apex of the emergence hood tower (Fig. 6). The emergence funnel is constructed with a 147 mm diameter glass funnel with either a 58 or 60° angle bowl and a stem of 3 cm. Commonly available funnels typically come with longer stems, hence the stem should be cut to the desired length. The funnel is mounted between two pieces of 23 x 23 cm plywood (1.9 cm thick). The basal section has a 14.6 cm diameter hole centered in it. The upper section has a 14.0 cm diameter hole centered in it which is cut at an angle to accommodate the shape of the funnel. The funnel is mounted between these two pieces of plywood, which are held together with four # 8, 3.2 cm screws. Silicone caulk is applied around the base of the funnel to assure a tight seal between the two sections and the funnel. Attached to the stem of the emergence funnel is a 102 cm, 1.3 cm ID Tygon® tube with nylon screen clamped over the distal end. This tube is commonly called the "emergence tube" and is suspended so that the distal end is always up and the amount of curvature allowed in the tube is varied according to adult emergence rate and frequency of adult collection. When adult collections are only conducted once a day, the emergence tubes should be pulled out straight to minimize the amount of compaction and resulting mortality that can occur in the tube.

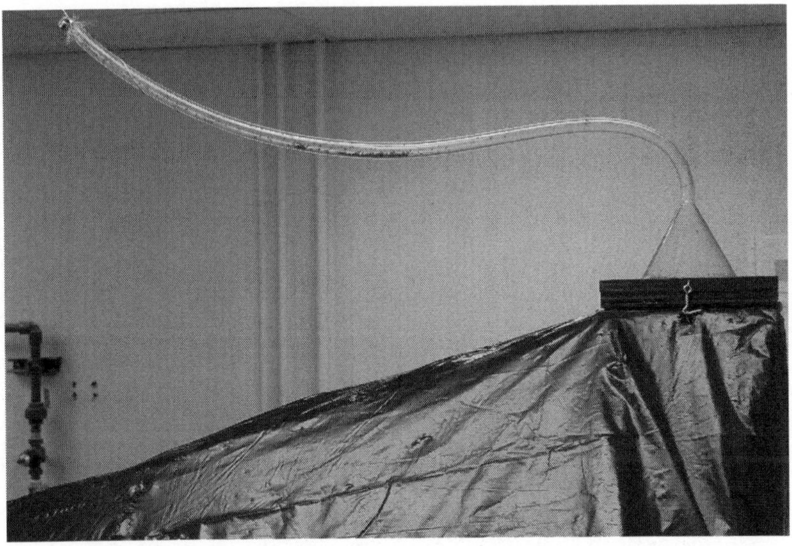

Fig. 6. Glass emergence funnel at the apex of the emergence hood with an adult tube attached.

Rearing Room Lighting

The emergence tube affixed to the funnel is positioned to attract as many flies as possible into the tube. Proper positioning of the emergence tube in relation to the room's lighting sources is critical to encouraging the largest number of flies to move quickly into the tube. Consequently, light placement and intensity are important. The colony is housed in essentially one large room (46 x 19′) with a partition in the middle. Each half of the room has eight recessed, 4-bulb, 34-watt (122 cm) fluorescent lights overhead and panel of eleven incandescent lights down one side of the room. The lighting and emergence tubes are positioned to facilitate phototactic behavior and efficient adult collection in the emergence tubes. All lights are on 24-hr timers (Dayton Electric Mfg., Chicago, IL) with a 16:8 (L:D) photo period and a 45-minute, three-phased crepuscular period to simulate dawn and dusk. In order to maximize adult collections and survival, the lighting system timers are set so that the overhead fluorescent lights go off first and come on last each morning. This lighting system is inexpensive when compared to commercially-available, crepuscular-lighting simulators. In addition, the overall light cycle is arranged so that the evening crepuscular period occurs between 1600–1730. Hence, few flies are encouraged into the emergence tubes after the late afternoon adult collection takes place. This schedule minimizes the length of time adults are confined in the emergence tube, thereby reducing adult mortality.

Mating and Adult Handling

Typically, adults will be emerging from 3 to 5 of the 9 rearing units in operation. Mating occurs in both the emergence tubes and a mating tube. Since adults are collected from all tanks from which they are emerging, the precise number of generations produced in our system each year is difficult to determine. Flies are typically collected from all emergence tubes containing flies at 0800, 1100 and 1600 hrs. When the adult flies are collected once daily, collection would ideally occur at approximately 1600 hrs. Adult collections are conducted every day of the year.

While it seems insignificant, the manner in which the emergence tube is suspended is important to adult survival, particularly on days when flies are collected only once. Under normal circumstances, emergence tubes are suspended to have a slight dip where the flies will eventually accumulate. This technique increases contact among the flies and likely improves mating as proposed by Raybould and Grunewald (1975). However, when adult emergence is high, or collections are made only once a day, the tube should be pulled out as straight as possible to reduce compaction and increase survival. Emergence tubes are suspended with coarse wire (18 gauge), which is attached to the ceiling over each emergence funnel. The coarse wire can be adjusted to provide the proper tube suspension depending on daily emergence rates.

After collection from the various rearing units, the flies are combined into a smaller, 1 cm ID Tygon® tube that is 45–60 cm in length to enhance mating and fertilization (Fig. 7). This tube also has nylon screening clamped over one end and is clamped closed on the other. Flies are enclosed in this mating tube at a density of 10–15 flies/cm (Fig. 8). The crowding of black flies in confined areas has been shown to increase mating behavior and fertilization rates (Simmons and Edman 1978, 1982, Bernardo 1986).

During the 45-minute mating period, the tube is gently shaken against a hard surface and air is drawn through it at least three times to insure the survival of the densely confined flies and to stimulate mating behavior. Simmons and Edman (1982) reported that "activity and movement are important to arouse males, since hitting the side of the cage to disturb the flies often resulted in a male clasping a female and attempting to mate". While this observation described behavior of members of the *S. damnosum* complex, the technique appears to be equally effective and important for *S. vittatum* IS-7. The emergence and mating tubes typically become fouled after 2–3 d of use. Tubes are cleaned by pulling a stiff bristle brush thru the tube after soaking the brush in a 1% detergent (Alconox®, Alconox Inc., White Plains, NY) solution. After cleaning, the tubes are rinsed with hot water and thoroughly drained to promote rapid drying.

After 45 minutes in the mating tube, the adults are placed in 473 ml, cylindrical paper cans (Neptune Paper Products Company, Inc., Newark, NJ). The paper cans are modified as adult holding containers by removing the center of the lid and replacing it with nylon screen (Fig. 9). The screen is

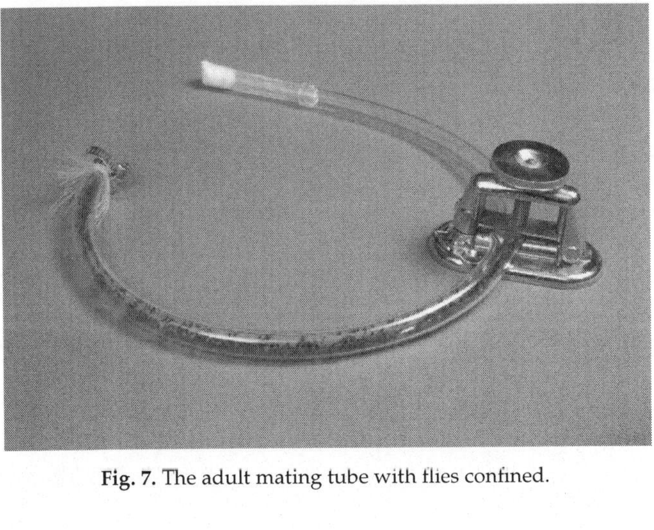

Fig. 7. The adult mating tube with flies confined.

Fig. 8. Close-up of black flies confined in mating tube.

Fig. 9. The adult storage container.

glued (Elmer's Glue-All, Elmer's Products Inc., Columbus, OH) to the rolled edge of the container lid. Four cotton balls (Fisherbrand Absorbent Cotton Balls-medium) are unrolled to create two cotton pads that are placed on the screened end of the can. One pad is moistened with a 10% sucrose solution and the other is moistened with distilled water. To prevent fly mortality, it is important that liquid neither drips nor condenses in the adult container. Waxed containers are not recommended due to condensation issues.

In addition to the screening, a 2.5 x 2.5 cm hole is cut in the side of the container. Two 3.5 x 3.5 cm pieces of dental dam (Colteen Whaledent, Cuyahoga Falls, OH.), each with 2.5 cm slits centered in them are secured to the container with duct tape. The two pieces of dental dam are secured in such a manner that the slits are 90° to each other. The dental dam is flexible enough to permit the adult mating tube to be inserted into the adult container, allowing the flies in the mating tube to be gently shaken into the container (Fig. 10). In theory, the dental dam would seal upon itself and prevent flies from escaping. To assure no flies escape the adult container, a cotton ball is usually placed over the dental dam and attached with a small (~5 cm) piece of tape.

After the flies have been gently transferred into the adult containers and the cotton pads moistened, the adult containers are placed in sealed-plastic containers and stored at 5°C until the following week. These conditions are similar to those suggested by both Rutschke and Grundewald (1984) and Ham and Fleming (1988) where high humidity and low temperatures were used to extend the survival time of adult black flies. In our system, the adults are typically stored for only one week. As previously mentioned, operational schedules and protocols were developed to ensure that large numbers of late instar larvae were available during the middle of the workweek. Consequently, adults are removed from refrigeration on Monday of each week, allowing egg development to resume in preparation for being placed in the oviposition chamber on Tuesday. Upon removal from cold storage, the cotton pads that are placed on the screened end of the adult container are checked for ample moisture. The pads should be moist enough that light finger pressure will squeeze excess moisture from the pad. If this is not the case, the pads are replenished with 10% sucrose solution. It is important that the flies have access to the sucrose solution when not under refrigeration.

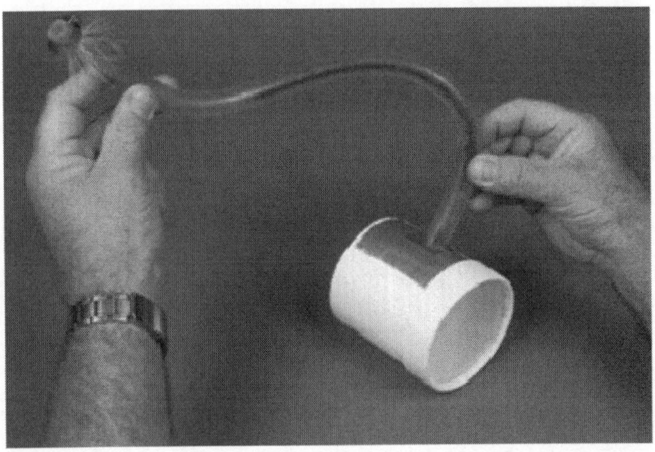

Fig. 10. Transferring adults from the mating tube to the adult storage container.

Oviposition

After cold storage, the adult flies are placed in a pre-oviposition chamber where they are warmed to room temperature. The pre-oviposition chamber consists of a 150-liter, plastic garbage can with a lid. Tap water (2–3 L) is placed in the chamber to create a humid environment to enhance adult survival. Foam insulation is placed in the bottom of the chamber to prevent the adult containers from coming in direct contact with the water. After approximately 24 hrs in the pre-oviposition chamber, the adults are ready for release into the insect cages of the ovipositon chamber. The time spent in the pre-oviposition chamber allows the flies to warm to 20–22°C, acquire a sucrose meal to maintain energy levels and resume egg development which was suppressed by cold storage. *Simulium vittatum* is autogenous and does not require a blood meal to produce their first batch of eggs. Consequently, we have not developed systems to conduct blood feeding. Cupp and Ramberg (1997) cover this topic thoroughly.

The oviposition chamber (Fig. 11) consists of a 37.8 liter aquarium with a 1.9 cm-plywood frame (65 x 36 x 41 cm) built around it. A description of the frame is as follows: two, 11.5-cm diameter holes are centered in the top of the frame. Nylon screen is tightly stapled over each of these holes and a 2.0 cm wall of silicone is formed around the edges. Off-set from each of these larger holes, a 3.8-cm diameter hole is drilled, 7.6 cm from the edge of the larger hole. This small hole provides access for tubing to provide water via a recirculation pump to the oviposition surface. A 46-cm fluorescent light is mounted flush with the back, bottom edge of the oviposition platform. Placing the light source below the oviposition platform causes the light to shine into the insect cage from below the egg-sheets, thereby taking advantage of the phototaxis of the adult flies to draw them to the egg-sheets. On the bottom of the oviposition platform, underneath each of the 11.5-cm holes, is a U-shaped funnel holder. The funnel holder is cut from 1.9-cm plywood so that a 147-mm diameter funnel with a 58 or 60° bowl is

Fig. 11. The oviposition chamber with one insect cage in place.

held tightly against the bottom of the oviposition platform. The U-shape allows the funnel to be slid into place and allows for easy disassembly for cleaning. Fitting the funnel tightly against the bottom of the oviposition platform insures all water that has been pumped over the egg sheets returns to the aquarium.

After the chamber is assembled, approximately 19 liters of tap water are added to the aquarium. A pump (Little Giant, Model P-AAA-WG, Oklahoma City, OK) is placed in the aquarium with a 1.27 cm female-to-male fitting attached to the pump outlet. The fitting allows a 1.11 cm ID Tygon® tube to be connected to the pump outlet with a hose clamp. The 10–12 cm piece of tubing attaches to a Y-fitting that splits the pump flow, providing water for each egg sheet. Attached to the distal end of the Y-fitting is a 10–12 cm piece of tubing which attaches to a plastic gate valve. Attached to the distal end of the gate valve is another 10–12 cm piece of tubing which attaches to a piece of glass tubing that is 7 mm OD and 10–12 cm long. The glass tubing is inserted through the center of a rubber stopper that fits in a 3.8 cm hole in the top of the oviposition deck. A hole is drilled through the center of the stopper that is slightly smaller than the tubing to make this fitting. A 7–8 cm piece of tubing is connected to the top of the glass tubing. A 10 mL glass pipette that has been heated and bent into a U-shape is then fitted into the other end of this tubing to complete the water delivery system for the oviposition chamber. All connections are held tightly in place with hose clamps. Constructed and adjusted properly, this system will deliver a constant trickle (75–125 ml/min.) of water to the center of the 11.5 cm hole.

The actual oviposition substrate is an 11.5 cm disc of pale green cloth (Appendix) that is placed on top of the screen that covers the 11.5 cm hole in the oviposition platform. On top of this disc is an 11.5 cm piece of 1.4 cm OD tubing which is also covered with the pale green cloth (Fig. 12). The combination of these two substrates provides an area simulating a small splash zone and areas wet with only a thin film of water. To prevent water from pooling on top of the egg-sheet itself, slits (2–3 cm) are made at 90° angles around the edge of each egg-sheet and at 90° angles in the center.

Eggs will be deposited on both pieces of cloth. Other species may prefer different colors of cloth (Cupp and Ramberg 1997). Once the water flow has been adjusted to a trickle onto the center of the cloth and tubing, an insect cage (30.5 x 61 cm) with weather stripping (high density foam, 1.3 x 1.9 cm) around its bottom edge is placed on the top of the oviposition deck to enclose the adults around the oviposition substrate (Fig. 11). Adults collected the previous week and enclosed in the preoviposition chamber for 24 hrs are then released into the insect cages. Upon releasing the flies into the insect cages, the entire apparatus is covered with black cloth to exclude light from the oviposition deck. The fluorescent light is turned on, lighting

Fig. 12. Egg-sheets after oviposition with "strings" of eggs. Dead flies have been removed to better view the eggs.

the oviposition deck from beneath and attracting the adults to the pieces of cloth (oviposition substrate). The system takes advantage of the black fly's strong phototactic response and the female's desire to lay eggs on green substrates (Peschken and Thorsteinson 1965, Golini and Davies 1987) bathed in a film of water. This system also operates in a constant stage of low light, thereby simulating twilight conditions, a period that has been reported conducive to black fly oviposition behavior by (Simmons and Edman 1980, 1982, Ham and Bianco 1983). The system is operated for 48 hrs, with the egg-sheets typically removed on Thursday afternoon. Because our objective is to rear as many larvae as possible each week, two oviposition chambers are operated. By providing more surface area for oviposition, we produce as many viable eggs as possible from each week's adult population.

After 48 hr in the oviposition chamber, a high percentage of the adults will have died. The egg-sheets are carefully removed from the oviposition deck and gently rinsed in the recirculating reservoir to remove loosely accumulated dead adults (Fig. 12). Dead flies on the egg-sheets represent a source of contamination to a new rearing. However, it should be noted that some dead adults will be entangled in the egg masses. After gently rinsing the egg-sheets, any areas that are void of eggs are carefully trimmed off with scissors. Egg-sheets are placed in Petri dishes and moistened with 5–10 mL of distilled water. Another 24 hr is allowed for egg development to continue and on Friday, five d after the oviposition process began, the egg-sheets are pinned to the top of the runway.

Water Temperature Regime

Water temperature during the early stages of each rearing is critical to the rate of larval development. Each rearing is initiated by filling the rearing unit to approximately 75% capacity with approximately 450 liters of 26–28°C tap water. After filling the rearing unit, the pump is turned on and the flow valve is opened to allow a high rate of flow for 30 minutes prior to "pinning" the egg-sheets on the runway. This period allows an initial volatilization of residual chlorine from the tap water. The use of warm water helps to stimulate egg hatching and early larval development. Because the rearing rooms are normally operated between 21–23°C, the rearing units are covered with a sheet of plastic for the first 4–6 d to help maintain an above ambient temperature condition for as long as possible. After 10–14 d, the compressor is turned on and the water temperature is gradually reduced to 15–16°C by day 17 of the rearing. The water temperature is maintained in this range for the remainder of the rearing.

Attaching the Egg-sheets

Upon initiation of each rearing cycle, the egg-sheets are carefully attached to the upper portion of the runway with disposable pushpins (Fig. 13). Proper attachment of the egg-sheets to the runway surface is a critical step in the rearing process. The initial flow used to moisten the eggs and to provide first instar larval habitat is approximately 2.5 L/min. This flow is maintained for the first 2–4 d. During this time it is important that the lip where the water originates is level, as is the runway. If either of these parts of the rearing unit is not level, the initial flow will be confined to only a small portion of the runway. As a result, inadequate moistening of the egg-sheets will occur and the amount of first instar habitat will be significantly

Fig. 13. Pinning new egg-sheets, notice low initial flow rate.

reduced. It is also important that the egg-sheets overlap slightly, with the upper portion of each successive egg-sheet being slightly overlapped by the sheet above it. This attachment technique will allow water flowing from the upper egg-sheet to continue flowing over each successive sheet. Water flow in the rearing unit is gradually increased after the first 2–4 d. Each day thereafter, the flow is gradually increased until by days 10–14 the flow is at full capacity for the rearing units (33.2 L/min.). Full capacity is reached when the water flow is as high as can be maintained without spilling over the back of the upper chamber.

Feeding

Feeding is initiated the first day that the egg-sheets are pinned to the runway. While many different types of food sources have been used to successfully rear black flies (Cupp and Ramberg 1997), a ratio of 1:1 of Purina Advanced Nutrition Rabbit Chow (18% minimum crude protein) and soybean meal (47% minimum crude protein) has proven very effective in this system. These two ingredients are added dry, in small volumes (200 cc) to a household blender and ground to a fine and uniform consistency. Approximately 100 cc of dry ground food is placed on a 53 μm sieve (USA Standard Testing Sieve No. 270) and washed with seven liters of tap water to produce a food slurry containing approximately 2.0 g of food/liter (Bernardo et al. 1986). The total volume of food slurry added to each rearing unit varies due to the number of larvae present and the age and condition of the rearing. On d 0–2, seven liters of the food slurry is added to the rearing unit. This material is poured directly into the lower reservoir through the catch screen at the end of the runway.

Microorganisms including nematodes, bdelloid rotifers, gastrotrichs, ciliated protozoa and bacteria (Bernardo et al. 1986) develop quickly during the early stages of each rearing. As a result, feeding is usually stopped for d 3–6 while larval growth occurs and the microorganisms developing on the larval substrates are consumed. By day 7, the larvae have typically reduced the microorganism populations, attached to the screen substrates and the runway itself is generally well established. If too much food is provided during the early stages of a rearing, a persistent film of microorganisms and food material will develop and the larvae will attach to this film rather than the screen substrate. When this occurs, the larvae are susceptible to washing off the runway when the film becomes dislodged.

On day 7, the automated feeding system is assembled and activated. This system consists of a rectangular (30.5 x 30.5 x 45.7 cm) polypropylene tank with a capacity of 41.6 L. This tank holds the food slurry and is refrigerated at 5°C to prevent spoilage of the food slurry (Fig. 14). Each food tank has two submersible pumps which keep the food slurry from

Fig. 14. Refrigerator with automated food system, adult storage containers and solutions for adult maintenance.

settling and pump the food into the rearing system. The stirring pump is larger (642.6 L/min capacity), and the feeding pump is a smaller version (453.6 L/min). Both pumps operate on ten minute timers with five second-trippers. The timers are adjusted so that the stirring pump is activated for 30 seconds and then the feeding pump comes on for a five-second burst. Attached to the feeding pump is a 0.95 cm ID Tygon® tube. This tube is approximately 100 cm long and runs from the food pump through the side of the refrigerator to the chambers at the head of the runway. The chambers adjacent to the spillway lip are covered with either plywood or Plexiglas® covers. Two small screw eyes inserted into these covers are used to anchor the food line in place. In addition, a jumbo hosecock clamp (Fisher Scientific) is attached to the chamber cover approximately 4 cm from the edge of the center chamber where water flows over the spillway lip. This hosecock clamp is used to regulate the amount of food being introduced into the rearing unit by the automated feeding system. Two important aspects of this portion of the rearing process involve not allowing the tip of the feeding tube to touch the water surface. If the end of the feeding tube becomes submersed in the tank water, a siphon will be formed and rearing unit water will be siphoned out of the rearing unit and back through the feeding system. It is also preferred that the feeding tube be positioned so the food is injected as far away from the spillway lip of the upper chamber as possible. This positioning allows the food material to mix with the water that is being pumped over the runway, thereby exposing more larvae to a more uniform food source.

The amount of food introduced into the rearing system varies with the number and size of the larvae present and the overall condition of the runway. High populations of microorganisms will require lower feeding levels as previously stated, until the microorganisms are reduced. Typically, 7–10 L of food slurry is introduced into the rearing system on d 6–10. After day 10, feeding is increased to 28 L of food slurry per day. This level of feeding is continued until a significant portion of the larvae have pupated. When a majority of larvae have pupated, the feeding rates can be reduced to 7–10 L of food slurry per day. However, most rearing units will have feeding systems in place throughout the entire rearing cycle. When feeding is discontinued, the automated feeding system is disassembled and thoroughly cleaned with hot, soapy water. Cleaning the feeding tubes requires forcing a soapy, stiff-bristled brush through the hose from each end and then rinsing the tube thoroughly with hot water.

Cleaning the Rearing Units

With nine rearing units in operation, an individual rearing typically develops for 60 d. After this period, the unit is drained and the larval substrates are removed. The runway is then scrubbed to remove the majority of the pupal cocoons and refilled with approximately 640 L of water and 3.8 L of 5.25% sodium hypochlorite (household bleach). The pump is turned on as soon as approximately 10 cm of water and the bleach have been added to the tank to expose all inner surfaces to a high concentration of the sanitizer. The pump is operated at full capacity for 24 hrs. The next day the unit is drained again, the Plexiglas® top removed and the entire inside of the water reservoir is scrubbed with Tilex® (The Clorox Company, Oakland, CA). All inner surfaces and the upper chambers of the Plexiglas® top are also cleaned thoroughly with Tilex®. The rearing unit is then reassembled and filled with water again and 11.3 L of household bleach are added. The rearing unit is again run at full capacity for 24 hrs. The unit is then drained and thoroughly rinsed with water. Upon completion of the final rinse the unit is vacuumed dry and covered with plastic sheeting in preparation for the next rearing.

Rearing Trouble Shooting

Rearing black flies on a large-scale, using this system, should only be attempted in a rearing room with floor drains. The possibility of water spills and water on the floor is constant and the consequences of large spills can be catastrophic in some situations. In addition, it is strongly recommended that two wet/dry vacuums be maintained as standard equipment, as water on the floor is inevitable. The wet/dry vacuums are very useful for routine cleaning and control of other arthropod pests including spiders, midges, roaches, etc.

The management and uniformity of a single room would be preferred for a rearing system as described herein. Unfortunately, securing nearly 1000 sq. ft. in such a configuration is difficult under most circumstances. As a result of the heat produced from all of the components (lights, compressors, pumps) necessary to operate this system, the cooling system associated with the rearing space is extremely important. The system should be dependable, easy to maintain and of proper capacity for the size of the room involved and the equipment contained therein.

The durability of the items described for this rearing system has been remarkable. The six original rearing units with which our colony was developed have been running nearly constantly for twenty years with minimal mechanical problems. Tank leakage has been our most common problem, typically arising during the cleaning cycle. Leaks are located and patched with fiberglass cloth and resin. The use of these components on the inside of the reservoir would be a significant upgrade from the epoxy resin alone that was used on the original units. After years of service, the circulation pumps will eventually fail. The pumps are relatively easy to replace with the Iwaki Model MD-30RT. As with the leaks, pump problems typically occur during the cleaning process when water levels are changing regularly. The most difficult aspect of rearing unit repair deals with the compressor and cooling system. Each unit has a compressor and cooling system which will eventually malfunction. At this point, a qualified HVAC technician will be required for repairs.

Although the Cornell Automated System has been described several times in the literature, no one has ever addressed the level of trained personnel required to maintain this system on a year-to-year basis. The University of Georgia laboratory typically operates in association with two full-time employees, two graduate students, and three or four part-time employees. While not all of these employees are working with the colony at one time, a group of cross-trained employees is extremely important as the colony is maintained 365 d a year. Mokry et al. (1981) touched on the topic when they stated that the rearing process "may prove too difficult and time-consuming for all but a few specialized research establishments". Unfortunately, this statement appears accurate as the University of Georgia black fly rearing and bioassay laboratory is currently a "one-of-a-kind" operation.

Bioassay Techniques

The eggs, larvae, pupae and adults produced in this system have all been used in research projects at one time or another. The larvae are the most commonly used stage in our laboratory. The orbital shaker bioassay developed at Clemson University (Figs. 15 and 16) has been the basis for a variety of subsequent experimental endeavors. One aspect of this work involved using larval black flies as environmental monitoring organisms. As a result, several studies were published using a longer exposure time and evaluating chemicals that could be found in streams as a result of urban runoff. Hyder et al. (2004) used a 24-hr orbital shaker toxicity test to demonstrate that the two cytospecies of *S. vittatum* exhibit equivalent susceptibility to the organophosphate-insecticide chlorpyrifos. Overmyer et al. (2002) used a 48-hr, orbital shaker acute toxicity test to evaluate the susceptibility of black flies to lawn-care insecticides. This study demonstrated the versatility of the orbital shaker, exposing larvae to three of the most

Fig. 15. An orbital shaker loaded with 250 ml flat-bottomed extraction flasks.

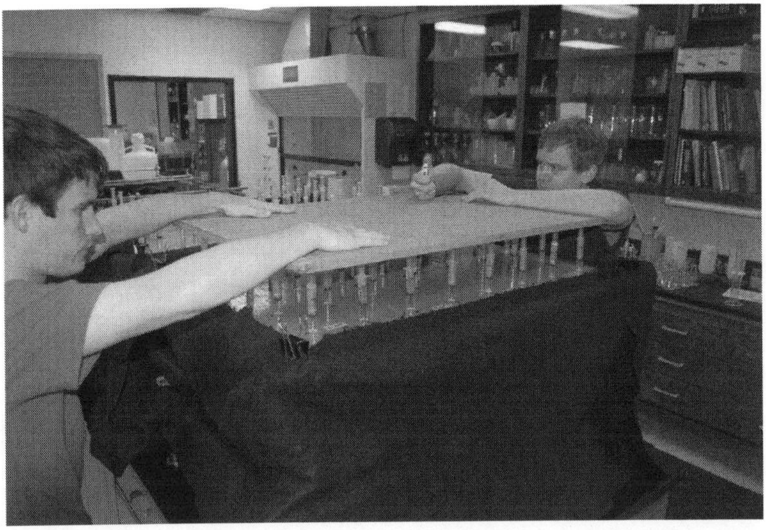

Fig. 16. Using dosing board to depress the plunger of forty syringes simultaneously and initiate a *Bti* exposure.

commonly used insecticides (carbaryl, chlorpyrifos, and malathion) for 48 hrs to determine the LC_{50} while producing little untreated control mortality. This work was followed by Overmyer and Noblet (2003) who demonstrated that different suspended solids in the larval media will significantly affect the bioavailability of various pesticides. The development of two new classes of pesticides, the phenylpyrazoles and the neonicotinoids, led to additional orbital shaker bioassay evaluations. Overmyer et al. (2005) determined that fipronil was more toxic than imidacloprid to colony larvae in a 48-hr orbital shaker bioassay. Overmyer et al. (2007) followed this work, demonstrating that the enantiomers of fipronil did not produce significantly different levels of mortality in a 48-hr orbital shaker bioassay using the colony larvae.

When evaluating the efficacy of *Bti* produced insecticide proteins, a shorter exposure time is more representative of operational black fly suppression treatments. In operational situations, the river is treated at a specified concentration over a specified time. As a result, the larvae are exposed to the insecticidal proteins for a short time as the proteins drift downstream past the larvae. Stephens et al. (2004) evaluated the effects of various algae species when mixed in the larval media of the bioassay on *Bti* induced mortality. This work demonstrated that various algae suspended in the larval media can reduce the activity of *Bti* on black fly larvae. Overmyer et al. (2006) demonstrated that by increasing the dose or the length of time the larvae are exposed to the *Bti* insecticidal proteins, mortality can be restored compared to a medium that has no algae present.

Recent experimentation conducted in our laboratory has focused on the effects of suspended solids in the larval habitat on *Bti* insecticidal protein induced mortality in larval black flies (Table 1). This data demonstrates how a change in *Bti* insecticidal protein concentration will affect larval mortality and the effect that suspended solids can have on this relationship in a 5-hr bioassay.

Iburg et al. (2010) described a controlled current toxicity test (CCTT) in which colony larvae were maintained in a larval media containing antibiotics for up 72 hrs. In these experiments, the larvae were maintained on gyrotory shakers for 43 and 67 hr, prior to being exposed to *Bti* insecticidal proteins for 10 minutes (Table 2). After the larvae were exposed to the insecticidal proteins the shakers were turned off, stopping the current in the flasks. The larvae were held in a stationary environment with no current, until mortality was evaluated at either 48 or 72 hr post-initiation. The results of these experiments demonstrated that anthropogenic concentrations of antibiotics in the larval media had no effect on *Bti* insecticidal protein toxicity

Table 1. Influence of 100 ppm ground organic matter on the activity of *Bti*-produced insecticidal proteins (IPs) on black fly larvae with IPs at 0, 0.375, 0.750, and 1.5 ppm.

***Bti* ICP concentration (ppm)**	**Larval mortality[1] (%)**	**Larval mortality[2] (%)**	**t-test re[3] P value**
0.000	2.1 ± 0.8^{a}	0.9 ± 0.6^{a}	0.2803
0.750	81.8 ± 2.4^{b}	26.3 ± 2.5^{b}	< 0.0001
1.500	86.3 ± 2.5^{bc}	42.0 ± 3.1^{c}	< 0.0001
3.000	93.4 ± 1.5^{c}	75.0 ± 2.5^{d}	< 0.0001

[1]All values in this column are the mean ± SEM of 10 replicate vessels containing 30 larvae each. The larval medium was deionized water.
[2]All values in this column are the mean± SEM of 10 replicate vessels containing 30 larvae each. The larval medium was deionized water contain 100 ppm of larval food.
[3]This analysis was conducted to compare the means in each row.
[a-c]Values with different superscripts within a column differ significantly ($P \leq 0.05$).

Table 2. The response of *S. vittatum* larvae to *Bacillus thuringiensis* subsp. *israelensis* (*Bti*) insecticidal proteins (IPs) after exposure to sulfamethoxazole (SMZ): trimethoprim (TMP) for 48 and 72 hrs. From: Iburg et al. (2010).

Treatment	**Mortality at 48 hrs (%)**	**Mortality at 72 hrs (%)**
Control	3.4 ± 0.8^{a}	2.5 ± 1.2^{a}
SMZ:TMP only[1]	5.0 ± 1.4^{a}	4.0 ± 1.0^{a}
Bti IP only[2]	75.5 ± 2.6^{b}	81.8 ± 5.7^{b}
SMZ:TMP [1] + *Bti* IP[2]	92.8 ± 2.3^{c}	93.5 ± 2.4^{c}

[a,b,c]Values (mean ± standard error of the mean) with different superscripts differ significantly within a column ($p \leq 0.05$).
[1]The concentration of SMZ:TMP in the larval medium was 25.0 mg/L:5.0 mg/L.
[2]The concentration of *Bti* IPs was 1.08 ITU/200 ml.

to larval black flies. To the contrary, exposing the larvae to the mixture of sulfamethoxazole and trimethoprim actually increased the effectiveness of the *Bti* insecticidal proteins in this case. This change was likely related to the suppression of folic acid synthesis by the larvae, thereby rendering the larvae more susceptible to *Bti* IP toxicity. This finding indicates that insect nutrition studies could be another option for the CCTT technique.

The development and maintenance of the black fly colony negates the need to obtain eggs and larvae from field sites, thereby significantly reducing the potential for variability within experiments. In addition, the colony allows experimentation to be scheduled as needed to meet the goals of the researchers. The aforementioned experiments demonstrate the tremendous scientific resource the University of Georgia black fly colony provides and the wide variety of experiments that can be conducted with this resource.

Conclusion

Black flies continue to be one of the world's most important groups of insect pests. As a result, the need to continue to learn more about their biology and control is quite apparent. The University of Georgia black fly colony has entered its 29th year, striving to do just that. With diligent oversight and attention to detail, we hope to use this resource to provide specimens for a wide variety of research projects for years to come.

The three most important factors affecting larval development in this system appear to be larval number, food availability and water temperature. Colbo and Porter (1979) reported that if suitable current is provided and water purity is maintained that these are the two main factors regulating black fly development. These researchers also reported that food deprivation resulted in loss of synchronous development. After many generations in our laboratories, we are confident that this occurs. While our system continues to produce large, robust larvae and adults, the length of development is variable and time to 50% emergence is extended as predicted by Bernardo et al. (1986) for high larval levels in this system. While large volumes (>28 L) of food slurry are added to this system, the amount of food actually available to the larvae may not be sufficient for some rearings. During rearings with *Simulium vittatum* IIIL-1, runways estimated to support 70,000 larvae exhibited extended developmental periods.

Over the past few years, the capacity of our rearing system has been pushed further than we had ever imagined possible. Today we routinely produce 250,000–300,000 larvae per rearing on approximately 5,040 cm^2 of runway surface. As a result of these large rearings, we regularly add more food material than reported in previous publications. However, due to the number of larvae, our system is still offering significantly less food/larvae (0.19–0.22 mg/larvae/day) as compared to Bernardo et al. (1986) who proposed maximum developmental conditions when larvae were fed 4 mg/day. One factor offsetting this difference is that our food solution is higher in protein than that reported by Bernardo et al. (1986) due to higher levels of protein in our soybean meal (47% as opposed to 34%). Hudson (1994), found larvae reared in this system have large fat reserves and developed through the normal seven instars for *S. vittatum*, thereby indicating a quality diet. However, some of our largest rearings, which approach 350,000 larvae, are probably on the verge of overloading this system. Colbo and Thompson (1978) reported that emerging adults excrete stored waste products, which can, in turn, inhibit the development of remaining larvae. Grunwald (1978) reported that a large volume of water per larvae is required for closed system laboratory rearing. The smallest volume/larvae that our system operates at is approximately 2.0 ml/larvae with no observable ill effects. On numerous occasions, rearing units have experienced reduced water levels for a variety of reasons and no ill effects were observed by the increased concentration of waste materials. Therefore, we believe that adequate food supply is our most important limiting factor. Whether it is food deprivation or excess nitrogenous wastes, our largest rearings have prolonged emergence spread over several weeks. Conversely, larvae reared at lower densities with fewer larvae always develop faster and emergence is more synchronous. Along these lines, we now often reduce larval density on d 5–10 in an effort to increase the rate of development for the remaining larvae. As early as 1966, Muirhead-Thompson discussed the fact that successful black fly rearing appeared to be related to the "meticulous care in providing measured quantities of a suitable food". Based on our experience, this fact appears to hold true over 45 years later.

Because the colony has been in operation since 1981, the question of inbreeding often arises. While this is definitely a concern for any prolonged colonization of a species in the laboratory, the fact that substantial numbers of robust larvae and adults are still being produced on a routine basis indicates that the genetic diversity has been preserved up to this point. Raybould and Boakye (1986) stated that large scale rearing is required to avoid inbreeding. The sheer numbers of specimens being produced in this system likely minimizes the rate of inbreeding. However, Brockhouse and Adler (2002) determined that while all sex chromosomes and major autosomal polymorphisms were still present in the colony specimens, some changes in the extent of sex linkage and frequency of autosomal inversions has occurred. As related to the primary purpose of the colony, comparisons between field and laboratory testing have demonstrated that the colony material continues to be very useful in predicting the effectiveness of *Bti* formulations in the field. Larval feeding related experiments have also proven instrumental in understanding the many complex issues related to *Bti* induced mortality and suspended solids in the lotic environment.

Edman and Simmons (1985) reported that genetic diversity can be reintroduced into a colony by introducing wild females which will mate with laboratory reared males more readily than laboratory females with wild males. While this has been considered for our colony, the introduction of natural pathogens into a system that is working very effectively at achieving its objective is a serious concern. Specimens from this system continue to be critically important in the development of improved *Bti* formulations and experimentation related to environmental contaminants, larval feeding and oviposition behavior. Consequently, as long as bioassays and experimentation is consistent and repeatable, major changes in the system are not anticipated. It should be stressed that the area of inbreeding in laboratory reared black flies is relatively unstudied and ripe for further basic research.

K.W. Cummings (1986) stated of black flies "we can't kill them and we can't grow them". As a result of the work of many individuals, the prior statement is incorrect on both fronts. The University Georgia black fly colony is long-standing proof that, with careful attention to details, black flies can be cultured successfully. Furthermore, the use of laboratory-reared larvae can be a valuable tool, both in formulation development and in various bioassessments. Effective black fly control programs ranging in scale from the former WHO onchocerciasis program to localized suppression programs of various sizes are now routinely conducted worldwide. Black fly population suppression is now a realistic goal and laboratory reared black flies have played a key role in achieving this reality.

Acknowledgements

We would like to thank Valent BioSciences Corporation and formerly Abbott Laboratories for their financial support of our research and for many years of collaborative support from their research staff. This unique University-Industry partnership has resulted in global advances in vector control programs that could not have occurred otherwise. We would also like to thank the many staff, graduate and undergraduate students who have worked in our laboratory maintaining and operating the colony and the bioassays systems. Our lives are all richer because of these enthusiastic and vibrant young people. We wish them all the very best! Lastly we want to thank Mr. Joe Iburg for his diligent efforts in maintaining the colony in recent years and Dr. Roger Wyatt for his scientific and editorial contributions to our laboratory group.

Appendix*

Advanced Nutrition Rabbit Feed (18% Protein), Purina Mills, Inc., P.O. Box 66812, St. Louis, MO 63166-6812

Dayno Aqua-Lab, Models 503 (302 L.) and 703 (643 L.) Dayno Manufacturing, Inc., 165 Commercial St., Lynn, MA 01905 (617-598-5566)

Dayton Time Switch, Model 2E024B (24 hr timer) and Model 2E355A (10 minute timer) Dayton Electric Mfg. Co. Chicago, IL 60648

Dental Dam, Catalog #H00540 (dark colored, heavy gauge) Colteen Whaledent, 750 Corporate Dr., Mahwah, NJ 07430 (800-221-3046)

Eggsheet cloth, Symphony Mint, 65% poly, 35% cotton blend, Jo-Ann's Fabric Shops

Fisherbrand Absorbent Cotton Balls (Medium), Fisher Scientific Company, Houston, TX 77038

Intermatic Repeat Cycle Timer, Model C8815 (10 minute timer), Intermatic Incorporated, Spring Grove, IL 60081

Iwaki Magnetic Pump, Model MD-30RT-115NL, Iwaki Co., LTD Tokyo, Japan, available through Cole-Parmer (www.coleparmer.com)

Little Giant Pump, Model P-AAA (food and oviposition pump), Model 2E-38N (food stirring pump), Little Giant Pump Co., Oklahoma City, Oklahoma.

Nylon Screen, Nitex Brand, 600 micron, 245 micron thread diameter, 50.5% open area, Small Parts, 15901 SW 29th Street, Suite 201, Miramar, FL 33027 (800-220-4242)

Paper cans, pint (adult containers), All Purpose, Liquid Tight Paper Food Containers, Neptune Paper Products inc., Newark, NJ 07102, sold by SSWLE Corporation, 168 W86th St, New York, NY 10024, (917-575-2788)

Soybean meal (47% protein), supplied by the University of Georgia poultry feed mill, also available through Gold Kist Feed, Gold Kist Inc., Atlanta, GA

Tilex Instant Mildew Remover, The Chlorox Company, Oakland, CA 94612

Possible Sources of Aquatic Rearing Units

Aqua Center, Leland MS (800-748-8921)
Aquatic Ecosystems, Inc., Apopca, FL (800-422-3939)
Aquaculture Supply, Dade City, FL (352-567-8540)
Eagar Aquaculture, North Salt Lake City, UT (800-423-6249)

References

Barton, W.E., R. Noblet and D.C. Kurtak. 1991. A simple technique for determining relative toxicities of *Bacillus thuringiensis* var. *israelensis* formulations against larval black flies (Diptera: Simuliidae). J. Am. Mosq. Control Assoc. 7: 313–315.

Bernardo, M.J., E.W. Cupp and A.E. Kiszewski. 1986. Rearing black flies (Diptera: Simuliidae) in the laboratory: Colonization and life table statistics for *Simulium vittatum*. Ann. Entomol. Soc. Am. 59: 610–621.

Bradley, G.H. 1935. The hatching of eggs of the southern buffalo gnat. Science 82: 277–278.

Brenner, R.J. and E.W. Cupp. 1980. Rearing black flies (Diptera: Simuliidae) in a closed system of water circulation. Tropenmed. Parasitol. 31: 247–258.

Brockhouse, C.L. and P.H. Adler. 2002. Cytogenetics of laboratory colonies of *Simulium vittatum* cytospecies IS-7 (Diptera: Simuliidae). J. Med. Ent. 39: 293–297.

Colbo, M.H. and G.N. Porter. 1979. Effects of the food supply on the life history of Simuliidae (Diptera). Can. J. Zool. 57: 301–306.

Colbo, M.H. and B.H. Thompson. 1978. An efficient technique for laboratory rearing of *Simulium verecundum* S. and J. (Diptera: Simuliidae). Can. J. Zool. 56: 507–510.

Cupp, E.W. and F.B. Ramberg. 1997. Care and maintenance of black fly colonies. pp. 31–40. *In*: Crampton, Beard and Louis (eds.). The Molecular Biology of Insect Disease Vectors.

Dalmat, H.T. 1955. The black flies (Diptera: Simuliidae) of Guatemala and their role as vectors of onchocerciasis. Smithsonian Misc. Coll. 125: 1–425.

Edman, J.D. and K.R. Simmons. 1985. Rearing and colonization of black flies (Diptera: Simuliidae). J. Med. Entomol. 22: 1–17.

Edman, J.D. and K.R. Simmons. 1987. Maintaining black flies in the laboratory. pp. 305–314. *In*: Kim and Merritt (eds.). Black Flies: Ecology, Population Management and Annotated World List. 528 pp. The Pennsylvania State University, University Park, United States.

Emery, W.T. 1913. Morphology and biology of *Simulium vittatum* and its distribution in Kansas. Kansas Univ. Sci. Bull. 8: 323–362.

Fredeen, F.J.H. 1959. Rearing black flies in the laboratory (Diptera: Simuliidae). Can. Entomol. 91: 73–83.

Golini, V.I. and D.M. Davies. 1987. Oviposition of black flies. pp. 261–275. *In*: Kim and Merritt (eds.). Black Flies: Ecology, Population Management and Annotated World List. 528 pp. The Pennsylvania State University, University Park, United States.

Grunewald, J. 1978. Die bedeutung der stickstoff-exkretion and ammoniak-empfindlichkeit von Simuliiden-Larven (Diptera) fur den Aufbau von Laboratoriumskulturen. Z. Angew Entomol. 85: 52–60.

Gray, E.W. 2010. The University of Georgia black fly colony. Wing Beats 21: 22–25.

Hall, R.E. and J.J. Harrod. 1963. A method of rearing *Simulium ornatum* var. *nitidifrons* (Diptera: Simuliidae) in the laboratory. Hydrobiologia 22: 197–201.

Ham, P.J. and A.E. Bianco. 1983. Maintenance of *Simulium Wilhelmia lineatum* Meigen and *Simulium erythrocephalum* de Geer through successive generations in the laboratory. Can. J. Zool. 62: 870–877.

Ham, P.J. and S.D. Fleming. 1988. A method for long-term storage of living black flies at 4°C. Annals of Tropical Medicine and Parasitology 82: 319–320.

Hudson, S.B. 1994. Factors affecting the survival of black fly larvae (Diptera: Simuliidae) when exposed to *Bacillus thuringiensis* serovar *israelensis* in the orbital shaker bioassay. M.S. thesis, Clemson University, South Carolina.

Hyder, A.H., J.P. Overmyer and R. Noblet. 2004. Influence of developmental stage on susceptibilities and sensitivities of *Simulium vittatum* IS-7 and *Simulium vittatum* IIIL-1 (Diptera: Simuliidae) to chlorpyrifos. J. Env. Toxicol. Chem. 23: 2856–2862.

Iburg, J.P., E.W. Gray, R.D. Wyatt and R. Noblet. 2010. Influence of selected antibiotics on the response of black fly (*Simulium vittatum*) larvae to insecticidal proteins produced by *Bacillus thuringiensis* subsp. *israelensis*. J. Env. Toxicol. Chem. 29: 1849–1853.

Mokry, J.E., M.H. Colbo and B.H. Thompson. 1981. Laboratory colonization of black flies. pp. 299–307. *In*: M. Laird (ed.). Black flies. The Future of Biological Methods in Integrated Control. Academic Press, New York, United States.

Muirhead-Thompson, R.C. 1966. Black flies. pp. 127–144. *In*: C.N. Smith (ed.). Insect Colonization and Mass Production. Academic Press, New York, United States.

Overmyer, J.P., K.L. Armbrust and R. Noblet. 2002. Susceptibility of black fly larvae (Diptera: Simuliidae) to lawn-care insecticides individually and as mixtures. J. Env. Toxicol. Chem. 22: 1582–1588.

Overmyer, J.P., B.N. Mason and K.L. Armbrust. 2005. Acute toxicity of imidacloprid and fipronil to a nontarget aquatic insect, *Simulium vittatum* Zetterstedt cytospecies IS-7. Bull. Environ. Contam. Toxicol. 74: 872–879.

Overmyer, J.P. and R. Noblet. 2003. Influences of a laboratory diet and natural seston on the bioavailability of carbaryl, chlorpyrifos, and malathion to black fly larvae (Diptera: Simuliidae) in an acute toxicity test. Arch. Environ. Contam. Toxicol. 45: 209–215.

Overmyer, J.P., D.R. Rouse, J.K. Avants, A.W. Garrison, M.E. Delorenzo, K.W. Chung, P.B. Key, W.A. Wilson and M.C. Black. 2007. Toxicity of fipronil and its enantiomers to marine and freshwater non-targets. J. Environ. Sci. Health. 42: 471–480.

Overmyer, J.P., M.S. Stephens, E.W. Gray and R. Noblet. 2006. Mitigating the effects of the green alga *Scenedesmus quadricauda* on the efficacy of *Bacillus thuringiensis* var. *israelensis* against larval black flies. J. Amer. Mosq. Control Assoc. 22: 135–139.

Peschken, D.P. and A.J. Thorsteinson. 1965. Visual orientation of black flies (Simuliidae: Diptera) to colour, shape and movement of targets. Entomol. Exp. Appl. 8: 282–288.

Puri, I.M. 1925. On the life history and structure of the early stages of Simuliidae (Diptera, Nematocera). Parasitology 17: 295–369.

Raybould, J.N. 1981. Present progress towards the laboratory colonization of members of the *Simulium damnosum* Theobald complex. pp. 307–318. *In*: M. Laird (ed.). Black flies. The Future of Biological Methods in Integrated Control. Academic Press, New York, United States.

Raybould, J.N. and D.A. Boakye. 1986. Temporary small-scale colonization of the Beffa form of the *Simulium damnosum* Theobald complex (Diptera: Simuliidae) in Africa. J. Am. Mosq. Control Assoc. 2: 14–17.

Raybould, J.N. and J. Grunewald. 1975. Present progress towards the laboratory colonization of African Simuliidae (Diptera). Tropenmed. Parasitol. 27: 155–168.

Rutschke, J. and J. Grunwald. 1984. A simple apparatus for maintaining black fly adults (Simuliidae) in the laboratory. Mosquito News 44: 461–465.

Simmons, K.R. and J.D. Edman. 1978. Successful mating, oviposition and complete generation rearing of the multivoltine black fly *Simulium decorum* (Diptera: Simuliidae) in the laboratory. Can. J. Zool. 56: 1223–1225.

Simmons, K.R. and J.D. Edman. 1982. Laboratory colonization of the human onchocerciasis vector *Simulium damnosum* complex (Diptera: Simuliidae), using an enclosed, gravity-trough rearing system. J. Med. Entomol. 19: 117–126.

Stephens, M.S., J.P. Overmyer, E.W. Gray and R. Noblet. 2004. Effects of algae on the efficacy of *Bacillus thuringiensis* var. *israelensis* against larval black flies. J. Am. Mosq. Control Assoc. 20: 171–175.

Tarrant, C.A., G. Scoles and E.W. Cupp. 1987. Techniques for inducing oviposition in *Simulium vittatum* (Diptera: Simuliidae) and for rearing sibling cohorts of Simuliids. J. Med. Entomol. 24: 694–699.

3

Maintenance of Ticks in the Laboratory

Daniel E. Sonenshine

ABSTRACT

Methods for breeding and maintaining colonies of ticks in the laboratory for biological investigations are described. This chapter reviews the types of laboratory facilities that are needed for breeding ticks. Techniques for feeding ticks on different types of small and medium-sized mammals and/or birds are described. Procedures for restraining and/or containing small laboratory animals also are noted. Finally, methods for storing non-feeding ticks for ecdysis, oviposition or long term survival are described. Selected figures are provided to illustrate the procedures or equipment used in the tick breeding/maintenance program. Attention also is given to the need for strict adherence to local and national regulations concerning animal welfare. Finally, the dangers inherent in rearing ticks that can bite humans are discussed and containment procedures for minimizing such risks are noted.

Key Words: Tick rearing, hard ticks, soft ticks, *Ixodidae, Dermacentor variabilis, I. scapularis, Aragasidae, Ornithodoros, O. tholozani, O. turicata, O. parkeri, Boophilus microplus, B. annulatus,* blood feeding

Department of Biological Sciences, Old Dominion University, Norfolk, VA 23529, USA.

Introduction

Ticks are notorious as vectors of disease-causing agents to humans and animals. Ticks transmit numerous protozoan, viral, bacterial (including rickettsial) and fungal pathogens. For these reasons, it is no surprise that ticks are the subject of numerous scientific investigations, including studies in the laboratory on their biology, morphology, and ability to transmit pathogenic organisms. In order to study ticks in the laboratory, the researchers must be acquainted with the methods used for rearing and maintaining ticks. This chapter addresses the various procedures, facilities and equipment needed for the creation of colonies of different tick species. Procedures and facilities for colonizing species of either hard ticks (Family *Ixodidae*) and soft ticks (Family *Argasidae*) are described. This chapter also addresses the importance of proper animal care, in accordance with local and national animal welfare regulations. For a more detailed review of the literature on tick rearing methods, tick biology and its role in developing rearing and maintenance methods, the reader should consult Biology of Ticks, Volumes 1 and 2 (Sonenshine 1993).

Laboratory Facilities for Rearing Ticks

Ticks are dangerous and should be treated accordingly. Before bringing living ticks into one's laboratory, a suitable area should be designated for the tick colony. This area should be dedicated to the rearing and colony maintenance effort, i.e., no unrelated activities should be allowed in this space. If a small colony of a single species and only intermittent feeding is planned, a corner of the laboratory may be sufficient. A small workbench will be needed for periodic feeding of the ticks on small laboratory animals. If large colonies and/or several tick species will be needed, a dedicated room or rooms with specialized facilities should be planned. This should include one or more dedicated workbenches for infesting animals with ticks, a supply cabinet and an incubator for maintaining the ticks during their lengthy non-feeding periods. Incubators are expensive and considerable thought should be given to obtaining the optimum size and capabilities needed for the facility. Each incubator must be able to maintain optimum humidity and temperature. Ideally, the facility should have two incubators, in order to hold some ticks in a low temperature incubator (e.g., ca. 10°C) for long term storage versus others in a warm incubator (e.g., ca. 25°C) for molting or oviposition.

Feeding Ticks in the Laboratory

Ticks must consume blood for their development and reproduction. Normally, this is done by exposing the ticks to an animal host for attachment and blood feeding. There is no single method for feeding all types of ticks; rather, methods vary from species to species as well as between investigators and laboratories. Before beginning any studies, the investigator must determine his or her institutional policy for assuring humane care of laboratory animals. In the United States, some species of laboratory animals, e.g., rabbits, are subject to inspection by personnel of the United States Department of Agriculture, to insure that activities are in accordance with animal welfare regulations. Many institutions, e.g., universities, hospitals, or research organizations, have an Institutional Animal Care and Use Committee (IACUC) which establishes policies and controls the use of animals. At such institutions, qualified investigators must submit a "protocol" to the committee for review and approval before they can commence using animals for feeding ticks. Such protocols must include a description of their procedures, including restraints, tranquilizers, anesthetics and, if appropriate, euthanasia, as well as justifying the need for the particular vertebrate species and number required. The protocol should also describe precautions that (1) minimize the risk of accidentally infesting the investigator as well as his/her technicians and/or students while feeding ticks; and (2) preventing escape of ticks (few situations are more embarrassing for the tick breeder than complaints about escaping human-biting ticks!). If known, human-biting ticks are being used, the investigator and/or laboratory personnel should wear protective clothing. A white lab coat or white coveralls and disposable plastic ("surgical") gloves should be worn when handling such ticks. For additional personal protection, double-sided tape can be placed over the wrists and cuffs if a long-sleeved garment is worn. Similarly, the trouser legs should be tucked into the socks and taped around the legs to prevent escaped ticks from entering the clothing from the laboratory floor.

Infestations of animals with ticks should be done on a clean, white laboratory bench free of clutter. The surface should be smooth, i.e., free of cracks, crevices or other places where ticks can hide. Strong plastic sheets (e.g., Formica^{Tm}) or stainless steel provides a suitable surface for this purpose. This will make it easier to observe any ticks crawling away from the host before they can escape into the laboratory environment. Additional precautions may be required when working with disease-carrying ticks, e.g., specially designed infestation tables surrounded by chemically treated moats and metal walls. This will prevent loss of ticks in the event of unexpected, violent host behavior. If work is planned with highly infectious microbes, the investigator must first consult with experts knowledgeable in the design of the air flow systems, filters and controlled access (e.g., Biosafety level-3 facilities) needed to meet stringent government regulations for such facilities.

Feeding Ixodid Ticks

Ixodid ticks feed slowly, requiring several d or even weeks to complete their blood meal. Moreover, many ixodid ticks are specific for certain kinds of vertebrate hosts, e.g., larvae and nymphs of the American dog tick, *Dermacentor variabilis* feed readily on rats (*Rattus norvegicus*) but feed poorly on laboratory mice (*Musmusculus*). In contrast, larvae of the lone star tick, *Amblyomma americanum*, feed poorly on rats or mice but will feed readily on young chickens. Rearing methods are specific for each tick species, especially in the choice of hosts. Thus, knowledge of the preferred hosts and feeding habits of the ticks to be reared is essential for developing a successful colony. Persons planning to rear ticks should consult with a specialist before proceeding with the rearing program.

To assist the novice investigator in developing methods for rearing ixodid ticks, several examples follow:

Feeding Dermacentor variabilis. For this species, I use the methods described in previous publications (Sonenshine 1968, Sonenshine et al. 1976). *Dermacentor variabilis* requires a combination of hosts to feed the different life stages. Rats, guinea-pigs or similar-sized small rodents are used for the immatures, while rabbits, guinea pigs or, if available, even larger hosts (e.g., dogs) are used for the adults. Larvae feed best if they are exposed to daylight or intense artificial light prior to releasing them on a host. This step is important if the ticks were kept in a dark incubator or in dim light since *D. variabilis* larvae will diapause under such conditions. The larvae should be protected from desiccating by confining them in a humid container. The exposure period should be increased gradually over several d. Following illumination, the larvae are released onto the rats. To minimize grooming and enhance attachment, the rats are restrained by enclosing them in narrow, cylinder-shaped wire screen cages (Fig. 1). Vegetable foods, e.g., sliced carrots or celery strips may be used to distract the animals and occupy their attention while the ticks attach. Larvae are released onto the head and neck of each rat with the aid of a camel's hairbrush (Fig. 2). Following their release, the cage containing the rat is wrapped with paper or cloth towels (to minimize tick escape), secured with rubber bands and left for 2–3 hr to allow the ticks time to attach. Following attachment, the towels are removed. Paper towels may be discarded in a secure waste can; cloth

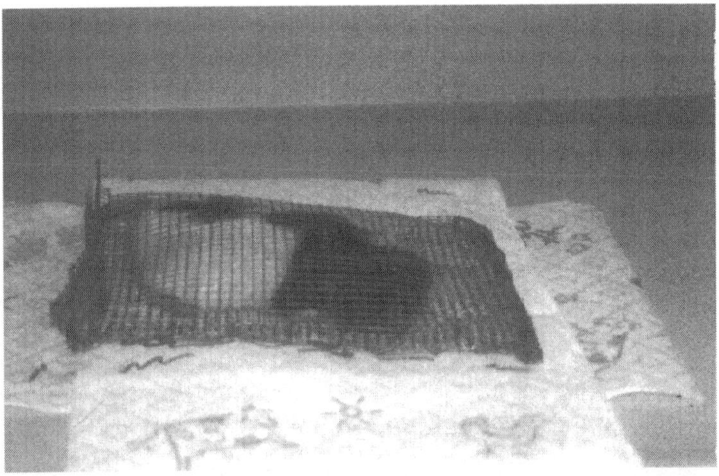

Fig. 1. Rat confined in funnel-shaped, wire screen (1/4 inch wire mesh) cage in preparation for infestation with larvae. One end of the funnel is opened, the rats are allowed to walk or run into the cage, and the end is closed by hand or with a pair of pliers.

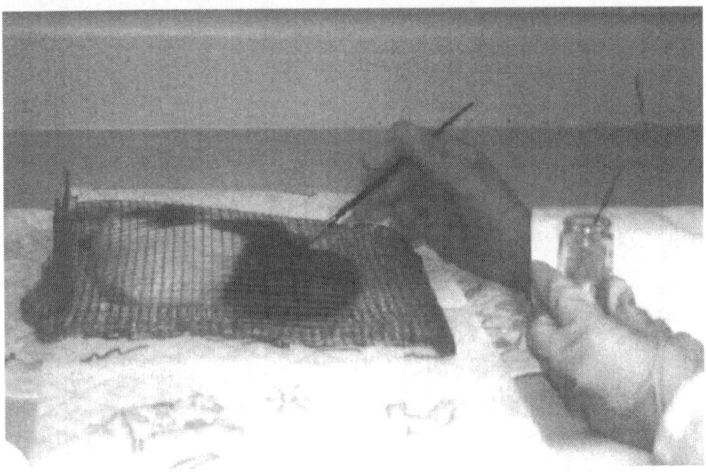

Fig. 2. Technician infesting a caged rat with larvae, using a camel's hair brush.

towels can be sealed in a plastic bag, washed with a mild, insecticidal soap, dried and stored for reuse. The tick-infested rats are released and returned to their standard cages for housing during the feeding period. The ticks are allowed to feed while the animals are confined in their standard cages (Fig. 3). To collect the engorged ticks after feeding, the cage floor should contain a litter pan separated from the animal by a wire screen grate. This will allow fed ticks to drop from the host and fall through the grate.

Although commercial litter may be used during the first 2 d after tick attachment, this must be replaced with white paper towels to allow

Fig. 3. Tick-infested rats confined in hanging cages over waste trays covered with paper towels. Engorged larvae or nymphs drop from the host, collect on the paper towels, and are removed with an aspirator. Double-side tape around the edges of the tray prevents tick escape. Reprinted with the kind permission of Oxford University Press.

collection of the engorged larvae after they fall from their hosts. Double-sided tape should be placed around the edges of the waste tray to prevent tick escape. Collection is done with an aspirator. The aspirator is made from a vial fitted with a rubber or cork stopper containing two openings. Glass tubes are inserted into the openings. One glass tube is connected by plastic tubing to a vacuum pump, while the other is connected to a short plastic tube that serves as the suction tube to collect the ticks (Fig. 4). Although the uppermost paper towels will be contaminated with host urine and feces, the sheets below will be clean, facilitating recovery of the engorged larvae. Since the fed ticks seek shelter after dropping from their hosts, most specimens will be found on the clean under sheets and metal surface of the waste tray. The contaminated uppermost sheets should be discarded, but the lower sheets often may be reused. Larvae trapped on the tape can also be collected, but this must be done carefully with fine forceps to avoid puncturing their delicate bodies. Collections should be made frequently, e.g., twice/d, so as to collect the larvae before they become trapped in the tape.

Some workers prefer water-filled trays instead of paper to trap engorge larvae. This method captures all of the detached ticks and there is no danger of losing specimens buried under fecal contamination or glued to tape. However, it is much more labor intensive since the tiny ticks must be picked

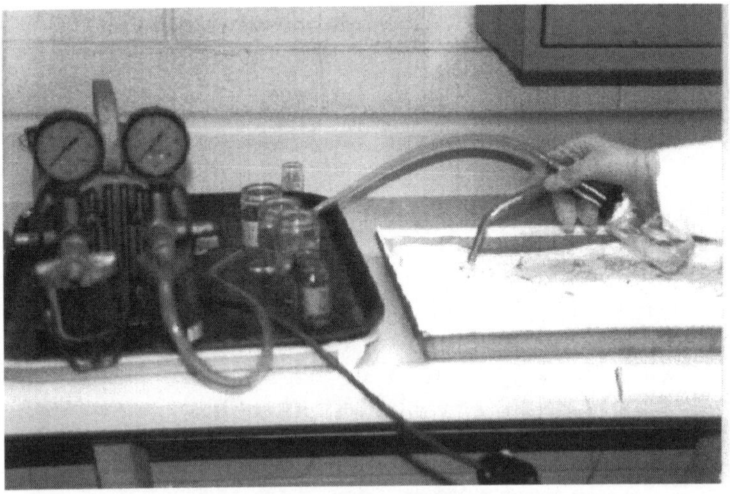

Fig. 4. Photograph showing technician collecting detached, fully engorged larvae from a paper-covered tray with an aspirator. The aspirator consists of a suction pump with adjustable gauges for controlling the vacuum, flexible tubing, a glass or plastic container with an inlet and an outlet port, and a flexible tube for sucking up the ticks. One piece of flexible tubing extends from the container to the inlet side of the vacuum pump; the other extends from the container and serves as the suction end of the aspirator.

out of the water, contaminated with fur, feces and urine, by hand. To simplify the process, the contents may be passed through a large, paper-filled funnel to remove the water. Then, the paper is spread onto a flat surface, allowed to dry and the larvae that crawl from the debris are collected. However, this is still tedious, since the larvae crawl from the moist mass at different rates. When maximum containment is required, an apparatus similar to that described by Endris et al. (1986) may be useful. Nymphal *D. variabilis* are fed in the same manner as the larvae, using the same hosts. However, the nymphs are considerably larger and, consequently, will consume much more blood. A typical *D. variabilis* nymph (~ 250 mg) will increase its body weight more than 40 X during blood feeding, mostly during the last 1–2 d (Sonenshine unpublished). Obviously, great care must be taken to avoid killing these small animal hosts due to exsanguination by the voracious nymphs.

Adult ticks are best reared on larger animals such as dogs or rabbits, although many other hosts may also be used. For investigators without ready access to domestic animals such as cattle, goats or sheep, the "laboratory rabbit" (i.e., the hare, *Oryctolagus cuniculus*) provides a convenient alternative. These animals can be housed in small cages, so that many individuals can be kept in one room of the animal facility. Rodents such as guinea pigs may also be used.

Since these laboratory animals are not natural hosts of *D. variabilis*, the adult ticks must be confined on the host to enhance feeding. Examples of containment devices are surgical stockinet sleeves (usually used for plaster casts), ear bags or capsules. For cattle or other ungulates, the host must first be restrained (e.g., nose ring, head gate, etc.). The stockinet sleeve is applied to the skin on the flanks, although any suitable area may be used. If the hair coat is very dense, it may be useful to first shave a circular area (6-an 8-cm diameter) with an electric razor. Then, a suitable piece of stockinet should be cut, e.g. (7.5 cm diameter and approximately 15 cm long) should be cut, applied firmly to the skin and glued with adhesive cement (cyanoacrylic cements are ideal for this purpose) (Figs. 5, 6). After the cement has dried, the investigator should check to insure that the sleeve is tightly bound to the skin, without gaps from which ticks might escape. Once it is secure, the ticks can be released within the sleeve. This can be done by inserting the vial containing the ticks into the sleeve with the protective gauze cap towards the skin and removing the lid. The vial is inverted inside the sleeve and held with one hand, while the cap is removed with the other hand, both operations done while holding the vial from the exterior of the stockinet. Next, the stockinet is twisted around the vial and sealed with adhesive tape. This insures that all ticks remain confined within the sleeve. Stockinet sleeves are convenient, require no specialized hardware, and rarely cause irritation or other adverse reaction to the host. After a suitable interval as

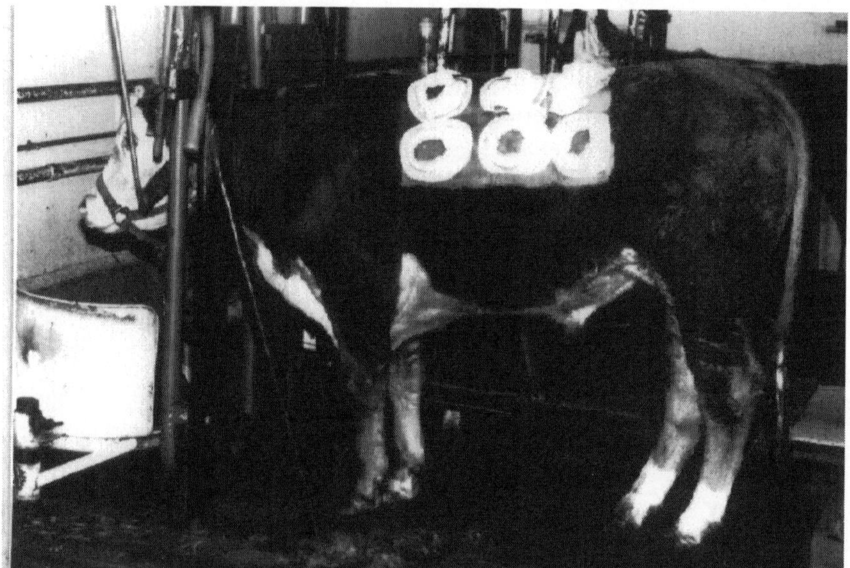

Fig. 5. Steer restrained by a head gate with stockinet sleeves cemented to the shaved skin of the flanks. Additional sleeves are secured on the opposite side of the animal.

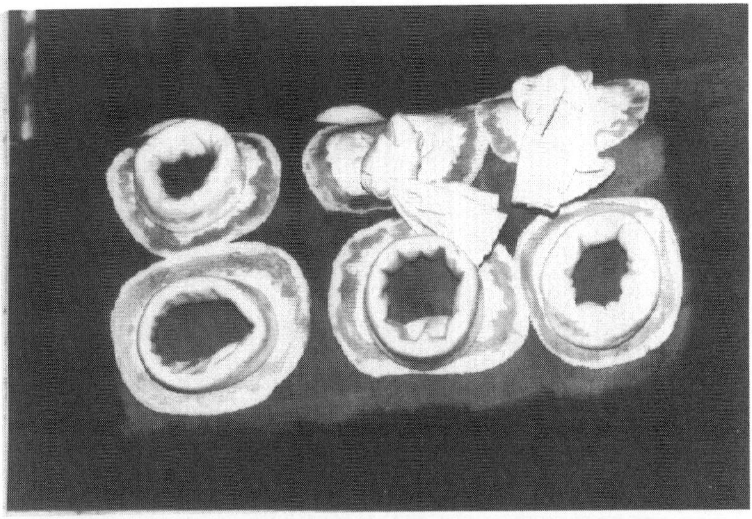

Fig. 6. Enlargement of the same animal shown in the previous figure, showing details of the stockinet sleeves. Two of the sleeves in the top row have been infested with ticks. After releasing the ticks, the protruding end of each sleeve is tied into a knot. The sides of the remaining sleeves are folded and are ready to receive ticks. Photographs furnished courtesy of Dr. J. Matthew Pound, Knipling Bushland US Livestock Insects Laboratory, USDA, Kerrville, Texas. Reprinted with the kind permission of Oxford University Press.

determined by the investigator (e.g., 12–24 hr), the ticks within the sleeve can be examined by removing the tape, folding the sleeve against the host's skin and counting the attached ticks. Unattached ticks can be removed at this time. Stockinet sleeves can also be applied to smaller animals such as laboratory rabbits (Fig. 7). Before proceeding, the animals should be tranquilized (PromAce^{Tm}, Acepromazinemaleate, Aveco, Fort Dodge, Iowa, USA) and the mid-section shaved to facilitate tick attachment.

In contrast to the sleeves applied to a small area on the flanks of ungulates, the entire mid-section of the rabbit's body is enclosed in stockinet. For this purpose, holes are cut near the front and back end of a 10–12 cm length of stockinet, the sleeve is inserted over the animal's head, and the forelegs inserted through the holes.

Adhesive tape is applied to the anterior end of the sleeve, just behind the shoulders. The stockinet is pulled back around the mid-section, the ticks are released onto the body, and sleeve pulled back towards the hindquarters. Next, the hindlegs are inserted through the remaining holes in the stockinet. The posterior end of the sleeve is sealed with adhesive tape. To prevent the rabbit from damaging the stockinet, a collar made of cardboard or polyethylene tubing can be installed around the animal's neck (Watts et al. 1972) (Fig. 8). Tick attachment and feeding success can be monitored by periodically removing the adhesive tape around the rear section and

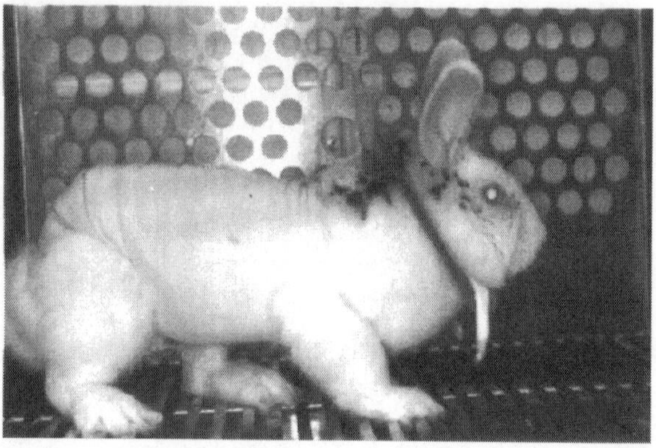

Fig. 7. A New Zealand white rabbit with a stockinet sleeve around the midsection of the body. Note the holes for the fore-and hindlegs. An elizabethan collar was used to minimize disturbance of the device. Reprinted with the kind permission of Oxford University Press.

Fig. 8. A New Zealand white rabbit with an Elizabethan collar to restrain the animal and prevent it disturbing the attached ticks. Reprinted with the kind permission of Oxford University Press.

observing the ticks. Ear bags also are commonly used for feeding ticks on animals with large ears, such as rabbits (Fig. 9). The simplest ear bags can be made with stockinet sleeves, sealed to the skin of the head with cyanoacrylic cement. After the ticks are released inside the bag, the opposite end is sealed with tape. Reusable ear bags may be fabricated with strong cloth (e.g., denim) and flat acrylic sheets were shaped to cover the top of the lids. Additional zippers, Velcro strips or clasps to seal them. These devices are washed after each application and reused.Another type of containment device is the capsule. Capsules are sturdier and more rigid than stockinet or cloth sleeves, providing greater security for valuable specimens or disease-infected ticks. Capsules usually are made of metal or plastic, with

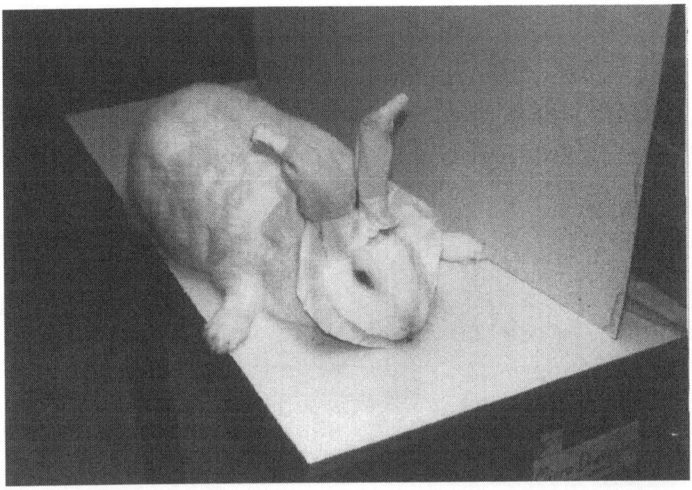

Fig. 9. A New Zealand white rabbit (*Oryctolagus cuniculi)* with ear bags and an elizabethan collar. Reprinted with the kind permission of Oxford University Press.

a removable lid. Some investigators cut off the neck of screw-cap plastic bottles to make inexpensive capsules which can be taped to the shaved flanks of suitable animals. Prior to installation, the rabbit is tranquilized and the midsection shaved with electric clippers. Following attachment of the capsule, the ticks are released and the lid screwed into place. Observations can be made directly through the clear plastic without disturbing the animal. For more detailed observations I fabricated capsules from acrylic plastic and used the university shop to cut uniform sections from different diameter tubing. Threads were cut in the tubes so that one section would thread into the other, one side forming the base while the other formed the lid. Circular sections of strips of circular plastic or a polyethylene tube was used to cushion the bottom section when applied to rabbits (Fig. 10). The bottom section also can be buffed or tapered to avoid sharp edges. To avoid irritation to the rabbit's skin, adhesive tape is applied liberally around the bottom and sides, so that only this softer material is placed against the animal's skin. Adhesive tape is used as described above to attach the capsule to the midsection of a rabbit (Fig. 11). removal of ticks, the lid is unscrewed. When installed correctly, these devices are well tolerated by the rabbits and there is no evidence of pain or discomfort. Ticks attach readily within the capsular area, where they remain confined. In my experience, these devices are the most secure and convenient means for feeding many adult ticks on rabbits. However, it is important that the capsules be installed properly. Careless installation may lead to sores, abscesses or other adverse skin reactions and require the capsule's early removal.

Fig. 10. A disassembled plastic capsule used to confine feeding ticks on a rabbit showing the base with threads around its margin and the threaded lid. Strips of adhesive tape are used to attached the device to the animal.

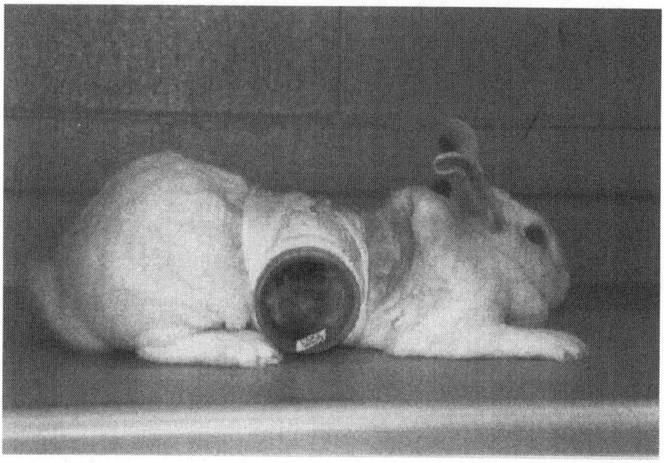

Fig. 11. A New Zealand white rabbit with a capsule taped to its midsection. With this technique, the ticks remain completely confined within the capsule.

Feeding capsules offer other advantages over ear bags or stockinet sleeves, e.g., 1) volatiles secreted by feeding ticks can be captured for chemical analysis (Apps et al. 1988, Diehl et al. 1991) or 2) the capsules permit observation of tick behavior during feeding in response to different external conditions Wooten-Saadi et al. (1991). Some tick species require different combinations of hosts than that illustrated by the American dog tick. An example is the rabbit tick, *Haemaphysalis leporispalustris*; immatures of this species feed on a wide variety of avian hosts, especially ground feeding birds, while the adults feed exclusively on rabbits. Similarly, larvae of the lone star tick, *Amblyomma americanum*, shun mice, rats or other small rodents but feed readily on avian hosts; nymphs and adults attack virtually all kinds of medium-sized or large-sized mammals.

To rear *A. americanum*, I use newly-hatched chicks for the larvae, and laboratory rabbits (*O. cuniculus*) for the nymphs and adults. Chicks can be housed in tall cylindrical glass or plastic "mouse jars" (e.g., Nalgene animal cages, VWR, Columbia, Maryland USA) on wire mesh screens. Double-sided tape is placed around the top of the cylinders to prevent tick escape. The space below the screens is filled with litter for waste collection but this is replaced with paper towels when larvae engorge and begin to drop off. Infestation with larvae can be done by releasing them directly onto the birds. Fed larvae are recovered by siphoning them from the paper towels with a vacuum aspirator. Following incubation and molting of the fed larvae, the nymphs can be fed on rabbits under stockinet sleeves in the same manner as described above for *D. variabilis*.

The black-legged tick (or deer tick), *Ixodes scapularis*, can be fed on rabbits or a combination of small rodent and rabbit hosts. Albino laboratory mice (*Mus musculus*) are convenient hosts for larvae. In my laboratory, the mice are tranquilized (Acepromazine), then placed in wire mesh baskets mounted on cork or rubber stoppers for support. A sliding aluminum or galvanized metal top is used to seal the cage. Care should be exercised during the infestation phase, since *I. scapularis* larvae will attack humans. Disposable plastic ("surgical") gloves should be worn, with the ends taped around the investigator's wrists. If a long-sleeved shirt or laboratory jacket is worn, the sleeves should be taped around the forearms as additional barriers to tick bite. After infestation, food and water are supplied for the mice. Each cage with its tick-infested mouse is placed in a metal or glass tray (a baking pan can be used for this purpose) with double-side tape around the edges to prevent tick escape. Litter for waste collection is replaced with paper towels when the larvae engorge and fall from the mice. Following feeding, fed larvae are vacuumed from the paper towels with a vacuum aspirator as described for *D. variabilis*. Hamsters or gerbils may be substituted for white mice, with similar success. Nymphal and adult *I. scapularis* can be fed on laboratory rabbits under stockinet or in capsules as described above.

Some economically important tick species will only feed on large animals, e.g., cattle. An example is the cattle tick, *Boophilus microplus* or *B. annulatus*. For feed cattle ticks on large domestic animals, restraints such as those used by livestock handlers are necessary. These devices are usually found in barns or other animal buildings. A common restraint used with such animals is the stanchion, used to secure the head and neck. Once the cow or bull is restrained, parts of the body can be shaved, stockinets sleeve glued to the shaved skin as described previously, and ticks released within.

Argasidae

Argasid ticks have different feeding requirements than ixodid ticks. Almost all argasids feed rapidly; often in as little as 15–30 min (larvae of many bat-parasitizing argasids require several d for a complete blood meal). Typically, argasid ticks are secretive and feed best in semi-dark or totally dark conditions. These habits simplify the feeding process, since the hosts need not be restrained for long periods.

Before attempting to feed the ticks, information about the preferred hosts must be obtained. Several cave, burrow or nest-infesting *Ornithodoros* species, e.g., *Ornithodoros tholozani, O. turicata,* or *O. parkeri,* are relatively easily to colonize using laboratory rodents such as mice, rats, hamsters or guinea pigs. These ticks survive best if allowed to burrow into litter, sand, or other material in which they can hide (this will be discussed further in section D (Maintaining Ticks for Oviposition, Development and Long Term Survival). I use albino laboratory mice (*M. musculus*) to feed my colony of *Ornithodoros parkeri*. First, the mice are tranquilized with Acepromazine. When the mice stop moving, they are placed in the enclosure containing the ticks. The latter can be a wooden or plastic box, or a glass dish. The interior is partially filled with a mixture of fine sand, saw dust and wood shavings containing the numerous ticks. All stages, including larvae, the several nymphal stages and the adults, occur in the litter material. Within a few min after the resting mice are introduced, hungry ticks begin to emerge from the litter and attach to various parts of the bodies of the mice.

To enhance feeding, the enclosure is covered and the lights are dimmed or turned off. Tick feeding can be observed using red light (using a red filter similar to that used in photographic dark room lamps). Within 20–30 min, ticks can be found feeding on virtually all parts of the animal's bodies. Some can be observed secreting coxal fluid while feeding; others wait until after they drop off to eliminate their coxal fluid. Feeding is usually completed within one hr, although a few specimens may remain attached for much longer periods. If it is necessary to collect the ticks for post-feeding observations such as molting or oviposition, feeding can be done in a separate container lined with clean white paper towels. Tranquilized mice are placed in acontainers. Then, hungry ticks are removed from the colony (use caution to avoid being bitten or allowing ticks to escape), placed in vials and transferred to the mice.

Small feeding containers, sufficient for a single mouse or rat, can be made from a wide-mouth jar (e.g., mason jar or "museum" jar). A hole is cut in the lid and the opening sealed with fine mesh cloth (e.g., nylon). Litter is placed in the bottom of the jar and this is covered with a circle of wire mesh screen cut so that it just fits into the jar. Ticks can be placed in the litter and allowed to maintain themselves between feedings. When the ticks are to be fed, a tranquilized mouse is placed in the jar on top of the wire mesh screen. Hungry ticks emerge, feed and drop off to shelter in the litter in the same manner as described for the larger containers described above (Endris et al. 1986). For bird ticks, baby chicks can be substituted for mice. In this case, a tall battery jar is preferred instead of the shallow mason jars. Otherwise, the procedures are very similar.

Capsules may be used to feed individual argasid ticks (Endris et al. 1986). Plastic vials small enough (~ 5 cm long x 2 cm wide) to be taped to the body of a mouse or rat are used. A plaster of paris/charcoal mixture, which turns dark gray when thoroughly moistened, can be used to provide a humidity source during feeding. The gauze or nylon mesh-covering opening may be placed in the lid or cut out of the side of the vial. When attached to the body of the tranquilized host, hungry ticks will feed directly through the cloth barrier. The advantage of these capsules is that no specimen handling is required for feeding. The ticks remain confined in the same vials before and after feeding, facilitating observations on molting, oviposition or other biological processes. As noted earlier, dim light (e.g., red filtered lamp) or total darkness may be necessary to feed the ticks. Controls to set the temperature, humidity and light: dark cycle should be positioned on the exterior of the incubator, so that changes in these conditions can be made without disturbing the colonies necessary for the ticks to feed.

Membrane feeding has been used for fast feed argasid ticks as an alternative to animal hosts. This procedure may be attractive in order to avoid delays in obtaining approval for use of expensive, highly regulated animals. It is also useful for adding known doses of chemicals or microbes to the blood meal. However, as a routine means of maintaining the colony, membrane feeding is cumbersome and can present serious risks to the ticks fed in this manner. Use of blood from incompatible hosts, unsatisfactory membranes, membrane leakage, inadequate pressure, poor (or no) blood circulation and poor temperature regulation can lead to poor feeding success and, in some cases, even tick mortality. Consequently, membrane feeding is best reserved for use in experiments (e.g., adding chemicals or microorganisms to the blood) rather than as the primary method of colony maintenance.

Maintaining Ticks for Oviposition, Development and Long Term Survival

To maintain ticks in the laboratory during the long intervals between feeding, the investigator or technician must know how to (1) prevent the ticks from escaping; (2) separate the different life stages; and (3) how to provide adequate moisture (humidity) and temperature to enhance tick development, oviposition and survival.

Organizing the Tick Colony

To prevent their escape, ticks should be kept in sealed glass or plastic vials, with gauze or nylon-covered openings for air exchange. The vials should be kept in a designated area of the laboratory designed in a manner similar to an insectary. If available, a small room should be set aside for the tick colonies. A numbering system for the vials with each life stage, e.g., L-1 for vial no. 1 of larvae, N-1 for nymphs, etc., helps to organize the colony. It also makes it easier to keep track of the number of specimens in the colony. This will be discussed further elsewhere in this chapter.

Incubation and Incubator Facilities

To maximize colony production, ticks should be maintained under optimum temperature and humidity conditions for their development, oviposition or long term survival. Requirements for different species reflect their natural habitats and seasons of activity. Survival is enhanced at temperatures lower than those needed for development or oviposition.

Except for a few species adapted to semi-desert conditions, most tick species require high humidity during their non-feeding periods. The simplest method for providing moisture is to keep the ticks in closed containers (e.g., desiccator jars) over water or a saturated salt solution (Fig. 12). Solutions of different salts will result in a specific relative humidity so long as the air space above the solution is relatively small (Winston and Bates 1960). Vials with ticks can be stored in such containers for brief periods (i.e., 2–3 weeks). However, mold tends to form at the higher humidity. In addition, several

Fig. 12. Glass desiccator jar for holding ticks in a closed, constant humidity environment.

hr are required to achieve a stable atmosphere in the humidity container after the lid has been closed, limiting the number of times that the lid can be removed to retrieve or introduce samples. Thus, closed containers are not suitable for long term maintenance.

For investigators who plan to develop and maintain large colonies for many years, commercial incubators that can control both temperature and relative humidity are essential. The incubator provides a designated, controlled environment where conditions are uniform and optimized for each species. In addition to temperature and humidity control, the incubator should be fitted with a powerful ventilating fan (sufficient for 5–10 air changes per hr) and internal lighting. Adequate ventilation is important to suppress mold formation, especially at high humidity. Optimizing the d light cycle is important since some tick species are light sensitive and will diapause when kept under dark conditions. A typical "reach-in" incubator is comparable to a floor model refrigerator (Fig. 13). If cost is a limiting factor, a commercial incubator that only controls temperature may be purchased. The tick breeder has several options for boosting the relative humidity. One

Fig. 13. A reach-in controlled environment chamber showing details of the shelves and containers for storing vials of ticks. Tick specimens are held in glass vials grouped in wire screen baskets; vials of the same type, e.g., engorged *D. variabilis* larvae are held in the same basket. All of the baskets of a given species are placed on the same shelf. Air flowing via large (15 cm diameter) ducts circulates between the chamber and the conditioner at the rear of the instrument.

of the simplest methods is to fill a large tray with wet sand and store the tick vials on the sand. A wire screen can be installed over the sand-filled tray to keep the vials dry. Using several such sand-filled trays in a small, closed environment will raise the relative humidity. In addition, a small circulating fan may be placed in a corner of the incubator to provide air circulation. The cooling system of the incubator should be capable of removing the limited amount of heat produced by the fan motor. However, the humidity will dissipate quickly each time the incubator is opened and several hr may be required to restore equilibrium. To check the relative humidity, a simple wet bulb/dry bulb indicator may be installed inside. This indicator can be constructed using two thermometers; the mercury end of the wet bulb is surrounded with cloth inserted into a beaker of water while the dry bulb thermometer is left exposed to the incubator air. Conditions within the incubator should be checked regularly (e.g., once/d) and water added to the open pans or sand as needed. An alternative to using moist sand is to store the tick-containing vials in sealed glass (or plastic) desiccators over a saturated salt solution as described above. Plastic is preferred over glass to minimize condensation. By using many such desiccators, each for a different life stage and species, one can avoid altering the relative humidity where the ticks are stored each time the incubator is opened. Ticks of the same life stage can be held in separate, small vials with appropriate labels, within each desiccator. A second incubator is needed if the investigator plans to hold some specimens of the tick colony for very long periods, e.g., 4–6 months. In this case, tick survival is enhanced at cool temperatures. With a two-incubator capability, the tick breeder can maintain one incubator at relatively warm temperatures, typically from 21°–27°C, as the primary environment for rapid development of the immatures and for oviposition. The other incubator is maintained at a lower temperature, e.g., from 10°–15°C to reduce metabolic activity and foster tick survival. In both cases, humidity should be adjusted to the optimum for the tick species being maintained. Some investigators lock the incubators to forestall undue tampering with the ticks and minimize the risk of tick escape. In situations, where the tick colony is expected to become very large, or ticks will be infected with transmissible microorganisms, a walk-in controlled environment room is recommended (Fig. 14).The walk-in room can be fitted with a work bench for handling the specimens and numerous shelves can be installed to store the ticks during the non-feeding periods. The room may also serve as the site where feeding of infected ticks can be done in isolation from all other facilities so as to prevent spread of tick-borne infections. Commercial walk-in controlled environment rooms suitable for this purpose typically contain an interior work space ranging from 1.9 to 2.5 m long by about 1.3 to 1.5 m wide. The controlled environment rooms must have the same temperature and humidity control capabilities as the smaller chambers. Such devices,

Fig. 14. A walk-in controlled environment room with an external control panel providing programmed or continuous temperature and relative humidity regulation. Photo courtesy of Parameter Generation and Control, Black Mountain, North Carolina, USA. Reprinted with the kind permission of Oxford University Press.

with their external controllers, can be very expensive. To save money, some institutions prefer to modify a small room and convert it into a controlled environment. Regardless of the type of incubator chosen, the ticks must be confined within glass or plastic vials and labeled appropriately. In my laboratory, ticks are confined within reusable glass vials with plastic lids. Holes are cut in the lids and a piece of fine mesh nylon or cotton cloth is placed under the lid. As noted earlier, plastic is preferred, especially if the ticks are to be held at low temperatures (e.g., in a refrigerator at 4–6°C), to minimize condensation (condensation will kill the ticks). Ticks of the same species and life stage are stored together in groups of vials, and the vials are labeled accordingly.

Care is taken to avoid overcrowding, and relatively small numbers of ticks are kept in each vial, e.g., 100–200 fed larvae, or 50–100 fed nymphs. Fed ticks often adhere to one another due to the secretion of cuticular waxes. Therefore, overcrowding may result in suffocation of the individuals furthest from the top of the mass of fed ticks. As noted above, vial labels should contain sufficient information for identifying the ticks, specifically the species, life stage, and date fed. We use a three letter labeling system, wherein the first two letters indicate the genus and species name while the third indicates the life stage (e.g., DVL for *Dermacentor variabilis* larvae, DVN for nymphs, etc.) (Fig. 15). We also include the date collected; if collections are done regularly during tick feeding, this will be the same as the date fed. Thus, if 10 tubes of larvae were collected on a single d, the tubes would be labeled DVL 1 through DVL 10, each with the same date. Wire baskets, plastic boxes or some other non-paper container can be used to consolidate vials of the same type (Fig. 16). Each container should be labeled to facilitate identification. If possible, all stages of the same species should be grouped together on the same shelf in the incubator, but this will depend upon the size of the incubator and the number of tick species in colony. When colony size is small, several species may be stored on the same shelf, but in different baskets. It is extremely important to maintain a precise separation to protect against mixing specimens of different species. Once the colony becomes contaminated with specimens of another species, it is difficult and time consuming to separate them again. Each investigator also will have to decide on the optimal conditions to be maintained in the incubator. In my laboratory, I maintain the ticks being bred for experiments, mostly *Dermacentor variabilis*, at 27° ± 1°C, 92 ± 2% relative humidity and a 16:8 hr Light:Dark (L:D) daily cycle. Different conditions may be required to optimize the conditions for other species. Unfed ticks not needed for immediate use should be transferred to a cooler environment for long-term survival.

Fig. 15. Tick vials for maintaining fed ticks in incubators during their non-parasitic phases. The glass vials contain tick specimens of the same species and life stage. Note the gauze cover held in place by a perforated plastic "snap-cap" lid. Coded labels indicate the species, life stage, dates of drop off from the hosts and other information as needed.

Fig. 16. Wire screen basket for holding vials of ticks to be stored in the incubator for development or reproduction. The baskets serve as organizers to facilitate grouping of similar tick material for later use.

Record Keeping

Although not essential, a log book showing the status of the tick colony or colonies is recommended if ticks will be needed frequently for experiments. This is especially important for large facilities where the tick breeding program must furnish numerous specimens on a regular basis. The record system may consist of a paper log book with hand written entries, or a computerized log. It is unrealistic to attempt counting individual ticks. However, the number of vials of each life stage can be noted and an estimate of the total number of ticks of that stage given in the summary column. Depositing uniform numbers in the vials during tick collecting activities makes it easier to estimate totals, e.g., 200–300 fed larvae or 100 fed nymphs per vial. It is especially important to record the numbers of replete, egg laying females since these ticks are the stock for renewing the next generation of the colony. The log should record for each vial of a given life stage the dates fed and, subsequently, the dates when molting occurred. For replete females, the numbers of ovipositing females, date when they dropped off, and the dates when oviposition and hatching began also should be recorded.

Security Considerations

Accidents are common in laboratories. Where ticks are concerned, laboratory accidents can have serious consequences beyond the specific damage to supplies or equipment in the tick breeding facility. Tick escapes present the most serious concern. If ticks capable of biting humans escape and are found where other people, unrelated to the tick breeding program, can encounter them, complaints may be expected.

To avoid such accidents, containment procedures should be instituted and rigorously enforced. While ticks are feeding on their animal hosts, confining the animals within wire screen enclosures prevents dispersing loosely attached ticks when the animals shake themselves. Collecting trays used to capture ticks that fall from their hosts after feeding should be covered around the margins with double-sided tape or a sticky barrier. Inexpensive masking tape, with the strips rolled on themselves to present a sticky surface, makes a suitable barrier. To insure that no ticks escape from the tick feeding room, additional sticky tape barriers can be installed around each wall near the junction with the floor, under the door and around the door frames. This will prevent escaped ticks from finding hiding places in tiny cracks or crevices in the walls or ceiling where they might not be noticed. The walls of the tick breeding room should be sealed and painted. Finally, thorough and frequent sanitation is essential to maintain clean conditions where ticks cannot breed.

Another opportunity for ticks to escape occurs when people transport them from one location to another, e.g., from the tick breeding room or area to the research laboratories. Accidents that occur in crowded institutions such as universities can cause serious embarrassment. Accidental escape of exotic ticks into the surrounding countryside would be especially serious, presenting the local authorities with a difficult and expensive containment/eradication problem. Accidents are likely to occur when technicians, students or other workers are exiting doorways, or transporting ticks on staircases or crowded passageways. Ticks may be spilled from unprotected containers, poorly sealed vials, or released accidently by dropping and breaking their glass vial containers.

To avoid such problems during transport, secure containers should be prepared and the tick vials installed in the containers before making the trip. Such containers can be simple boxes which will not break if dropped and will contain the ticks if the vials are broken or the vial covers dislodged. Portable carts with 2 or 3 shelves should be used if additional supplies must be carried along with the ticks.

References

Apps, P.J., H.W. Viljoen and V. Pretorius. 1988. Aggregation pheromones of the bont tick, *Amblyomma hebraeum*: identification of candidates for bioassay. Onderstepoort J. Vet. Res. 55: 135–37.

Diehl, P.A., P. Guerin, M. Vlimant and P. Steullet. 1991. Biosynthesis, production site and emission rates of aggregation-attachment pheromone in males of two *Amblyomma* ticks. J. Chem. Ecol. 17: 833–847.

Endris, R.G., T.M. Haslet, M.J. Monahan, W.R. Hess and L.C. Rutledge. 1986. Techniques for mass rearing soft ticks (Acari: Argasidae). J. Med. Entomol. 23: 225–229.

Sonenshine, D.E. 1991. Biology of Ticks. Oxford University Press, New York. 447 pp.

Sonenshine, D.E. 1993. Appendix 1. pp. 372–412. *In*: Biology of Ticks. Oxford University Press, New York. 465 pp.

Sonenshine, D.E. 1968. Mass rearing of radioisotope tagged larval ticks for ecological investigations (Acarina: Ixodidae). J. Econ. Entomol. 64: 1423–29.

Sonenshine, D.E., R.M. Silverstein, E. Plummer and J.R. West. 1976. 2,6-dichlorophenol, the sex pheromone of the Rocky Mountain wood tick, *Dermacentor andersoni* Stiles and the American dog tick, *Dermacentor variabilis* (Say). J. Chem. Ecol. 2: 201–209.

Watts, B.P., Jr., J.M. Pound and J.H. Oliver. 1972. An adjustable plastic collar for feeding ticks on ears ofrabbits. J. Parasitol. 58: 1105.

Winston, P.W. and D.H. Bates. 1960. Saturated solutions for the control of humidity in biological research. Ecology 41: 232–234.

Wooten-Saadi, E., C.H. Cole and G.I. Garris. 1991. A light-impermeable feeding chamber for ticks (Acari: Ixodidae). Exp. Appl. Acarol. 28: 111–113.

4

Maintenance of the Triatomine Bugs *Rhodnius prolixus* and *Triatoma dimidiata* under Laboratory Conditions

R.V. Durvasula,[1,*] *J. Taneja,*[1] *K. Cobb*[2] and *E.M. Dotson*[2]

ABSTRACT

The triatomine bugs *Rhodnius prolixus* and *Triatoma dimidiata* are vectors of Chagas disease. They have been reared in laboratories to study their biology and their interaction with *Trypanosoma cruzi*. The bugs are easy to rear and handle and can be maintained in clean plastic containers. They are hematophagous throughout their life cycle and require infrequent blood-meals to survive and reproduce. The most commonly employed method of feeding has been on an immobilized live animal, but an alternative method of feeding human or animal blood via a membrane is widely utilized. Here we describe methods

[1] Center for Global Health, University of New Mexico School of Medicine, Albuquerque, NM USA.

[2] Division of Parasitic Diseases, Centers for Disease Control and Prevention, Atlanta, GA, USA.

* Corresponding author

for containment, rearing and feeding of these insects. Additionally we include information on maintaining bugs infected with *T. cruzi* and aposymbiotic bugs that lack gut-associated symbiotic bacteria.

Key Words: Triatomine bugs, *Rhodnius prolixus, Triatoma dimidiata*, laboratory maintenance, *in vtiro* blood feeding

Introduction

Triatomine Systematic

Triatomine bugs are hemimetabolous insects with piercing/sucking mouth parts. They belong to the order Hemiptera which consists of two suborders, Homoptera and Heteroptera. All Homopteran and many Heteropteran bugs feed on plant fluids. Some Heteropterans are predaceous, feeding on the body fluids of other arthropods and vertebrates. Two families of the latter suborder contain blood sucking insects. One of them, the Cimicidae, or the bed bug family, are all hematophagous. The other, the Reduviidae, is commonly known as the assassin bug family. Most Reduviidae are predators of other insects, but the subfamily, the Triatominae, have adapted to feed on vertebrate blood. There are more than 140 known species of Triatominae, *R. prolixus* and *T. dimidiata* being two of them.

Distribution

The triatomine bugs are mainly distributed in the American continent and some Caribbean islands. In addition, a few species have also been reported from Africa, Asia and Australia (Lent and Wygodzinsky 1979, Kalshoven 1970, Else et al. 1977, Monteith 1974). *Rhodnius prolixus* is mainly distributed in the northern part of South America (Venezuela, Colombia, Guyana, French Guyana, Surinam, and in some Central American countries such as, Guatemala, Honduras and El Salvador, although its range in Central America has been drastically reduced by recent control programs. *Triatoma dimidiata's* range includes Ecuador, Peru, all of Central America and the south of Mexico.

Habits and Habitats

All triatomine bugs are sylvatic in origin. Their natural habitats are primarily associated with the nests or burrows of a wide variety of small mammals and birds and occasionally with lizards (Lent and Wygodzinsky 1979). Some species, however, have become closely associated with human domestic and peri-domestic environments. They usually inhabit animal houses such as stables and corrals, chicken and pigeon coops, guinea-pig and rabbit houses. Human habitations become infested only if they have unplastered or cracked walls, thatched roofs or other corners and crevices where the bugs can easily find refuge. The sylvatic epitopes of *R. prolixus* include palm tree crowns and bird nests, whereas can be found in rock and wood piles, tree bark and caves. Both *R. prolixus* and *T. dimidiata* are frequently inhabiting domestic and peridomestic sites.

Most triatomines are nocturnal in habit. They are found aggregated in their refuges during the day becoming active at night, when they search for a host if in need of a blood-meal. Only the adult stages of these bugs are winged; the nymphs are wingless and incapable of flying. The adults of some triatomine species are known to fly over 100–200 km distances when dispersing from one habitat to another (Schofield et al. 1991). Such dispersive flights mainly occur at high temperatures (ca. 30°C) and conditions of low nutritional status (Lehane et al. 1992). If sufficient food sources are available the bugs generally do not leave a habitat. A more common means of dispersal is passive transport by man via roofing materials such as palm fronds and straw and in furniture, clothes and other household goods. Migratory birds can also serve as dispersal agents for the eggs and young stages of some nest-dwelling species such as *R. prolixus* (Lent and Wygodzinsky 1979, Schofield 1979).

Life Cycle

The triatomines have a typical hemimetabolous life-cycle, from eggs through five nymphal instars to adult males and females. All nymphal and adult life-stages of triatomine bugs are obligate hematophagues; most species feed at night on their sleeping hosts. They have a relatively long life-cycle and low rate of reproduction. Fertilized females lay a few eggs at a time and each female can lay 100 to 600 eggs during her adult life of 3 to 12 months, depending on species and quantity of blood ingested. The eggs hatch within 10 to 30 days after oviposition and the first instar nymphs are ready for their first blood-meal 2 to 3 days later, but can survive 2 to 3 weeks without feeding. The blood-meals are large; each nymph can ingest blood that is almost 8 to 9 times its own starved weight. The later nymphal instars can withstand starvation for weeks or even months. Usually a single replete blood-meal of 20 to 30 min' duration is sufficient to initiate molting to the next nymphal stage, but 2 to 5 smaller meals are often taken by second to fifth instar nymphs before molting (Lent and Wygodzinsky 1979).

Triatomines as Vectors of Chagas Disease

Medical and Economic Importance

Many triatomine bugs, including *R. prolixus* and *T. dimidiata*, are vectors of American trypanosomiasis, or Chagas disease. Chagas disease is caused by a protozoan parasite, *T. cruzi*. The insect's bite is almost painless because of the anesthetizing action of its saliva and the bug usually defecates either while feeding on the host or shortly after the blood-meal. Because the infective form of *T. cruzi* is transmitted only through the feces, the time span between the blood meal and the first subsequent defecation is epidemiologically significant. This time span varies among different species of bugs. Those species that defecate immediately after feeding are more efficient vectors than those that do not (Zeledon and Rabinovich, 1981). The trypanosomes penetrate host tissues through skin abrasions, mucosal membranes, or the conjunctivae. The parasite passes between its two hosts, triatomine bugs and vertebrates, and undergoes a complex series of morphological and physiological changes (Garcia and Azambuja 1991). The epimastigote and amastigote stages are the multiplicative forms found in the insect vector and within vertebrate host cells, respectively. The trypomastigote stage represents the highly infectious non-replicative form found as metacyclic trypomastigotes in the hindgut of the insect and as blood-form trypomastigotes in the bloodstream of infected vertebrate hosts. All blood feeding stages can acquire the parasite and once infected, the bugs can transmit the parasite the remainder of their life cycle.

Chagas disease is a serious public health problem in Central and South America. Over half of the known species of Triatominae have been naturally or experimentally infected with *T. cruzi* and because of their similar behavior and physiology all species are regarded as potential vectors. This disease occurs mainly in rural areas with poor housing conditions and especially affects children and young adults, resulting in chronic infection with high morbidity. The World Health Organization estimates that over 10 million persons are infected and 65 million are at risk in the endemic regions (WHO, TRS Report 2008).

Neither vaccine nor effective chemotherapy exists for chronic Chagas disease, leaving vector control as the only alternative. Insecticidal control is the only method that has been extensively applied. The Southern Cone initiative to eradicate these bugs from South America has made much progress but an additional, more environmentally acceptable, economically viable and long-lasting control method still remains to be developed.

Rearing and Maintenance

Triatomine bugs have long been reared in laboratories all over the world to study their biology as well as their interaction with *T. cruzi*. *Rhodnius prolixus* has been one of the model insects in studies of insect physiology (Wigglesworth 1972) mainly because of its size, blood feeding habit, long life-cycle and ease of rearing and handling. The rearing and maintenance of large numbers of *R. prolixus* and *T. dimidiata* are relatively simple. The rearing containers do not require frequent cleaning. Rather, these bugs thrive in a soiled environment. The bugs are quite hardy making it easy to handle them and, since they are walking insects, they pose a low threat of escape, especially after a blood-meal. Though the source of blood and feeding conditions have to be correct for them to initiate feeding and fully engorge, it is relatively easy to feed them, as they require a blood-meal as infrequently as once a month. Thus, it is possible to rear colonies with minimal expenses and man-power. To obtain best results the bugs should be maintained in a closed environment with favorable conditions, fed with blood from a suitable host source and kept undisturbed with limited handling.

Containment

Insectary

The insects should be contained in an insectary room or incubators designed to prevent escape of insects. Furthermore, environmental factors such as temperature, humidity and light conditions should be regulated and maintained. Our insectary set-up closely follows the, Arthropod-bome Infectious Diseases Laboratory (AIDL) standards (Higgs and Beaty 1996), as described below.

An appropriate insectary consists of:

1. limited-access, sealed room(s), preferably without windows, low ceilings and with multiple-door access to the rearing and handling/manipulation areas.
2. The entrance to the rearing room should preferably be via a series of doors that constitute sequential barriers to accidental escape of insects.
3. All doors should open inward and close automatically; this will push any escaping insect back into the room.
4. Two contiguous doors must never be opened simultaneously. It is essential that a rear door is closed before opening one in front.

5. In the rearing room, all surfaces should be finished with a material that can be easily cleaned and disinfected. Certain surfaces, such as walls, ceilings, shelves and counter-tops, should be painted white so that any escaping insect can be easily located and killed or caught with a fine mesh to prevent escape.
6. Minimal equipment and supplies should be kept in the insectary, to limit potential hiding places for insects.
7. Extra precautions should be taken if infected insects are kept. The infected insects should be doubly contained (in a container within a container) and preferably kept in an area separate from the stock colonies.
8. Appropriate biohazard information notices should be prominently displayed on the entrance door of the room containing infected insects.

Temperature, Humidity and Light Requirements

The optimum temperature requirements of the two species are quite different. *R. prolixus* does best at higher temperatures (ca. 27°C); *T. dimidiata* prefers lower temperatures (ca. 23°C). The conditions in our insectary are maintained at an intermediate temperature of 23–25°C because *T. dimidiata* is more sensitive to temperature fluctuations than *R. prolixus.* The relative humidity conditions for rearing are not as critical as temperature. Excess humidity (>90%) can cause fungal growth in overcrowded colonies. At low humidity (<40%) the life span of a bug can be reduced; both extremes should be avoided. In our insectary the relative humidity varies between 60% to 80%. Several activities of the triatomine life are synchronized with the natural light-dark cycles of day and night (Ampleford and Steel 1982, Saunders 1982, Ampleford and Davey 1989, Lazzari 1992). The best way of conditioning is to mimic the natural light-dark cycle. Sometimes, for convenience, the cycle can be shifted or even reversed. When it is not possible to set a light-dark cycle, the bugs can be kept in complete darkness, though this might affect certain physiological and behavioral phenomena.

Rearing Containers

Any clean plastic container can be used to rear these bugs. Glass containers should never be used as they can break while handling. These bugs normally show thigmotactic behavior, preferring to aggregate and hide in cracks, crevices and corners (Lorenzo Figueiras et al. 1993) and such crevices should be provided for them in the rearing containers. We use plastic 1 liter containers (Nalgene, USA) to maintain the insect colonies (Fig. 1). The central disc of the lids is removed by cutting it out. A nylon mesh cut to size and secured with the hollow rim of the lid is used to cover the top of the containers. Finer mesh (–800 holes/sq. cm.) is used on containers with small nymphal instars.

Special care is required with the first instar nymphs as they can escape if the mesh size is not small enough. A larger mesh size (100 holes/sq. cm.) is used with bigger nymphal instars and adults, especially with *T. dimidiata,* adults of which have difficulties piercing their proboscis through a finer mesh when feeding. The mesh should be regularly checked for any damages or holes through which the bugs could escape; if any are found it should be discarded and replaced with a new one immediately. To avoid such a problem, a metal mesh cut to size can also be used. The containers are further prepared by washing and drying them. It is important to wash new containers to eliminate chemicals that might be toxic to these insects. Filter paper discs (Whatman no. 1) cut to size and accordion folded are placed vertically in the containers as a hiding substrate. A circular disc of filter paper is also placed at the base of the container since, after a blood-meal, most bugs descend to the bottom of the container and defecate profusely; the feces can be absorbed by the filter paper (Fig. 1). The prepared containers are usually autoclaved before their first use. Autoclaving is not critical; we follow this procedure to make sure that we are not introducing bacterial flora when setting up experiments with aposymbiotic bugs (see

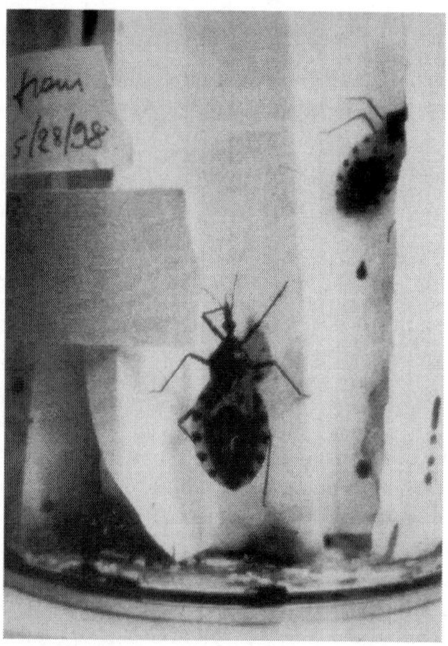

Fig. 1. *Triatoma dimidiata* adults on an accordion folded filter paper in a plastic container.

below). To prevent accidental escape, all containers with insects are kept on partly submerged supports in trays filled with soapy water or ringed with petroleum jelly (Fig. 2).

A stock colony can be started either by taking insects from another established colony or from field insects of one strain collected from a single colony. Ideally, a colony should be started from a single female of known strain, especially if the insects are to be used for genetic studies. In the 1 liter jars we can maintain up to 200 bugs, including 15 to 20 pairs of adult *R. prolixus* or 10 to 15 pairs of *T. dimidiata*, the latter being larger in size. The containers should not be cleaned too often, as these bugs thrive in their own dirt. Young nymphs have a special need for the feces of other bugs to survive. They exhibit coprophagy, i.e., they ingest feces to obtain essential gut symbionts (Baines 1956, Schaub et al. 1989). Therefore, any container with young nymphs must contain at least one filter paper disc heavily soiled with the feces of older bugs. Though these bugs live under crowded conditions, care should be taken not to exceed the above limits. If overcrowding occurs, a few adults should be transferred to a different container to start a new colony.

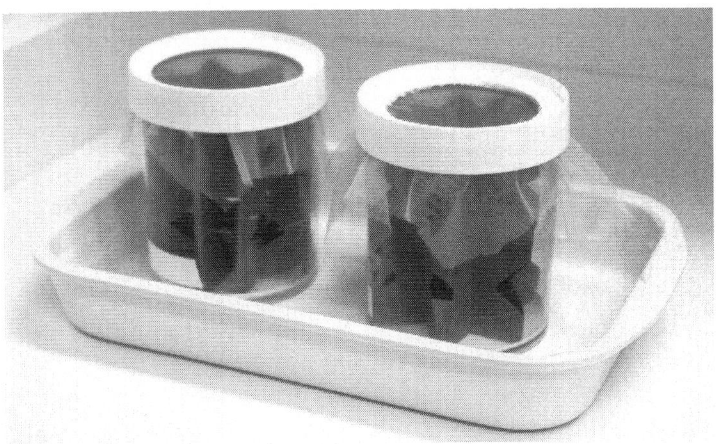

Fig. 2. Bug containment in trays filled with soapy water to prevent escape.

Handling and Manipulation

These bugs can be described well as walking, rather than flying insects. Though adults are winged, they rarely attempt flight under laboratory conditions. This makes it easier to handle them as the threat of escape is low. Precautions should still-be-taken and containers with bugs should always be opened over a container coated with fluon or glass containers with straight vertical walls. All stages of *T. dimidiata* and all immatures of *R. prolixus* are incapable of climbing glass. Younger instar nymphs climb and cling to the mesh, therefore, the containers should be lightly tapped to bring all insects to the bottom before opening. The bugs are fast walkers and can climb the walls of plastic containers with agility, the adults more so than nymphs. In addition, warmth, CO_2 and other odors emanating from the person handling them can stimulate the hungry bugs to climb and start feeding on the available host. Any manipulations should thus be carried out carefully and quickly. Furthermore, it is a good idea to have a second person who can watch for escaping insects, particularly when an inexperienced person handles the colony. The bugs have a thick cuticle which also makes it easy to handle them, though excessive handling should be avoided as it can have adverse effects on their development and longevity. The best way to handle them is with soft entomological forceps. The younger stages can be picked up by the legs but the older nymphs and adults should be held more firmly at the pronotum or thorax. Handling of engorged insects should be avoided.

Blood Sources

All stages of these bugs are obligate feeders of vertebrate blood. They require large blood-meals, about 8 to 12 times their unfed weight, at infrequent intervals of about two weeks to a month for normal growth and reproduction. Though the bugs do best when fed regularly on live animals, this is not always feasible and ethical, particularly when rearing large numbers. An alternative method of feeding freshly drawn animal or human blood through synthetic membranes has proven to be quite successful. The quality and type of blood source should be carefully chosen, as it influences the development, reproduction and fecundity of the bugs to a large extent. The donor should preferably be a healthy adult free of any disease. Azambuja and Garcia (1997) report that human and rabbit blood give best results, while chicken, sheep and horse blood when used alone have deleterious effects (Gardiner and Maddrell 1972, Gomes et al. 1990). We have observed that human blood is best to establish *T. cruzi* infection in the bugs. Also, not all anticoagulants are considered suitable. Citrated blood is reported to be better than heparinized blood and oxalates or fluorides can be deleterious, hence should not be used as anticoagulants (Azambuja and Garcia 1997). In our laboratory, we use rabbit blood for feeding the bugs at two-four week intervals.

Feeding

The most widely used method of feeding these bugs has been the use of immobilized live animals such as a rabbit, mouse, guinea-pig, hamster, or chicken. Another method, the artificial feeding method, involves warm blood from humans or other vertebrates applied to a membrane. (Pimley and Langley 1978, Nunez and Segura 1987, Nunez and Lazzari 1990). Very often the artificial feeding method has been used for experimental purposes, since it facilitates the modification of blood composition. Using this technique, substances such as drugs, parasites, feeding stimulants, feeding deterrents and bacteria can easily be fed to bugs in different media such as blood, serum or saline. Several factors that might affect feeding and engorging by these bugs have been reported. Some of these include thermal, visual and olfactory stimuli; the composition, osmotic pressure and pH of the diet; length of time since previous meal; and the degree of abdominal stretch (Langley 1976, Friend and Smith 1977, Schofield 1979, Lazzari and Nunez 1989, Guerenstein and Nunez 1994). Some laboratories use a combination strategy, alternating between live animal and membrane feeding. For ethical reasons we try to feed our bugs using the artificial method of feeding most of the time. Live animal use can be reserved for boosting the colonies a couple of times per year.

Live Animal Feeding

Materials

A healthy vertebrate host, an anesthetic, syringe, needle (gauge, 26G1/2), an enclosed container for the vertebrate host, rearing containers with hungry bugs, restraining equipment if bigger animals are used.

Procedure

A hamster or guinea pig is injected with 10 mg/kg dose of Ketaset® (Fort Dodge, Iowa, USA) and placed in an enclosed container until the animal is fully anesthetized, i.e., it should not exhibit any voluntary movement in response to tactile stimuli. This can take up to 10 to 15 min; the animal then sleeps for one to two hr. The anaesthetized animal is then placed on top of a rearing jar containing hungry bugs. The body heat and odor emanating from the animal attract the bugs (Wigglesworth and Gillett 1934, Nunez 1982, Taneja and Guerin 1997, Taneja and Guerin 1998). The bugs then orient themselves in the direction of the food source and climb up the filter paper in the container and start feeding on the animal by piercing their proboscis' through the mesh. Feeding should be terminated within an hour before the animal shows any signs of consciousness. A maximum of about 50 bugs, including 10 adults and 10 fifth instars can be fed on one animal. Up to 100 bugs can be fed if the feeding is continued until exsanguinations. We achieve 90 to 100% feeding success (percent of colony insects that engorge) with this method.

Notes

1. When using bigger animals such as a guinea-pig or a rabbit for feeding the bugs, it is advisable to restrain them, despite anesthetic; this is not so important with smaller animals such as a mouse or a hamster.
2. Using rats for feeding is not a good idea. They are very active and aggressive, making it difficult to restrain them.
3. To facilitate feeding by the bugs on a bigger animal with a thick furry coat, an area on the skin surface can be shaved where the insect container will be attached, preferably the belly, usually, in an inverted position.
4. Irrespective of which animal is used for feeding, the feeding set up should not be left unattended over long periods of time. The status of feeding should be checked every 15 min. The animal could wake up ahead of time and start moving and struggling, thus dismantling the set up, or the animal could die. The bugs when fed on a dead animal either do not survive or do not develop normally.
5. Covering the bug container with a black cloth or a black container facilitates frequent checking on the sleeping animal's condition without disturbing the feeding bugs.
6. Animals used for feeding should be disease free and free of drug treatments, internal or external. A fully grown, healthy young adult is always preferable.

The animal used for feeding the bugs should preferably be closely related to their natural hosts. Feeding for prolonged periods of time on distantly related animals such as cows or sheep has been reported to have adverse effects on the bugs, i.e., slow development, low reproduction and high mortality rates (Azambuja and Garcia 1997).

Artificial Feeding

In this method, blood is drawn from live vertebrates and mixed with anticoagulants or subjected to defibrination to prevent coagulation. It is then fed to insects through membrane devices. Various membranes such as thin rubber, latex or other synthetic surgical gloves, condoms, or laboratory film (Parafilm, USA) can be used as feeding membranes. Several facile artificial feeding methods have been used in our laboratory. Three of these are explained below:

Water-Jacketed Feeding Chambers

Materials

Defibrinated or citrated sheep blood, or citrated human blood (0.019M sodium citrate), feeding chambers, synthetic membrane, adhesive tape, a stand with clamps, a water bath with circulation at ca. 40°C, rearing jars containing hungry bugs.

Procedure

A feeding chamber is custom designed from glass. It consists of a central reservoir of 10 ml capacity open at both ends, enclosed in a double-walled jacket. The lower end of the central reservoir is covered with stretched membrane which is then carefully secured with adhesive tape. Thus prepared feeding chamber is then held inverted over a decontaminating container (with 10% bleach) and 5 to 7 ml fresh blood is slowly poured into it from the top. The jacket of the feeding chamber has an inlet and an outlet, so that warm water from a water-bath can be constantly circulated around the central reservoir. This keeps the blood warm (ca. 37°C). Once the blood has reached the desired temperature the feeding chamber is placed on a rearing jar and secured with clamps (Fig. 3). The set-up is then placed in the dark so insects can feed undisturbed. This method gives a feeding success rate of approx. 55 to 70%.

Notes

1. Any unused, powder-free, disposable, synthetic gloves can be used as membrane. They should be sterile, preferably irradiated before use, then cut to size and utilized only once. The membrane is stretched so that most bugs can pierce their proboscis' through it. Care is taken to not tear the membrane.
2. Holding the feeding chamber over a bleach solution when pouring blood is a safety measure. In case there are any leaks, blood will fall in the decontaminating container.

Fig. 3. Water-Jacketed feeding chamber.

3. Air bubbles in the blood should be avoided from settling directly under the membrane. Presence of air bubbles can cause interruption in feeding of a bug before full engorgement.
4. Glass feeding chambers should be handled with utmost care, especially when filled with blood. The jacket around the central blood reservoir can be much bigger than the central reservoir itself. The warmth emanating from the circulating warm water will stimulate hungry bugs and attracting them to the blood.
5. Glass feeding chambers should be handled with utmost care, especially when filled with blood.
6. The jacket around the central blood reservoir can be much bigger than the central reservoir itself. The warmth emanating from the circulating warm water will stimulate hungry bugs and attracting them to the blood.
7. These water-jacketed feeders are heavy and should be used very carefully on the nylon mesh covering the cages to avoid tears.
8. After each feeding, the feeding chambers should be washed and autoclaved before re-use.

Using Conical Tubes as Feeding Chambers

This is a very simple, easy, and inexpensive method (Fig. 4).

Fig. 4. Conical tube as feeding chamber.

Materials

Defibrinated or citrated sheep blood or citrated human blood (0.019M sodium citrate), 15 ml or 50 ml sterile conicals (Coming, NY, USA), synthetic membrane, adhesive tape, stand with clamps, a steam/water heating device.

Procedure

Two to 5 ml of blood is poured in a sterile conical tube. The tube is covered with a stretched membrane which is then secured with adhesive tape. The blood is warmed by holding the tube for ten s in the steam from a small portable room humidifier or for five minin a 40°C water bath. The tube containing warm blood is then clamped in an inverted position on a rearing container. Most bugs climb up the filter paper in the container immediately and start feeding. The setup is then covered with black cloth, so the bugs can feed undisturbed in the dark. The tube containing the blood needs re-warming every 20–30 min. We achieve approx. 40–60% feeding success with this method.

Notes

1. Care should be taken when stretching the membrane, as too much stretch can cause tearing.
2. Both the conicals and the membranes should be disposed of after a single use. Better to use disposable materials than something that might need regular cleaning for re-use. Before re-use, items should be sterilized.
3. It is a good idea to check for leaks before feeding with inverted tubes, by inverting the tube over a decontaminating solution (10% bleach) in a beaker.
4. The blood should be heated to approximately 37°C only. Overheating should be avoided. It is better to re-heat blood at frequent intervals (approx. every 20–30 min).
5. It is better to use broad (50 ml) tubes rather than the narrow (15 ml) tubes. The larger surface area provides a better opportunity to most bugs that are motivated to feed. Two or more narrow tubes can be used for each container to increase the surface area available for feeding.
6. Air-bubbles should be avoided from settling on the membrane. The narrow tubes (15 ml) should contain a minimum of 2 ml and the broad (50 ml) ones 5 ml of blood when starting to feed. The blood level should be constantly monitored while feeding. It is good to check for the presence of any air-bubbles or air gaps every time the blood is re-heated. Overall, this is the simplest method but it requires repeated warming of blood and checks on the status of feeding.

Use of Membrane Covered Glass-Plate as a Feeding Chamber

Materials

A glass plate (25 cm x 30 cm) with ground edges, laboratory film (Parafilm®, USA) an electric hot plate/heating pad, blood, syringe, needle (gauge, 26G1/2).

Procedure

The glass plate is autoclaved. Under sterile conditions, one side of the plate is then covered with a membrane by carefully stretching laboratory film and sealing at the edges of the plate to form an enclosed chamber. Five to 7 ml of blood is then carefully injected into the chamber by inserting a fine needle through the membrane from a corner. The glass plate chamber with blood is then placed on a hot plate or heating pad set at 40°C. Once the glass plate and the blood have reached the desired temperature (37°C) the rearing containers with hungry bugs are placed inverted on the membrane through which they pierce their proboscis and feed on the warm blood underneath. In this set-up, the mesh on the rearing jars is secured with rubber bands instead of the hollow lids to avoid damage to the membrane (Fig. 5).

Fig. 5. Membrane covered glass plate as a feeding chamber.

Notes

1. The feeding chamber is a bit more difficult to set-up than the conicals, but once prepared properly, it is an efficient method of feeding many insects.
2. It is important to make sure that the Parafilm membrane is leak-proof. The blood should be filled slowly and carefully. If any leaks are detected, however small, they should be immediately fixed with some new membrane.
3. The needle used for injecting the blood in the chamber should be very fine. A bigger bore needle may tear the membrane.
4. The containers with insects should be placed very carefully and gently onto the feeding chamber. Any extra pressure could tear the membrane.
5. A decontaminating solution (10% bleach) should always be kept within reach in case of any relatively big leaks which cannot be repaired. In such a case, the feeding should be stopped and the plate setup

immersed in the decontaminating solution as neatly as possible. A new plate chamber then needs to be assembled.

6. Never fill too much blood in the chamber. Too much pressure on the membrane can cause a leak. It is a better idea to fill a sufficient quantity of blood in the chamber (1 to 2 mm thick layer) from which the bugs can feed. Gradually as the blood level goes down with the intake by the bugs some more blood can be added. This can be repeated as long as the bugs keep feeding.

Frequency and Duration of Feeding

If insects feed to full engorgement, monthly meals are sufficient for survival, development and reproduction. However, in most feeding regimes, not all bugs feed to full engorgement, thus making multiple feedings necessary.

Feeding on live animals

As the bugs feed more readily on a live animal, each rearing jar containing 10–50 bugs can be fed in ca. 20–30 min. The frequency of feeding can also be lower since the feeding success of the bugs is higher and almost all bugs that are hungry would take a full meal. We find that a feeding frequency of once a month is adequate, when using animals such as rabbits, guinea pigs or hamsters.

Membrane feeding

To achieve the success equivalent to live animal feeding, both the duration and frequency of membrane feeding need to be increased. To feed a rearing jar containing 10 to 50 bugs, a two to three hr duration gives quite good results. A frequency of feeding at weekly or biweekly intervals is necessary, since not all bugs feed at a given time and can be ready for a meal a week later. In addition, an increase in the feeding frequency is required for optimum egg production, to get the fifth instars to molt into adults and to establish infection with *T. cruzi.*

Infection with Parasites via Feeding

One of the main advantages of membrane feeding is that it permits the introduction of experimental compounds and organisms into the gut of the bugs. To infect the bugs with *T. cruzi*, we follow the protocol by Garcia and Azambuja 1997. Human blood is first complement-inactivated by heating it at 56°C for an hr. *T. cruzi* from liquid culture of epimastigotes maintained in our laboratory is then added to the heat-inactivated blood to get the desired titer (ca.10^4–10^5epimastigotes/ml). This blood containing parasites is then fed to the bugs through a synthetic membrane as described above. Extra care should be taken when handling the tubes containing *T. cruzi* infected blood. Double gloves are worn and the disinfectant solution is always kept within reach in case there is a leak or rupture in the membrane. It is important to avoid too much heating of the blood before feeding as 1) it can kill the parasites in blood and 2) hot blood can rupture the membrane and cause leakage.

Oviposition

The *Rhodnius prolixus* females glue their eggs onto a surface. In the accordion folded filter paper they usually glue them in the folds. In nature these bugs are often found associated with birds' nests; therefore, bird feathers (soft fluffy down or young feathers) can be provided to them in rearing containers. We use washed and dried chicken or pigeon feathers that have been sterilized with ethanol (Fig. 6). When these are put in the rearing containers, the bugs invariably prefer the feathers to paper for gluing their *eggs*. This can also stimulate a slight increase in the egg production.

Triatoma dimidiata females lay their eggs loose which can get collected at the base of the rearing container. One female can lay up to 40 eggs per blood-meal for both species. The incubation period of the eggs before they hatch into the first instar nymphs is seven to 14 d for *R. prolixus* and 14 to 21 d for *T. dimidiata*. Newly hatched nymphs of both species do not feed during the first three to seven d.

Fig. 6. Eggs of *Rhodnius prolixus*, laid on and glued to pigeon feathers.

Aposymbiotic Colonies

Both *Rhodnius prolixus* and *Triatoma dimidiata* harbor extracellular bacteria in their gut that serve symbiotic functions. We have shown in *R. prolixus* that such bacteria can be genetically altered and re-introduced into the insect to produce molecules toxic to *T. cruzi* (Durvasula et al. 1997). To maintain lines of aposymbiotic bugs, i.e., bugs lacking extracellular symbionts in their gut, we follow the protocol of Beard and Aksoy (1997). Briefly, freshly laid eggs are immersed for three min in a Clinidine Solution (CliniPad, USA) (Fig. 7). They are then rinsed in sterile PBS until all traces of iodine are gone (this usually takes five to seven rinses).

The rinsed eggs are transferred to a sterile rearing container that is housed within another sterile container (Fig. 8). Aposymbiotic nymphs are maintained strictly on membrane feeders, using only sterile membranes. This permits introduction of wild-type or genetically transformed bacteria during blood-meals. Aposymbiotic colonies have higher rates of mortality than wild-type colonies. With *R. prolixus* symbiotic bacteria should be introduced to nymphs by the second instar stage to prevent mortality.

Fig. 7. *T. dimidiata* eggs immersed in Clinidine Solution.

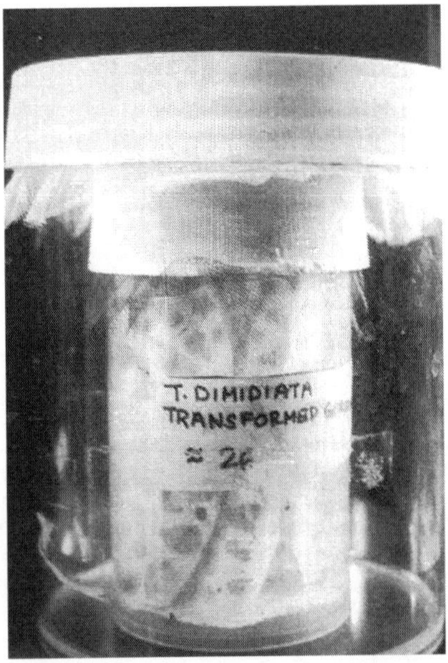

Fig. 8. Double sterile containers method for housing aposymbiotic bugs.

References

Ampleford, E. and K. Davey. 1989. Egg laying in the insect *Rhodnius prolixus* is timed in a circadian fashion. J. Insect Physiol. 35: 183–187.

Ampleford, E. and C. Steel. 1982. Circadian control of ecdysis in *Rhodnius prolixus* (Hemiptera). J. Comp. Physiol. 147: 281–286.

Azambuja, P. and E.S. Garcia. 1997. Care and maintenance of triatomine colonies. pp. 56–64. *In*: J.M. Crampton, C.B. Beard and C. Louis (eds.). The Molecular Biology of Insect Disease Vectors—A Methods Manual. Chapman & Hall, New York.

Baines, S. 1956. The role of the symbiotic bacteria in the nutrition of *Rhodnius prolixus* (Hemiptera). J. Exp. Biol. 33: 533–541.

Beard, C.B. and S. Aksoy. 1997. Genetic manipulation of insect symbionts. pp. 555–560. *In*: J.M. Crampton, C.B. Beard and C. Louis (eds.). The Molecular Biology of Insect Disease Vectors—A Methods Manual. Chapman & Hall, New York.

Durvasula, R.V., A. Gumbs, A. Panackal, O. Kruglov, S. Aksoy, R.B. Merrifield, E.F. Richards and C.B. Beard. 1997. Prevention of insect-borne diseases: An approach using transgenic symbiotic bacteria. Proc. Natl. Acad. Sci. 94: 3274–3278.

Else, J.G., W.H. Cheong, S. Mahadevan and L.G. Zarate. 1977. A new species of cave-inhabiting *Triatoma* (Hemiptera: Reduviidae) from Malaysia. J. Med. Entomol. 14: 367–369.

Friend, W.G. and J.J.B. Smith. 1977. Factors affecting feeding by bloodsucking insects. Ann. Rev. Entomol. 22: 309–331.

Garcia, E.S. and P. Azambuja. 1991. Development and interactions of *Trypanosoma cruzi* within the insect vector. Parasitol. Today 7: 240–244.

Garcia, E.S. and P. Azambuja. 1997. Infection of triatomines with *Trypanosoma cruzi*. pp. 146–158. *In*: J.M. C-rampton, C.B. Beard and C. Louis [eds.]. The Molecular Biology of Insect Disease Vectors—A Methods Manual. Chapman & Hall, New York.

Gardiner, B.O.C. and S.H.P. Maddrell. 1972. Techniques for routine and large-scale rearing of *Rhodnius prolixus* (Hemiptera, Reduviidae). Bull. Ent. Res. 61: 505–516.

Gomes, J.E.P.L., P. Azambuja and E.S. Garcia. 1990. Comparative studies on the growth and reproductive performances of *Rhodnius prolixus* reared on different blood sources. Mem. Oswaldo Cruz. 85: 299–304.

Guerenstein, P.G. and J.A. Nunez. 1994. Feeding response of the haematophagous bugs *Rhodnius prolixus* and *Triatoma infestans* to saline solutions: a comparative study. J. Insect Physiol. 40: 747–752.

Higgs, S. and B. Beaty. 1996. Rearing and containment of mosquito vectors. pp. 595–605. *In*: B.J. Beaty and W.C. Marquardt [eds.]. The Biology of Disease Vectors. University Press of Colorado.

Kalshoven, L.G.E. 1970. Observations on the blood-sucking reduviid *Triatoma rubrofasciata* (De Geer) in Java. Entomol. Berichten Amsterdam 30: 41–47.

Langley, P.A. 1976. Initiation and regulation of ingestion by hematophagous arthropods. J. Med. Entomol. 13: 121–130.

Lazzari, C.R. 1992. Circadian organization of locomotion activity in the haematophagous bug *Triatoma infestans*. J. Insect Physiol. 38: 895–903.

Lazzari, C.R. and J.A. Nunez. 1989. Blood temperature and feeding behavior in *Triatoma infestans* (Heteroptera: Reduviidae). Entomol. Gener. 14: 183–188.

Lehane, M.J., P.K. McEwan, C.J. Whitaker and C.J. Schofield. 1992. The role of temperature and nutritional status in flight initiation by *Triatoma infestans*. Acta Trop. 52: 27–38.

Lent, H. and P. Wygodzinsky. 1979. Revision of the Triatominae (Hemiptera: Reduviidae), and their significance as vectors of Chagas disease. Bull. Am. Mus. Nat. Hist. 163: 123–520.

Lorenzo Figueiras, A.N., A. Kenigsten and C.R. Lazzari. 1994. Aggregation in the haematophagous bug *Triatoma infestans*: Chemical signals and temporal pattern. J. Insect Physiol. 40: 311–316.

Monteith, G.B. 1974. Confirmation of the presence of Triatominae (Hemiptera: Reduviidae) in Australia, with notes on Indo-Pacific species. J. Aust. Ent. Soc. 13: 89–94.

Nunez, J.A. 1982. Food source orientation and activity in *Rhodnius prolixus* Stal (Hemiptera: Reduviidae). Bull. Ent. Res. 72: 253–262.

Nunez, J.A. and C.R. Lazzari. 1990. Rearing of *Triatoma infestans* Klug (Hemiptera, Reduviidae) in the absence of a live host: I. Some factors affecting the artificial feeding. J. Appl. Ent. 109: 87–92.

Nunez, J.A. and E.L. Segura. 1987. Rearing of Triatominae. *In*: R.R. Brenner and A.M. Stoka [eds.]. Chagas' Disease Vectors. CRC Press, Florida. 2: 31–40.

Pimley, R.W. and P.A. Langley. 1978. Rearing triatomine bugs in the absence of a live host and some effects of diet on reproduction of *Rhodnius prolixus* Stal (Hemiptera, Reduviidae). Bull. Ent. Res. 68: 243–250.

Saunders, D.S. 1982. *In*: Insect Clocks. Pergamon Press, Oxford.

Schaub, G.A., C.A. Boker, C. Jensen and D. Reduth. 1989. Cannibalism and coprophagy are modes of transmission of *Blastocrithidia triatominae* (Trypanosomatidae) between Triatomines. J. Protozool. 36: 171–175.

Schofield, C.J. 1979. The behavior of Triatominae (Hemiptera: Reduvidae): a review. Bull. Entomol. Res. 69: 363–379.

Schofield, C.J., M.J. Lehane, P. McEwan, S.S. Catala and D.E. Gorla. 1991. Dispersive flight by *Triatoma sordida*. Trans. R. Trop. Med. 85: 676–678.

Taneja, J. and P.M. Guerin. 1995. Oriented responses of triatomine bugs *Rhodnius prolixus* and *Triatoma infestans* to vertebrate odourson a servosphere. J. Comp. Physiol. A. 176: 455–464.

Taneja, J. and P.M. Guerin. 1997. Ammonia attracts the haematophagous bug *Triatoma infestans*: behavioral and neurophysiological data on nymphs. J. Comp. Physiol. A. 181: 21–34.
WHO Expert Committee. Control of Chagas Disease (WHO Technical Series Report). Geneva: World Health Organization 1991. 811: 1–95.
Wigglesworth, V.B. 1972. *In*: Principles of Insect Physiology. Chapman and Hall, London.
Wigglesworth, V.B. and J.D. Gillett. 1934. The function of the antenna in *Rhodnius prolixus* (Hemiptera) and the mechanism of orientation to the host. J. Exp. Biol. 11: 120–139.
Zeldon, R. and J.E. Rabinovich. 1981. Chagas' disease. An ecological appraisal with special emphasis on its insects vectors. Ann. Rev. Entomol. 26: 101–133.

5

Laboratory Rearing of Bed Bugs

Mark F. Feldlaufer,[1,*] *Harold J. Harlan*[2] and *Dini M. Miller*[3]

ABSTRACT

The resurgence of bed bugs *Cimex lectularius* L. in the United States and worldwide has resulted in an increase in research by university, government, and industry scientists. This research has primarily been directed at the biology and control of this blood-feeding pest. A need has subsequently arisen for producing sufficient biological material for research purposes and a variety of rearing methods are currently being employed. Colony rearing and maintenance of bed bugs, however, must be conducted carefully to both reduce the possibility of the researcher being bitten, and to prevent the unwanted and unwitting spread of bed bugs both within the laboratory environment, and beyond. We report rearing methodologies and procedures, including *in vitro* blood-feeding that maximize the production of desired numbers of specimens of consistent size and biological qualities of all bed bug stages in a controlled environment while minimizing the likelihood of escape.

Key Words: bed bugs, *Cimex lectularius*, rearing, *in vitro* feeding

[1] Invasive Insect Biocontrol and Behavior Laboratory, USDA-ARS, Beltsville (retired), MD 20705, USA.

[2] Armed Forces Pest Management Board, Silver Spring, MD 20910, USA.

[3] Department of Entomology, Virginia Polytechnic Institute and State University, Blacksburg, VA 24061, USA.

* Corresponding author

Mention of trade names or commercial products in this article is solely for the purpose of providing specific information or examples, and does not imply recommendation or endorsement by any of the authors or the United States Department of Agriculture, the Department of Defense, or the Virginia Polytechnic Institute and State University.

Introduction

The resurgence of the common bed bug, *Cimex lectularius* L. (Hemiptera: Cimicidae), is well-documented in the scientific literature as well as in the the lay media (Anderson and Leffler 2008, Wang et al. 2010, National Pest Management Association 2010). Explanations for this resurgence include but are not limited to, increased travel, changes in pest control practices for other urban pests, and insecticide resistance (Boase 2004, Harlan et al. 2008, Romero et al. 2007). While there still remains no scientifically valid evidence that the common bed bugor its tropical cousin *Cimex hemipterus* can act as a biological vector of human pathogens, mechanical transmission of certain viruses remains debatable (Blow et al. 2001, Jupp and Lyons 1987, Jupp et al. 1980, 1983, Silverman et al. 2001, Webb et al. 1989). Apart from disease transmission, the common bed bug is annoying, difficult to control, and their bites can cause allergic reactions. These reactions range from little if any irritation, to both immediate and delayed immune responses, to relatively severe allergic hypersensitivity (Leverkus et al. 2006, Goddard and DeShazo 2009, Reinhardt et al. 2009). In one report, an antigen prepared from the common bed bug was linked to bronchial asthma (AbouGamra et al. 1991), while in locations where very large infestations of bed bugs exist, their blood-feeding can result in victims presenting with anemia (Pritchard and Hwang 2009). A review of illnesses associated with the pesticides used to control bed bugs has recently been published (Centers for Disease Control and Prevention 2011).

The economic impact of bed bug infestations, though difficult to document, can be huge. Costs can include expensive detection methods, such as bed bug-sniffing dogs, intensive inspections and treatment applications that take hours to complete, the perceived need to replace bedding and furniture, and the potential for negative publicity or lawsuits generated when people encounter bed bugs in hotels or businesses (Miller 2007, Wenk 2007).

The bed bug resurgence has led to an increase in both basic and applied research directed at the biology and control of this pest. The increased research efforts have created a subsequent need for producing sufficient numbers of consistent quality biological specimens for empirical tests. The general topic of feeding common bed bugs on laboratory host animals has previously been addressed (Peterson 1959, Burden 1966). In this current article, we present additional rearing techniques for the common bed bug, including *in vitro* membrane feeding, that allow for the production of large numbers of consistently similar bed bug specimens, while minimizing the likelihood of bugs escaping into the surrounding environment.

We note that several other species of cimicids have also been successfully reared in the laboratory through at least one complete generation. Published examples have included the Mexican chicken bug, *Haematosiphon inodorus* (Duges), colonized by Lee (1955); the cliff swallow bug, *Oeciacus vicarius* Horvath colonized by Rush (1981), and more recently by Oesterle et al. (2010); an uncommon north American bat bug, *Cimex brevis* Usinger and Ueshima, colonized by Bower and Woo (1981); and the eastern bat bug, *Cimex adjunctus* Barber, colonized by Reeves (2000). In each of these cases, the rearing conditions were very similar to those typically used for the common bed bug, as detailed in this article, with the necessary variations of temperature, light-dark cycle, humidity and; host blood source for each of the respective species reared.

Field Collection Methods

Laboratory strains of common bed bugs *C. lectularius* have understandably originated from field collections. One widely-available strain, often designated as the 'Harlan strain', was established in 1973 from an infestation in a U. S. Army barrackson Ft. Dix, NJ (Bartley and Harlan 1974). The Harlan strain is considered to be sensitive to pyrethroid (e.g., deltamethrin) exposure (Romero et al. 2007). Another insecticide-susceptible strain that has often been used in laboratories originated from field collections in Gainesville FL (Yoon et al. 2006, Yoon et al. 2008). With the current resurgence of bed bug infestations nationwide, new field populations are constantly being introduced into the laboratory for research purposes, particularly in those studies involving insecticide resistance (see Moore and Miller 2006, Romero et al. 2007, Yoon et al. 2008, Zhu et al. 2010, Kilpinen et al. 2011).

Laboratory Maintenance

Materials and Methods

Containers and Harborages

All stages of bed bugs must be maintained in some sort of plastic or glass container that includes a harborage (usually constructed from filter paper or cardboard) where the bugs can aggregate and hide. Most containers also have a cap containing a fine mesh material that prevents bugs from escaping while allowing all stages of the bugs to feed on a suitable blood source. The senior author (MFF) maintains large numbers of bed bugs in 480-ml (pint), wide-mouthed glass canning jars (Jarden Corp., Rye, NY), available at most hardware stores. Lids are removed and discarded from the cap bands, and fine (100 mesh) nylon, available at most fabric stores, is cut to size (84-mm dia.) and attached to the inside of the cap band with a general purpose, silicone adhesive sealant (Fig. 1). After curing, the cap can

Fig. 1. Bed bug rearing components. From LEFT, Mason jar, cap with lid removed and nylon mesh, finished cap, filter paper harborage.

be screwed onto the jar forming a tight seal. The nylon mesh cap provides suitable containment and a matrix through which the bugs can feed.

Harborages for these particular jars are constructed of rectangular-cut (140 by 100 mm), fan-folded filter paper (Whatman International Ltd, Maidstone UK). The smaller dimension(100mm) of the harborage is important, since this becomes the height of the harborage when the filter paper is fan-folded along the 140 mm side. By not having the filter paper (four to six/jar) extend to the top of the jar (approx. 115 mm), bed bugs cannot easily escape when the cap of the jar is removed for either colony maintenance or removal of bugs for testing. One method of feeding (see below) involves inverting a colony jar over a blood source, so that the fan-folded filter paper drops onto the nylon mesh part of the cap affording the bed bugs ready access to the blood meal. This allows feeding with minimal handling and there is no need to remove the cap during the feeding process. As mentioned above, other jars and caps of various sizes, as well as harborages, can also be used to maintain bed bugs, though the principles regarding harborage and cap construction still remain the same.

Temperature/Photoperiod Requirements

Bed bugs are cosmopolitan pests, attacking in most environments where their human hosts are found. As such, bed bugs can be kept and reared at most ambient, indoor temperatures. Developmental and physiological studies where temperature has been a variable have utilized bed bugs kept at temperatures ranging from below 0°C–48°C for individual experiments (see Usinger 1966, and references therein; Benoit et al. 2009a, Reinhardt et al. 2010). When temperature is not a variable, a review of the recent literature indicated that laboratory colonies of *C. lectularius* are commonly kept in incubators at temperatures within this range, usually from 20°C–30°C (Montes et al. 2002, Siljander et al. 2007, Olson et al. 2009, Romero et al. 2009, Weeks et al. 2010, Polanco et al. 2011, Reis and Miller 2011, Suchy and Lewis 2011). In these instances, photoperiodic regimes have varied from 12–16 h of light to total darkness.

Rearing Techniques

Feeding

Bed bugs are obligate, blood-feeding insects. Each of five nymphal stages requires a blood meal to molt and reach the next nymphal or the adult stage, and adults require periodic blood meals to reproduce. Therefore, the blood-feeding of laboratory colonies is a critical component of any laboratory research program designed around bed bugs. In the past several years, researchers have employed a variety of blood sources for both the *in vivo* and *in vitro* feeding of bed bugs. A review of the literature in the past decade reveals that chickens (Pfeister et al. 2008), rabbits (Stutt and Siva-Jothy 2001, Reinhardt et al. 2003, Anderson et al. 2009), mice and pigeons (Araujo et al. 2009), and humans (Moore and Miller 2006, Siljander et al. 2007, Wintle and Reinhardt 2008) have been used as direct hosts for *in vivo* feeding of bed bugs.

In vitro feeding has utilized the blood (or blood products) of cattle (Montes et al. 2002), chickens (Montes et al. 2002, Romero et al. 2007, Siljander et al. 2007, Benoit et al. 2009b, Ryne 2009, Zhu et al. 2010), rabbits (Romero et al. 2009), sheep (Montes et al. 2002, Harraca et al. 2010, Weeks et al. 2010) and humans (Moor and Miller 2006, Yoon et al. 2008, Araujo et al. 2009, Olson et al. 2009, Feldlaufer et al. 2010, Seong et al. 2010). Most current *in vitro* feeding techniques are derived from methods developed by Montes et al. (2002), or modifications of a method originally developed for head lice (Takano-Lee et al. 2003a, 2003b, Yoon et al. 2006). Other *in vitro* methods have been used prior to this time though most were designed for either feeding individual bugs or for specific feeding experiments and not for colony maintenance (see Rivnay 1930, De Meillon and Goldberg 1947, Hall et al. 1979, Ogston and Yanovski 1982). During the *in vitro* feeding of bed bugs, the blood meal is warmed to 37–39°C by circulating heated water through a jacketed feeding vessel. The importance of warming the blood source to stimulate bed bug feeding has been previously noted (Montes et al. 2002). Jacketed beakers of various dimensions are available commercially (Fig. 2), and can be used either inverted over a colony of bugs, or right-side-up, where the colony of bed bugs is inverted over the jacketed beaker. For *in vitro* feeding we allow all stages of bed bugs to feed through a stretched Parafilm® M (Pechiney Plastic Packaging, Chicago,

IL) or Nescofilm® (Karlan Research Products Corp., Cottonwood, AZ) membrane. The senior author (MFF) uses human red blood cells fortified with plasma as previously described (Feldlaufer et al. 2010). The feeding setup is shown in Fig. 3. This particular jacketed beaker (Kimble Chase-Kontes Glass, Vinel and, NJ) has a relatively large surface area (approx. 14 cm^2) through which bugs can feed.

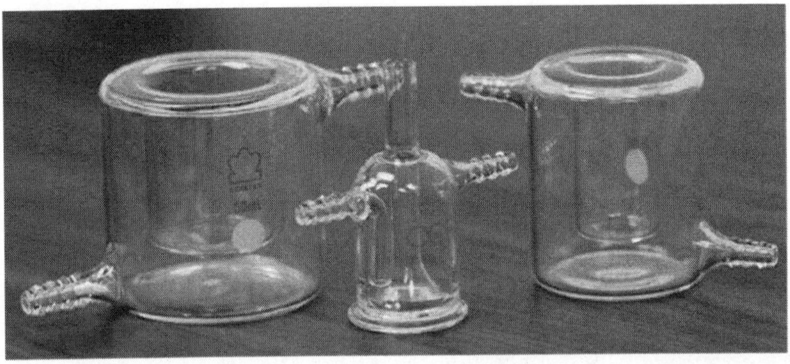

Fig. 2. Some commercially-available jacketed water beakers. The jacketed beakers on the LEFT and RIGHT are used with the bed bug colony inverted over the blood source. The jacketed funnel in the MIDDLE can be placed over an upright, bed bug colony. Note the various dimensions of the openings.

Fig. 3. LEFT-Jacketed beaker with blood covered by a Parafilm membrane. RIGHT-Bed bug colony inverted over blood source.

Information about other glass feeders, heating units, and systems is available online through the Malaria Research and reference Reagent Resource Center (2011). Regardless of the apparatus used, *in vitro* feeding has several benefits for the investigator. *In vitro* feeding eliminates the need for IRB (Institutional Review Board) approvals when humans are used as a direct source of blood, and the need for IACUC (Institutional Animal Care and Use Committee) approvals when laboratory animals are used as a direct blood source (Garber et al. 2010, Penslar and Porter 1993, Silverman et al. 2000, and information therein). Finally, whether an *in vivo* or *in vitro* method is employed for rearing bed bugs, all blood or blood products need to be kept, used and disposed of according to institutional procedures.

Other Considerations

Routine Colony Maintenance and General Laboratory Practice

The greatest risk of bugs escaping or investigators being bitten occurs during routine colony maintenance or when handling large numbers of bugs. Bed bugs do not move efficiently on slippery surfaces (Loudon and Boudaie 2009), so smooth surfaced enclosures are generally the foundation of all containment protocols. Glass Petri dishes lined with a polytetrafluroethylene product such as Fluon® (AGC Chemicals Americas, Inc., Exton, PA), polyethylene plastic Petri dishes, or other non-stick surfaces can all help confine bed bugs. Chilled rooms or surfaces slow bed bug movement and can help limit their potential for escape. The following general laboratory protocols provide examples of how bed bugs can be kept. All vessels containing bed bugs, including colony jars, and Petri dishes used in individual experiments are kept on commercially-available, non-stick cookie sheets (e.g., 337 by 235 mm), edged by a thin layer of petroleum jelly. Double-sided tape can also be used as a perimeter barrier. Depending on the number of bugs being reared and the frequency of feeding, mesh covers must be periodically replaced, since bed bug excreta can clog the mesh and reduce the surface area through which bugs can feed. Exuviae also must be periodically removed from the jars, particularly if jars are inverted during the feeding process. Exuviae can form an unwanted, physical barrier between the bed bugs and the blood meal. Removal of exuviae requires removing the cap of the jar, grasping the fan-folded harborages with long forceps, lifting the entire harborage out of the old jar, and placing it into a new jar. Most exuviae are left behind in the old jar, and a new cap can be placed on the new jar. Forceps and the old cap can be dipped into a beaker of hot water held at 55–60°C (Naylor and Boase 2010), immediately killing any bugs retained in the cap. Trying to harvest bed bugs remaining in the old cap is an easy way for the investigator to get bitten, or bugs to escape.

First-instar nymphs are very small, hard to see, and subject to movement by static electricity and air currents (Feldlaufer and Loudon 2011). Hot water (55–60°C) can be poured into old jars containing exuviae, killing any live bugs remaining in those jars.

At some point in time filter paper or cardboard harborages must be replaced, usually when they become so contaminated with excrement they take on a different color and texture. The decision to change harborages is somewhat subjective. To replace heavily soiled harborages, one can remove the old harborage with forceps, position the harborage matrix over a new jar, and then tap and 'gently shake' the old harborage matrix so that most of the bugs fall off of it, and down, into a clean jar. Then a new "clean" folded harborage (matrix) is placed into the jar. A camel's hair brush can also be used to gently dislodge bugs that still cling to the old harborage. This works well for populations that are easily reared and have many more individuals than are needed for study.

An alternative method relies on the bugs' tendency to climb upward when seeking a blood meal. A pre-folded section of new harborage material is place in contact with and above the old harborage, and bugs are allowed to climb onto it. It is then placed in a new jar. This is repeated for each harborage matrix sheet to be replaced (six sheets are used in the set-up illustrated in Fig. 1). In either case, old harborages are treated with hot water to kill any bugs that did not transfer. Insecticide-resistant strains usually have far fewer individuals, reproduce slower, and as many individuals as possible must be transferred to any replacement harborage. For such smaller populations or strains, the same earlier steps are followed, but before immersing the old, soiled harborage sheets in hot water, each one is placed in a 150x20 mm Petri dish on a chill table, carefully examined under at least 5X magnification, and all live bugs found removed by forceps to the new harborage. In addition, the contents of the old jar, including exuviae are dumped into a 150x20 mm petri dish on a chill table and all live bugs found are removed to the new container.

Removing bed bugs from a colony for individual experiments also poses a risk of bugs escaping and of the investigator being bitten. Besides working on non-stick surfaces, working on a chilled platform or in a cold room is useful to sort bed bug stages and sexes. Slowing the bugs' movement by cooling eliminates the need to use carbon dioxide, which is known to influence insect behavior in empirical studies (Badre et al. 2005). All bed bug sorting and manipulation are done on non-stick surfaces, with a beaker of hot water (55–60°C) nearby to clean forceps, brushes, or other tools used to manipulate the bugs. Personnel working with bed bugs can reduce the possibility of errant bed bugs becoming established by using plastic lawn/ patio chairs that have no cracks or crevices for bed bugs to hide in or become established. White chairs are best since all stages, except unfed first-instar nymphs, are more easily seen against light-colored backgrounds.

Personal Protection

In addition to the precautions described above, it is important to ensure bed bugs do not attach or crawl onto the researcher's apparel. Disposable, smooth polyethylene gowns with back closures (Poly Conversions, Inc., Rantoul IL) offer a protective barrier that bed bugs have great difficulty walking upon. Laboratory exam gloves pulled over the sleeves of the gown provide additional protection. Disposable polyethylene booties (Continental Plastics Corp., Delava WI) can also be used, if deemed necessary. Textured polyethylene coveralls or sleeves such as Tyvek® (DuPont™, Wilmington DE) offer less protection. Bed bugs can readily crawl on the textured surface, and the elastic closures are easily breached. Wearing clothes that can be left in the laboratory and laundered immediately after they are removed can also lessen the likelihood of transporting bed bugs to the researcher's home.

Conclusion

Bed bugs are relatively easy to maintain in the laboratory. Since blood feeding is a critical component of any bed bug rearing program, we have described a technique that has proven satisfactory, and have referenced other techniques. Our overall objective was to describe methods for rearing and manipulating bed bugs in a laboratory that will both minimize the risk of being bitten and the likelihood of bed bugs escaping. Many of these suggestions are based on the bed bugs' inability to move easily on smooth surfaces. While some of these procedures might be deemed "overcautious", they have been developed from research experience and the knowledge that once established bed bug infestations are difficult to control, even in the laboratory.

Acknowledgements

The authors wish to thank the following individuals for generously sharing their ideas, background information, experiences of rearing bed bugs, and technical suggestions: Jerome Goddard (Mississippi State University), Susan Jones (The Ohio State University), and Woodbridge Foster (The Ohio State University).

References

AbouGamra, E.M., F.A. el Shayed, T.A. Morsy, H.M. Hussein and E.S. Shehata. 1991. The relation between *Cimex lectularius* antigen and bronchial asthma in Egypt. J. Egypt Soc. Parasitol. 21: 735–746.

Anderson, A.L. and K. Leffle. 2008. Bed bug infestations in the news: A picture of an emerging public health problem in the United States. J. Environ. Hlth. 70: 24–27.

Anderson, J.F., F.J. Ferrandino, S. McKnight, J. Nolen and J. Miller. 2009. A carbon dioxide, heat and chemical lure trap for the bedbug, *Cimex lectularius*. Med. Vet. Entomol. 23: 99–105.

Araujo, R.N., F.S. Costa, N.F. Gontijo, T.C. Gonçalves and M.H. Pereira. 2009. The feeding process of *Cimex lectularius* (Linnaeus 1758) and *Cimex hemipterus* (Fabricius 1803) on different bloodmeal sources. J. Insect Physiol. 55: 1151–1157.

Badre, N.H., M.E. Martin and R.L Cooper. 2005. The physiological and behavioral effects of carbon dioxide on *Drosophila melanogaster* larvae. Comp. Biochem. Physiol. Part A. 140: 363–376.

Bartley, J.D. and H.J. Harlan. 1974. Bed bug infestation: Its control and management. Military Med. 139: 884–886.

Benoit, J.B., G. Lopez-Martinez, N.M. Teets, S.A. Phillips and D.L. Denlinger. 2009a. Responses of the bed bug, *Cimex lectularius*, to temperature extremes and dehydration: levels of tolerance, rapid cold hardening and expression of heat shock proteins. Med. Vet. Entomol. 23: 418–424.

Benoit, J.B., S.A. Phillips, T.J. Croxall, B.S. Christensen, J.A. Yoder and D.L. Denlinger. 2009b. Addition of alarm pheromone components improves the effectiveness of desiccant dusts against *Cimex lectularius*. J. Med. Entomol. 46: 572–579.

Boase, C.J. 2004. Bedbugs—reclaiming our cities. Biologist 51: 9–12.

Blow, J.A., M.J. Turell, A.L. Silverman and E.D. Walker. 2001. Stercorarial shedding and transtadial transmission of hepatitis B virus by common bed bugs (Hemiptera: Cimicidae). J. Med. Entomol. 38: 694–700.

Bower, S.M. and P.T.K. Woo. 1981. Development of *Trypanosoma* (*Schizotrypanum*) *hedricki* in *Cimex brevis* (Hemiptera: Cimicidae). Can. J. Zool. 59: 546–554.

Burden, G.S. 1966. Bed Bugs. pp. 175–182. *In*: C.N. Smith (ed.). Insect Colonization and Mass Production. Academic Press, Inc. New York, NY. USA.

Centers for Disease Control and Prevention. 2011. Acute illnesses associated with insecticides used to control bed bugs—seven States, 2003–2010. MMWR. 60: 1269–1274.

De Meillon, B. and L. Golberg. 1947. Preliminary studies on the nutritional requirements of the bedbug (*Cimex lectularius* L.) and the tick *Ornithodorus moubata* Murray. J. Exp. Biol. 24: 41–63.

Feldlaufer, M.F., M.J. Domingue, K.R. Chauhan and J.R. Aldrich. 2010. 4-oxo-aldehydes from the dorsal abdominal glands of the bed bug (Hemiptera: Cimicidae). J. Med. Entomol. 47: 140–143.

Feldlaufer, M.F. and C. Loudon. 2011. Undesirable dispersal of eggs and early-stage nymphs of the bed bug (Hemiptera: Cimicidae) by static electricity and air currents. J. Entomol. Sci. 46: 1–2.

Garber, J.C., R.W. Barbee, J.T. Bielitzki, L.A. Clayton, J.C. Donovan, D.F. Kohn, N.S. Lipman, P. Locke, J. Melcher, F.W. Quimby, P.V. Turner, G.A. Wood and H. Wurbel. 2010. Guide for the Care and Use of Laboratory Animals, 8th ed. The National Academies Press, Washington, DC.

Goddard, J. and R. DeShazo. 2009. Bed Bugs (*Cimex lectularius*) and clinical consequences of their bites. J. Am. Med. Assoc. 301: 1358–1366.

Hall, R.D., E.C. Turner, Jr. and W.B. Gross. 1979. A simple apparatus for providing blood diets at constant temperature and with different corticosterone levels to individual bed bug colonies (Hemiptera, Cimicidae). J. Med. Entomol. 16: 259–261.

Harlan, H.J., M.K. Faulde and G.J. Baumann. 2008. Bedbugs. pp. 131–151. *In*: X. Bonnefoy, H. Kampen and K. Sweeney (eds.). Public Health Significance of Urban Pests. World Health Organization, Copenhagen, Denmark.

Harraca, V., R. Ignell, C. Lofstedt and C. Ryne. 2010. Characterization of the antennal olfactory system of the bed bug (*Cimex lectularius*). Chem. Senses 35: 195–204.

Jupp, P.G. and S.F. Lyons. 1987. Experimental assessment of bedbugs (*Cimex lectularius* and *Cimex hemipterus*) and mosquitoes (*Aedes aegypti formosus*) as vectors of human immunodeficiency virus. Aids 1: 171–174.

Jupp, P.G., S.E. McElligott and G. Lecatsas. 1983. The mechanical transmission of hepatitis B virus by the common bedbug (*Cimex lectularius* L.) in South Africa. S. Afr. Med. J. 63: 77–81.

Jupp, P.G., O.W. Prozesky and S.E. McElligott. 1980. Absence of biological multiplication of hepatitis B virus in the common bedbug. S. Afr. Med. J. 57: 36.

Kilpinen, O., M. Kristensen and K.M.V. Jensen. 2011. Resistance differences between chlorpyrifos and synthetic pyrethroids in *Cimex lectularius* populations from Denmark. Parasitol. Res. 109: 1461–1464.

Lee, R.D. 1955. The biology of the Mexican Chicken Bug, *Haematosiphon inodorus* (Duges). Pan-Pacific Entomol. 31: 47–61.

Leverkus, M., R.C. Jochim, S. Schäd, E.-B. Bröcker, J.F. Andersen, J.G. Valenzuela and A. Trautmann. 2006. Bullous allergic hypersensitivity to bed bug bites mediated by IgE against salivary nitrophorin. J. Investigative Dermatol. 15: 91–96.

Loudon, C. and J. Boudaie. 2009. Walking with grappling hooks: bed bug locomotion on different surfaces. pp. 95. *In*: The 57th Annual Meeting of the Entomological Society of America; 13–16 December 2009, Indianapolis, IN. Entomol. Soc. Amer., Lanham, MD.

Malaria Research and Reference Reagent Resource Center. 2011. 2.4.8 Bloodfeeding: Membrane apparatuses and animals. http://www.mr4.org/AnophelesProgram/TrainingMethods/tabid/140/Default.aspx.

Miller, K. 2007. The cost of bedbugs. Business Week. November 8, 2007.

Montes, C., C. Cuadrillero and D. Vilella. 2002. Maintenance of a laboratory colony of *Cimex lectularius* (Hemiptera: Cimicidae) using an artificial feeding technique. J. Med. Entomol. 39: 675–679.

Moore, D.J. and D.M. Miller. 2006. Laboratory evaluations of insecticide product efficacy for control of *Cimex lectularius*. J. Econ. Entomol. 99: 2080–2086.

National Pest Management Association. 2010. The NPMA Releases First-Ever Comprehensive Global Bed Bug Study to Determine Extent of Resurgence. http://www.pestworld.org/bedbug.

Naylor, R.A. and C.J. Boase. 2010. Practical solutions for treating laundry infested with *Cimex lectularius* (Hemiptera: Cimicidae). J. Econ. Entomol. 103: 136–139.

Ogston, C.W. and A.D. Yanovski. 1982. An improved feeder for bloodsucking insects. J. Med. Entomol. 19: 42–44.

Olson, J.F., R.D. Moon and S.A. Kells. 2009. Off-host aggregation behavior and sensory basis of arrestment by *Cimex lectularius* (Heteroptera: Cimicidae). J. Insect Physiol. 55: 580–587.

Oesterle, P., N. Nemeth, G. Young, N. Mooers, S. Elmore, R. Bowen, P. Doherty, J. Hall, R. McLean and L. Clark. 2010. Cliff swallows, swallow bugs, and West Nile Virus: An unlikely transmission mechanism. Vector-borne Zoon. Dis. 10: 507–513.

Penslar, R.L. and J.P. Porter. 1993. IRB Guidebook. Office for Human Research Protections, U.S. Dept. Health & Human Services, Washington, DC. USA.

Peterson, A. 1959. Bedbug. pp. 117–188. *In*: A. Peterson (ed.). Entomological Techniques. How to Work with Insects. 9th ed. Edward Bros, Inc., Ann Arbor, MI.

Pfiester, M., P.G. Koehler and R.M. Pereira. 2008. Ability of bed bug-detecting canines to locate live bed bugs and viable bed bug eggs. J. Econ. Entomol. 101: 1389–1396.

Polanco, A.M., C.C. Brewster and D.M. Miller. 2011. Population growth potential for bed bug, *Cimex lectularius* L.: A life table analysis. Insects 2: 173–185.

Pritchard, M.J. and S.W. Hwang. 2009. Severe anemia from bedbugs. Can. Med. Assoc. J. 181: 287–288.

Reeves, W.K. 2000. Bionomics of *Cimex adjunctus* (Heteroptera: Cimicidae) in a maternity cave of *Myotis austraioriparius* (Chiroptera: Vespertilionidae) (South Carolina, USA). J. Entomol. Sci. 36: 74–77.

Reinhardt, K., D. Isaac and R. Naylor. 2010. Estimating the feeding rate of the bedbug *Cimex lectularius* in an infested room: An inexpensive method and a case study. Med. Vet. Entomol. 24: 46–54.

Reinhardt, K., P. Naylor and M.T. Siva-Jothy. 2003. Reducing a cost of traumatic insemination: female bedbugs evolve a unique organ. Proc. R. Soc. London (Series B-Biological Sciences) 270: 2371–2375.

Reinhardt, K., D. Kempke, R.A. Naylor and M.T. Siva-Jothy. 2009. Sensitivity to bites by the bedbug, *Cimex lectularius*. Med. Vet. Entomol. 23: 163–166.

Reis, M.D. and D.M. Miller. 2011. Host searching and aggregation activity of recently fed and unfed bed bugs (*Cimex lectularius* L.). Insects 2: 186–194.

Rivnay, E. 1930. Techniques in artificial feeding of the bed bug, *Cimex lectularius* L. J. Parasitiol. 16: 246–249.

Romero, A., M.F. Potter and K.F. Haynes. 2009. Behavioral responses of the bed bug to insecticide residues. J. Med. Entomol. 46: 51–57.

Romero, A., M.F. Potter, D.A. Potter and K.F. Haynes. 2007. Insecticide resistance in the bed bug: A factor in the pest's sudden resurgence? J. Med. Entomol. 44: 175–178.

Rush, W.A. 1981. Colonization of the swallow bug in the laboratory. Annals Entomol. Soc. Amer. 74: 556–559.

Ryne, C. 2009. Homosexual interactions in bed bugs: alarm pheromones as male recognition signals. Anim. Behaviour 78: 1471–1475.

Seong, K.M., D.Y. Lee, K.S. Yoon, D.H. Kwon, H.C. Kim, T.A. Klein, J.M. Clark and S.H. Lee. 2010. Establishment of quantitative sequencing and filter contact vial bioassay for monitoring pyrethroid resistance in the common bed bug, *Cimex lectularius*. J. Med. Entomol. 47: 592–599.

Siljander, E., D. Penman, H. Harlan and G. Gries. 2007. Evidence for male- and juvenile-specific contact pheromones of the common bed bug *Cimex lectularius*. Ent. Exp. Appl. 125: 215–219.

Silverman, J., M., A. Suckow and S. Murthy. 2000. The IACUC Handbook. CRC Press, Boca Raton FL.

Silverman, A.L., L.H. Qu, J. Blow, I.M. Zitron, S.C. Gordon and E.D. Walker. 2001. Assessment of hepatitis B virus DNA and hepatitis C virus RNA in the common bedbug (*Cimex lectularius* L.) and kissing bug (*Rodnius prolixus*). Am. J. Gastroenterol. 96: 2194–2198.

Stutt, A.D. and M.T. Siva-Jothy. 2001. Traumatic insemination and sexual conflict in the bed bug *Cimex lectularius*. Proc. National Acad. Sci. USA 98: 5683–5687.

Suchy, J.T. and V.R. Lewis. 2011. Host-seeking behavior in the bed bug, *Cimex lectularius*. Insects 2: 22–35.

Takano-Lee, M., R.K. Velten, J.D. Edman, B.A. Mullens and J.M. Clark. 2003a. An automated feeding apparatus for *in vitro* maintenance of the human head louse, *Pediculus capitas* (Anoplura: Pediculidae). J. Med. Entomol. 40: 795–799.

Takano-Lee, M., K.S. Yoon, J.D. Edman, B.A. Mullens and J.M. Clark. 2003b. *In vivo* and *in vitro* rearing of *Pediculus humanus capitas* (Anoplura: Pediculidae). J. Med. Entomol. 40: 628–635.

Usinger, R.L. 1966. Monograph of Cimicidae. Thos. Say Foundation, Vol. VII, Entomol. Soc. Amer., Lanham, MD.

Wang, C., K. Saltzmann, E. Chin, G.W. Bennett and T. Gibb. 2010. Characteristics of *Cimex lectularius* (Hemiptera: Cimicidae), infestation and dispersal in a high-rise apartment building. J. Econ. Entomol. 103: 172–177.

Webb, P.A., C.M. Happ, G.O. Maupin, B.J. Johnson, C.Y. Ou and T.P. Monath. 1989. Potential for insect transmission of HIV: Experimental exposure of *Cimex hemipterus* and *Toxorhynchites amboinensis* to human immunodeficiency virus. J. Infect. Dis. 160: 970–977.

Weeks, E.N., J.G. Logan, S.A. Gezan, C.M. Woodcock, M.A. Birkett, J.A. Pickett and M.M. Cameron. 2010. A bioassay for studying behavioural responses of the common bed bug, *Cimex lectularius* (Hemiptera: Cimicidae) to bed bug-derived volatiles. Bull. Entomol. Res. 1–8.

Wenk, T.M. 2007. Guarding against bedbugs. Lodging Law 8: 1–3.

Wintle, K. and K. Reinhardt. 2008. Temporary feeding inhibition caused by artificial abdominal distension in the bedbug, *Cimex lectularius*. J. Insect Physiol. 54: 1200–1204.

Yoon, K.S., J.P. Strycharz, J.R. Gao, M. Takano-Lee, J.D. Edman and J.M. Clark. 2006. An improved *in vitro* rearing system for the human head louse allows the determination of resistance to formulated pediculicides. Pesticide Biochem. Physiol. 86: 195–202.

Yoon, K.S., D.H. Kwon, J.P. Strycharz, C.S. Hollingsworth, S.H. Lee and J.M. Clark. 2008. Biochemical and molecular analysis of deltamethrin resistance in the common bed bug (Hemiptera: Cimicidae). J. Med. Entomol. 45: 1092–1101.

Zhu, F., J. Wigginton, A. Romero, A. Moore, K. Ferguson, R. Palli, M.F. Potter, K.F. Haynes and S.R. Palli. 2010. Widespread distribution of knockdown resistance mutations in the bed bug, *Cimex lectularius* (Hemiptera: Cimicidae), populations in the United States. Arch. Insect Biochem. Physiol. 73: 245–257.

6

Laboratory Maintenance of Phlebotomine Sand Flies

*Farida Mahmood** and *Michelle G. Colacicco-Mayhugh*

ABSTRACT

This chapter is intended to provide the reader with important aspect of sand fly colonization such as methods of collecting large number of sand flies from their natural breeding and feeding habitats. Biosafety precautions necessary while collecting the vectors species from endemic areas, their transportation from remote areas to the main laboratories, and later laboratory maintenance are emphasized. We have addressed major concerns in insectary safety to avoid escape and establishment of species in new nonnative habitats. Methods of *in vivo* and *in vitro* blood feeding and the equipment required for achieving the final goal of maintaining a successful colony are presented.

Key Words: *Leishmania, Lutzomyia, Phlebotomus, Sergentomyia,* sand fly colony, Sand fly rearing, emergence traps, light traps, Shannon traps, *in vitro* blood feeding

---

Department of the Army, Public Health Command Region-South, 2899 Schofield RD. JBSA Fort Sam Houston, TX 78234-6233, USA.

*Corresponding author

The views and opinions expressed in this chapter are solely of the authors and do not reflect the views of US Army Public Health Command Region-South or US Department of Defense. The specific materials described for maintenance of sand fly colonies were chosen based on availability, effectiveness and cost efficiency. The resulting usage is not to promote the individual companies mentioned; users can substitute similar equipment or substances manufactured by other companies. Mentioning of a proprietary product does not constitute an endorsement or recommendation by the authors or US Army Public Health Command Region-South or the US Department of Defense for its use.

Introduction

Sand flies belong to order Diptera, family Psychodidae, and subfamily phlebotominae. They lay eggs in terrestrial habitats that are rich in decomposed organic material such as dried leaves and feces of various animals including rodents. Sand flies have four larval instars, a pupal stage, and an adult stage in their life cycle. Adult sand flies are slow flying insects whose bodies are covered with long setae giving them a hairy appearance (Fig. 1). Although, most sand flies look similar in their external appearance, they show distinct morphological differences on closer examination. External morphological differences such as male claspers in combination with shape of internal organs like differences in shape of pharynx and, in case of females, the shape of their spermathecae (Fig. 2) are used to differentiate species. Conspecific species below the level of genera are identified by

Fig. 1. A female *Lutzomyia diabolica* collected in Bexar Co., Texas USA. Photographer Dr. Farida Mahmood.

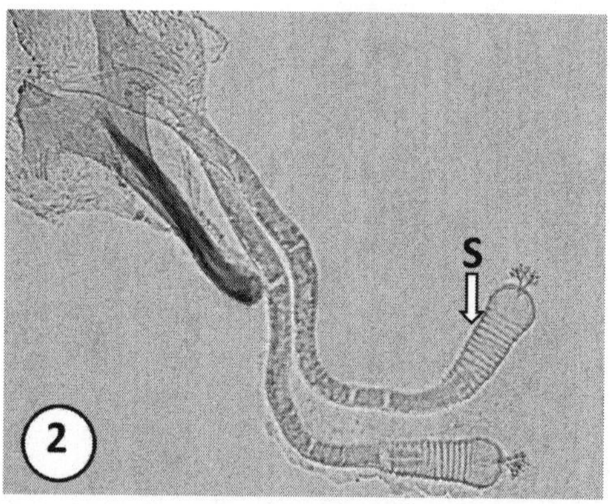

Fig. 2. Spermathecae (S) of a *Lutzomyia diabolica* female showing top shaped spermathecae. Photographer Dr. Farida Mahmood, PHCRS, JBSA Fort Sam Houston, TX.

associating wild caught males and females. Appendages on the head, its height from the vertex to the tip of the clypeus, together with its width are also used in taxonomic identifications. Internal structures of the head, such as the pharynx and pharyngeal teeth are also used in identification. Similarly, length and width of the wings are used in identification (Young 1979). Individuals interested in learning sand fly taxonomy are referred to start with current works by Young (1979), Young and Duncan (1994) and work of earlier taxonomists.

Males are identified using morphological differences in their bilaterally symmetrical genitalia and structures such as the spines on the style that are extremely important taxonomic characteristics (Young 1979). Many adult sand flies rest and blood feed in rodent burrows and rodent nests in xeric environments where availability of water drives distribution of many animal species (Figs. 3 and 4).

Fig. 3. A rodent burrow under rocks breeding and resting habitat of *Lu. diabolica* in Bexar County Texas. Photographer Dr. Farida. Mahmood, PHCRS. JBSA Fort Sam Houston, TX.

Fig. 4. A rodent nest breeding site under root of a tree in Bexar County, Texas. Photographer Mr. Walter Roachell PHCRS, JBSA Fort Sam Houston, TX.

Sand fly colonization is necessary for learning about their role in transmission of various pathogens. The first attempt at rearing sand flies in the laboratory was by Smith (1925) and led to later improvements by many leading sand fly researchers. Determining the role of different species as primary vectors of Leishmaniasis requires concomitant feeding of different species on blood with the same parasitemia under the same environmental conditions. Such studies can only be conducted if pure sand fly laboratory colonies are available.

Besides studying their biology, many other aspects of their physiology, genetics, life table characteristics, and differences in their vector competence also require availability of laboratory colonies. Information about the biology and life cycles of sand flies are important in building disease transmission models, taking in account the effect of temperature on their rate of development in nature and time of extrinsic incubation of phleboviruses and other pathogens including *Leishmania*.

In this chapter, we discuss the importance of sand flies as vectors of different diseases and the precautions necessary during their capture in the field, transportation and further colonization of wild caught flies in the laboratory. We also discuss methods of laboratory rearing of sand flies with examples of maintenance of *Lu. anthophora* (Addis) (Addis 1945) that was colonized from Brownsville, Texas, from *Neotoma micropus* and *Peromyscus leucopus* nests under a collection of refuse boards (Endris et al. 1984).

Medical and Veterinary Importance of Sand Flies

Phlebotomine sand flies (Diptera: Psychodidae) are the sole vectors of *Leishmania,* a group of parasites that causes a number of enzootic and zoonotic diseases collectively referred to as Leishmaniasis. Sand flies are widely distributed from approximately 50°N to 40°S in the world. Of the 700 sand flies species known to occur in the world, 70 species are proven or suspected vectors of Leishmaniasis (Lane 1993). At least 14 different species of *Leishmania* are known to cause various forms of this disease in humans. In the Old World, Leishmaniasis is transmitted by sand flies of the genus *Phlebotomus* Rodani and Bert that are mainly found in temperate areas including Central Asia, India, North Africa, and Mediterranean Europe. In the New World, Leishmaniasis is transmitted by sand flies of the genus *Lutzomyia* Franca and its distribution is mostly tropical except for autochthonous cases that were recorded in Texas (Lawyer 1984, McHugh et al. 1996, Washingtonpost.com 2007). Military service members from nonendemic areas are susceptible to leishmaniasis and other sand fly borne diseases when they deploy to endemic areas (Kern and Pedersen 1973, Magill et al. 1993, Ellis et al. 2007). From 1 October 2001 to 30 September 2006, leishmaniasis was diagnosed in approximately 1300 U.S. service members deployed in support of Operations Enduring and Iraqi Freedom (Anonymous 2007). As a result, some interest developed in determining the capacity of North American sand fly species to serve as vectors of Old World Leishmania species (Claborn et al. 2008); however, at present time the results of this research remain inconclusive.

Besides humans, other domestic mammalian species such as dogs (Boggiatto 2010), cats (Trainor et al. 2010), and horses (Reuss et al. 2012) have been found naturally infected with different *Leishmania* species in the United States of America. Species of *Leishmania* infecting wild animals have been isolated from armadillos, tree porcupines in the neotropics, and from rodents, and domestic guinea pigs. The distribution of various *Leishmania* species and their vectors are reviewed by Young and Lawyer (1987). Species of *Leishmania* infecting reptiles in nature were successfully fed to sand flies after mixing with reptilian blood in the laboratory and the parasites thrived successfully in sand flies (Adler 1964). Such studies to associate parasites and vectors under controlled conditions also require colonized sand flies.

Sand flies vector other diseases like *Bartonella bacilliformis,* commonly referred as Carrion's disease, which exhibits two clinical forms, Oroya fever and Verruga peruana. This disease is limited to Peru, Ecuador, and Colombia in the New World, with *Lutzomyia verrucarum* (Townsend) as the probable vector (Alexander 1995, Hertig 1939). Arboviruses belonging to three families (*Rhabdoviridae*, *Reoviridae*, and *Bunyaviridae*) are also transmitted by sand flies (Tesh 1988, Tesh and Modi 1983, 1984). Transovarian transmission of *Phleboviruses* has been shown for three *Lutzomyia* species (Endris et al. 1983, Tesh and Modi 1983). Viruses of seven New World serotypes were isolated from male sand flies (Tesh 1988), and *Phlebovirus* was also isolated from both sexes of sand flies in Panama (Tesh et al. 1974). In the Old World, viruses belonging to these families have been isolated from *Phlebotomus* sand flies. The *Phlebovirus* in the Old World is known as Papatacci fever, sand fly fever, and three day fever and is reported from India, China, Central Asia, and the Mediterranean region.

Since, sand flies are vectors of so many pathogens, any attempt to colonize them from wild caught live materials from an endemic area should be performed in an insectary with appropriate precautions in place and with an anteroom provided with a negative air pressure using fans to keep any escaped sand flies sitting on a worker to be blown off back in the insectary while exiting from the insectary.

Field Collection Methods of Sand Flies for Colonization

Materials

The following equipment and supplies are necessary for collection, colonization, and later maintenance of sand fly colonies.

1. Mechanical battery operated aspirators with collection tube and/or mouth aspirator with a HEPA filter (Model 612; John W. Hock, Gainesville, FL). Castro aspirator, and or reservoir type aspirators (Alexander 2000).
2. One Pint carton cages with organza netting on top and a side hole covered with 2 pieces of dental dam with a slit in the middle taped at right angles to each other on the hole for releasing sand flies in the carton.
3. One gallon ice cream carton cages with an organza netting at the top and a side hole with cotton stockinet sleeve for releasing adult sand flies in the cage.
4. Adult sand fly holding cages made from nylon cloth (18–20 cm^3) and supported with a steel or plastic frame (Bioquip®).
5. Nalgene Plastic containers 125–500 ml capacity with a layer of settled plaster of Paris poured in the bottom and completely dried before use (Fig. 5). These containers have 8 small holes in the bottom to allow absorption of water in the plaster of Paris.

Fig. 5. A sand fly oviposition/rearing container. Showing eggs and dead females on surface of wet plaster of Paris and live gravid females. Photographer Mr. Tobin Rowland from Vector and Parasite Biology, Entomology Branch, WRAIR.

6. Plastic vials (300 ml) with plastic screw lids with a circular hole cut out in the middle and covered with fine nylon or organza cloth for rearing eggs laid by gravid females.
7. Flash lights with or without red filter.
8. 25 ml vials with screened (18 mesh/cm) plastic snap lid and extra solid lids with tiny holes for air exchange. These tubes are used to raise larvae from single females.
9. CDC light traps with 1 gallon cooler for dry ice and a collection bag or plastic container with fine mesh netting attached to the bottom of the container.
10. Large Styrofoam box with tight fitting lid for transporting sand flies from field to laboratory.
11. Tyvek full body suite or white full sleeved shirt and full pants.
12. Rubber gloves.
13. EPA registered insect repellent endorsed by the CDC for efficacy against biting flies.
14. Dry ice.
15. 10–30% sugar solution in water and Karo® corn syrupin 1:1 ratio mixed with water. Instead of sucrose fructose can also be used as a source of energy.
16. Cotton balls.
17. Rubber bands.
18. Dry rabbit feces obtained from laboratory rabbits and dried under a fan.
19. Purina rabbit chow (complete diet 5315®).
20. An aged mixture (1:1) of rabbit feces and Purina rabbit chow as larval food (Young et al. 1981) .
21. Ketaset (VeterinaryProducts, Bristol Laboratories, Syracuse, NY 13201.
22. 16 Gauge disposable plastic syringes.
23. Gloves.
24. Environmental chambers with controls for temperature, photoperiod, and humidity.
25. Organza cloth.
26. Hot water circulator.
27. Chicken belly membrane, or Salted hog sausage casing.
28. Water bath circulator for maintaining 37°C water temperature in the upper glass blood feeder.
29. Glass blood feeder.

Methods

1. Larval Collection

Although, field collection methods for adults are very efficient, larval collection methods are proven to be time consuming and least productive due to cryptic terrestrial breeding habitats used by sand flies (Alexander 2000). Feliciangeli (2004) has provided an excellent review of natural larval breeding habitats of sand flies. She classified three types of breeding habitats including domestic, peridomestic, and sylvatic habitats that included rodent nests (Figs. 3 and 4), ant hills and termite nests beside other habitats. Feliciangeli (2004) discussed four methods for collecting immature stages including larvae and pupae. Out of the four methods used for finding breeding habitats only two: 1) visual searches of potential breeding sites such as leaf litter, organically rich soils due to the presence of animal feces in areas surrounding cow sheds, around chicken coops and in rodent burrows containing rodent feces, and 2) use of emergence traps (cages) on sand fly breeding ground provide direct access to adults that can be used for colonization. The number of sand flies collected by these methods is so small to make the efforts very time consuming, tedious and least productive and thus is not preferred methods used for establishing sand fly colonies.

2. Adult Collection

Collections of adult mated and often blood fed females for colonization is the best method for starting new colonies and adults can be collected using methods below:

a) CDC light traps (Sudia and Chamberlain 1962) supplemented with dry ice as CO_2 source to attract host seeking and gravid sand flies (Fig. 6).
b) Shannon traps used around animal dwellings including rodent (wild rats and gerbil) nesting. In South America Shannon traps are successfully used in conjunction with a black light to collect landing sand flies near animal sheds. Black Shannon traps were more attractive to sand flies compared to white Shannon traps (Galati et al. 2001).
c) Indoor resting sand flies can be collected using battery powered aspirators, a castro aspirator (Pessoa and Barretto 1948), or a reservoir aspirator (Barraud 1929). Collection of indoor resting sand flies using mouth aspirators from walls of animal shed like sheep housing areas and by using CDC light traps supplemented with dry ice as CO_2 are proven methods for collecting *Phlebotomous* and *Sergentomyia* sand flies

(Fig. 7). In rural areas, chicken coops with soil rich in organic material from chicken droppings are good breeding sites of sand flies (Fig. 8) and were also used successfully for collecting sand flies for colonization. At times caves occupied by animals also serve as breeding and resting sites for sand flies.

Fig. 6. *Lutzomyia diabolica* breeding habitat in Bexar County, Texas, showing a CDC light trap with dry ice and a collection container with fine mesh on bottom of the container. Photographer Dr. Farida Mahmood, PHCRS, JBSA Forts Sam Houston, TX.

Fig. 7. A sheep herd shed from a rural area used for collecting *Phlebotomous* and *Sergentomyia* sand flies by hanging light traps. Photographer Dr. Michelle G. Colacicco-Mayhugh.

Fig. 8. An indoor resting site. A chicken coop in a rural area from where *Phlebotomous* and *Sergentomyia* sand flies were collected using aspirators as well as CDC light traps. Photographer, Dr. Michelle G. Colacicco-Mayhugh.

d) In South America outdoor resting collections were made by disturbing, tree buttresses, and leaf litter under a screened frame folded like a tent (Alexander 2000), and under the rocks by disturbing sand flies using sticks and then collecting landing adults using mechanical battery powered aspirators.
e) Funnel traps placed on tree hole openings or on the opening of rodent nests.
f) Using animal baits together with CDC light traps.
g) Animal baited cone trap is an excellent method for collecting live unfed as well as blood fed flies (Montoya-Lerma and Lane 1996). Sand flies collected using mouth aspirators can be placed in 1 pint carton cages with screen tops on which cotton soaked with 10–30% sugar solution is placed to provide humidity as well as a source of energy during transportation to the laboratory for further colonization attempts. A counter flow geometry trap (MM-X model, American Biophyscis Corp., USA) was also used successfully by Mann et al. (2009) for collecting *Lu. shannoni* in Florida. Sand flies in CDC light traps should be collected early in the morning to avoid desiccation due to high environmental temperatures later in the day.

A review of Phlebotomine sand fly sampling techniques by Alexander (2000) provides an excellent source of field collection techniques and for finding sand fly breeding habitats or ecotypes present in an area.

Precautions during Collections

Collectors should wear a full body suit or a light colored full sleeved shirt and long pants during sand fly collections in areas where Leishmaniasis or other pathogens transmitted by sand flies are endemic. The bottom of the pant and cuffs of the shirts should be taped to prevent sand flies from crawling inside the collector's clothes. Sometimes sand flies landing on the collector can get entangled in the head hair and the head should be covered during collections or thoroughly checked before returning inside the vehicle. Sand flies in the field should always be collected using aspirators with filters because sand flies shed a lot of scales that may be contaminated with bacteria and fungal spores. Many bacteria and fungi grow in dark and moist areas like caves and basements and there is possibility of ingesting or breathing in the disturbed spore.

Transportation of Wild Caught Sand Flies

After field collection of sand flies using CDC light traps, a simple glass mouth aspirator with a HEPA filter can be used to transfer the flies into 120 ml specimen containers and 10–50 flies can be placed in each container that had water soaked plaster of Paris layer at the bottom (Fig. 5). During transportation, such containers can be directly placed on wet filter paper in a Styrofoam® box (Endris et al. 1982). Sand flies can also be transferred into cloth cages that are tied in a steel frame and a moist sponge placed inside the cage. The whole assembly can then be placed inside a large plastic bag that is tied close to prevent desiccation. A small cotton ball soaked in 30% sucrose and/or fructose should be added in the cage to provide energy source if the field laboratory is far from the sand fly collection sites.

Montoya-Lerma et al. (1998) described a transportation technique for sand flies that were collected from remote areas and later eggs were obtained from such sand flies in the field and sent to actual rearing facilities. *Lutozomyia evansi* were collected from Shannon traps at two villages situated in Caribbean Coast of Colombia. Collected sand flies were transferred into Barraud cages containing a wet sponge for providing high relative humidity. These cages were kept in polystyrene containers during transportation to the field laboratory where sand flies were fed on an anaesthetized hamster to provide a blood meal on the night of capture. Gravid females were allowed to oviposit in 300 ml Nalgene™ containers with damp plaster in the bottoms and tops covered with mesh cloths. These containers were kept inside a polycarbonate container. After most sand flies oviposited, the dead flies were removed and oviposition containers were placed in polystyrene boxes and transported to the permanent laboratories in Cali, Colombia, where larvae were reared according to Modi and Tesh (1983).

The above technique can be successfully adapted to transport field collected adult sand flies from xeric areas with high environmental temperatures and low humidity to avoid mortality. While transporting in Nalgene containers the plaster should be moistened with plenty of water and the extra water wiped of completely from the plaster of Paris and inside plastic walls of the containers before going to the field to avoid sand fly drowning. Sand flies stored in such containers would not die due to desiccation. A cotton ball soaked in 10–30% sugar water after removing extra sugar water to avoid dripping if placed on the screen mesh of transportation containers will result in increased survival of the females and thus a faster establishment of the colony. The temperature in the Styrofoam box can be lowered by placing water filled plastic bottle that was frozen in a refrigerator's freezer. The bottle can be wrapped in a towel to avoid direct contact of the containers with sand flies to wet ice otherwise a high mortality would occur in the field collected specimens.

Establishment of New Colonies of Sand Flies

During establishment of a new colony of a single species a lot of time and effort is required in the beginning. The patchy nature of the sand fly's immature developmental habitats could be exploited by collecting single species. Sand flies from such sites in the field can be released in a single cage and gravid and blood fed females aspirated out and isolated individually in 25 ml snap cap polystyrene vials with 2 cm of set plaster of Paris in the bottom that is soaked with water and remaining water dried completely before the release of fed females. A cotton swab soaked in 10–30% sugar solution or a drop of corn syrup Karo® is placed on top of the netting (18 mesh/cm) to provide sugar meal source.

The ovipostion containers (Fig. 5) are checked daily to remove dead females that have already oviposited on the plaster to avoid fungal growth that can hinder first instar hatching and growth. Dead females in good condition can be placed in a dilute soap solution in a cavity slide and gently swirled to remove the large number of scales. These females are washed in normal saline in a cavity slide before dissecting their spermathecae in normal saline for taxonomic identification of species as well as to check for the presence of internal parasites such as, nematodes, ascogregarines, fungi, and ecto parasites such as mites. Eggs of infected females should be discarded after freezing. Although time consuming, this early precaution will result in establishment of disease free sand fly cultures.

The second major problem while starting a new colony from field collected flies is high mortality after first oviposition (Killick Kendrick 1978, Schlein et al. 1990) and later development of fungi (Marchais et al. 1991) that causes higher larval mortality during development (Modi and Rowton 1999). This makes it necessary to remove the dead bodies of sand flies after oviposition. Many sand fly species do not blood feed readily in captivity and introduce another factor of mortality during colonization.

At 25–27°C sand flies first instar larvae hatch from eggs within 8–10 d after oviposition. Time taken for complete embryonic development depends on the insectary or environmental chamber's temperature. A high humidity in the oviposition container is necessary to keep eggs from drying out and few drops of water should be added after 2–3 d on the plaster of Paris or wet sponges should be placed in the plastic box containing the oviposition containers. A small amount of food prepared from aged rabbit feces is sprinkled on the hatched larvae. The larvae are watched daily, extra food is sprinkled every other day and water added with a syringe on the food. Another method of achieving moisture control is by drilling holes in the bottom of the larval rearing container before adding wet plaster of Paris in the larval rearing container and letting the container stand for few days to dry the plaster of Paris. In such mass oviposition and rearing containers the plaster of Paris is open to the outside at the bottom of the containers. These containers are soaked in a shallow water bath so that the plaster of Paris draws moisture up into the chamber. The moisture content is felt from the bottom of the container by feeling the dryness of the plaster of Paris from the bottom holes during the rearing process and more water is added by setting containers in the shallow water bath.

Unfed females are fed on a mice, golden hamster, or gerbil after anesthetizing them with Ketamine (0.2 ml/adult hamster). The blood feeding and rearing procedure are presented in detail under laboratory maintenance of sand flies and are same for field collected flies and work for establishing a new colony once the eggs are obtained from field collected gravid females and species identified using identification books such as the one by Young and Duncan (1994).

Preparation of Larval Rearing Diet

Most success in sand fly colonization was achieved after the development of a larval diet by Young et al. (1981) (Figs. 9 and 10) and is summarized below. Different types of larval foods are used in laboratories but the preferred diet was developed by Young et al. (1981). The diet consists of an aged mixture of air dried rabbit feces and Purina Rabbit Chow complete diet 5315®.

Fig. 9. Early stage in fermentation of sand fly larval rearing diet. Note dense growth of fungus mycelia. Photographer Mr. Tobin Rowland, Vector and Parasite Biology, Entomology Branch, WRAIR.

Fig. 10. Fermentation of larval rearing diet (late stage), showing degeneration of fungal mycelial mat. Photographer Mr. Tobin Rowland, Vector and Parasite Biology, Entomology Branch, WRAIR.

The mixture is 1:1 by weight or volume ratio of above components and is prepared by grinding the two components very fine in separate electric grinders. A coffee grinder can be used for this purpose. A thin layer of the above dry diet mixture is spread on the surface of an enameled or plastic photographic pan with smooth surface. The food is packed a little to a depth of 2–5 cm and is sprayed with distilled water to saturation. The pan with moist food is either covered with a lid or with a polythene sheet to keep the flies and other insects from laying eggs in the fermenting diet (small scale food preparation) or the trays can be placed in an acrylic plastic incubator (Fig. 11) during mass food preparation. The trays with the fermenting food mixture develop a thick layer of fungal mycelia that gives it a white mat like appearance (Fig. 9). The food should be checked every other day for dryness and more water sprinkled on its surface. After about 3 weeks the fungal growth dies off and turns into yellow to brownish color (Fig. 10). The procedure is continued until the food start smelling like humus. The whole procedure takes one month. The fungal inoculum can be obtained from spoiled bread or from older food for sand flies (Young et al. 1981).

Peterkova-Koci et al. (2012) compared the role of bacteria present in unsterilized and sterilized rabbit feces by autoclaving and concluded that bacterial by products act as oviposition attractant for *Lu. longipalpus* resulting in significantly more eggs collected in containers with regular unsterilized food compared to autoclaved food and larvae grew faster in

Fig. 11. Acrylic-plastic incubator cabinet for aging larval diet. Photographer Mr. Tobin Rowland, Vector and Parasite Biology, Entomology Branch, WRAIR.

unsterilized food. Additives, like beef liver powder, were later added to the food prepared by Young et al. (1981) method to increase nutritional components of the food (Modi and Tesh 1983). Beef liver powder was added after the fermentation of rabbit feces and rabbit chow to the humus form. They did not provide the ratio of beef liver powder added to the fermented diet. After addition of the liver powder the mixture should be stirred thoroughly and aged for 2 more weeks before grinding it and feeding to the larvae. Modi and Tesh (1983) used this technique for successfully raising 4000–5000 adult *P. papatasi* and *Lu. longipalpis*. The most important parameter is the time required for complete aging of the larval diet to a humus form that is no longer prone to fungal growth. Food can be tested before use in rearing the sand flies by using some food in an empty rearing container and providing moisture and keeping it under the humidity and temperature used for rearing in an incubator or an insectary.

Further advances in larval food preparation included development of acrylic plastic incubator cabinets or Lexane® cabinets (Fig. 11) that allowed control of humidity and biologically safe environment (Lawyer et al. 1991, Modi and Rowton 1999). Although the above larval food works for most colonized sand fly species and is used in many laboratories actively involved in research (Modi and Rowton 1999), research is continued for improvement and development of new larval rearing diets and is mainly mentioned here to be experimented as alternative rearing diets for species that would not develop in the above food. Souza et al. (1999) used a mixture of liver powder, 14 amino acids, pollen grains, fish food, and vitamins, in equal amounts and readers are referred to their publication for trying addition sources of various nutritional components.

The aged rabbit feces and rabbit chow diet (Young et al. 1981) was successfully used by the first author to maintain a 12 years old colony of *Lu. anthophora* (Addis) (Addis 1945) originally colonized from Texas (Endris et al. 1982), for studying different biochemical aspects of its physiology (Mahmood 1990, Mahmood and Borovsky 1992, 1993, Mahmood et al. 1992).

Insectary

A sand fly rearing insectary can be created in a room without windows, so that a proper photoperiod is maintained using simple timers that are connected to a florescent bulb. The choice of the photoperiod depends on the researcher and can be adjusted so light period is during working hour. Care should be taken not to use a photoperiod that would initiate diapause in adults or larval stage. Walls of the insectary should be devoid of holes and cracks and should be painted with white enamel paint to allow detection of escaped adults. The ceiling should not be very high and a worker should be able to catch the escaped adults while standing and using either a filtered mouth or mechanical aspirator. Air conducting inlets should be fitted with screen (18 mesh/cm) or a simple screen can be made using organza cloth fitted to a frame that can be fitted with duct tape on the outside of the inlets. A separate air conditioning unit should be used for maintaining the temperature in the insectary or portable heater and humidifiers can be used to maintain temperature and humidity in this room. The room should be provided with a suitable sized sink for washing rearing containers and pans used for food preparation. An insectary should either have double curtains or negative air pressure curtain at the door opening from the insectary to an anteroom before opening to the main corridors. No two doors should be opened simultaneously in an insectary. It should have just the most essential furniture and equipment to maintain the colony thus minimizing places for hiding accidentally escaped adults. Most sand fly colonies can be maintained properly in environmental growth chambers or incubators and thus eliminate the need for controlling temperature, humidity and light regimes in the insectary.

During large scale rearing of sand flies more space may be needed and movable racks or shelving may be required to place adult cages, ovipostion containers in plastic boxes and larval rearing containers. An insectary should have movable shelving for such occasions. These racks may be necessary if more than two or three different species are raised in one insectary. An ample supply of colored masking tape should be at hand to color code different species of sand flies or different treatments of an experiment may require color codes.

Laboratory Rearing

The reach-in environmental chambers or incubators that are provided with temperature, light and humidity controls present the best method of laboratory rearing (Figs. 12 and 13). Since adult sand flies are very small in size, a large number of flies can be maintained in small cages and humidity can be maintained by using a wet cotton pad or water soaked paper towels on top of the screen if a gallon cage or nylon cloth cage is used for adult maintenance. Such cage can be placed inside transparent plastic bags with wet sponges and closed to maintain high humidity inside the plastic bag containing the sand fly cage.

Three different types of cages are typically used for maintenance of adults and include: 1) square nylon holding cages in wire frames or plastic frames (Modi and Rowton 1999); 2) cages made of Lexane (31 cm^2) (Fig. 14) with bottom and back lined with plaster of Paris. The top and front sides of these cages have holes. The side-hole has a cotton stockinet sleeve (36 cm long x 15 cm wide) attached to it with a front plastic panel. Similar cages

Fig. 12. Environmental chamber containing sand fly larval rearing containers **(LC)**. **Fig. 13.** Environmental chamber containing oviposition box **(OVB)** with ovipostion containers in the bottom two shelves and adult sand fly cages **(SC)** in the upper two shelves. Photographer, Mr. Tobin Rowland, Vector and Parasite Biology, Entomology Branch, WRAIR.

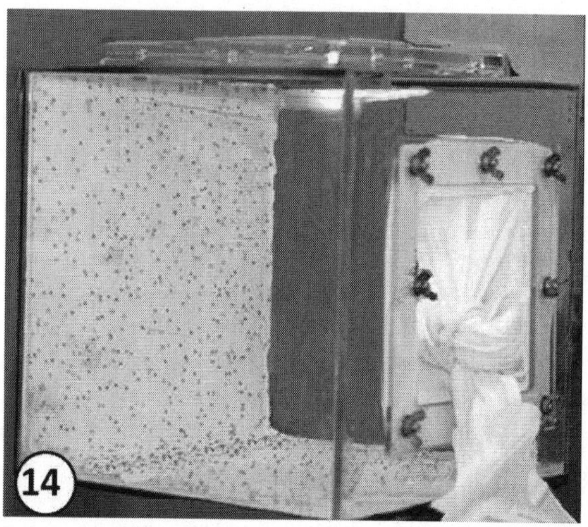

Fig. 14. A Sand Fly adult rearing cage showing bottom and back of the cage covered with a layer of plaster of Paris.

were prepared from glass aquaria in the beginning (Endris et al. 1982) but were later replaced with the Lexane cages due to safety concerns; 3) cages made from 1 gallon ice cream carton cardboard cages with top replaced with organza cloth netting and a side hole with stockinet cotton sleeve. Since sand flies are not fast fliers accidental escape of adults can be avoided by using a double cage system while releasing the adults in a smaller cage.

Oviposition and Larval Rearing

The oviposition containers can be divided into two categories: 1) twenty-five mL tubes used for isolation of single flies and 2) larger oviposition containers that later serve as larval rearing containers (Figs. 5 and 15–17).

Fig. 15. Sand fly oviposition containers in a plastic box that has water soaked piece of sponge to provide humidity. Photographer, Mr. Tobin Rowland, Vector and Parasite Biology, Entomology Branch, WRAIR.

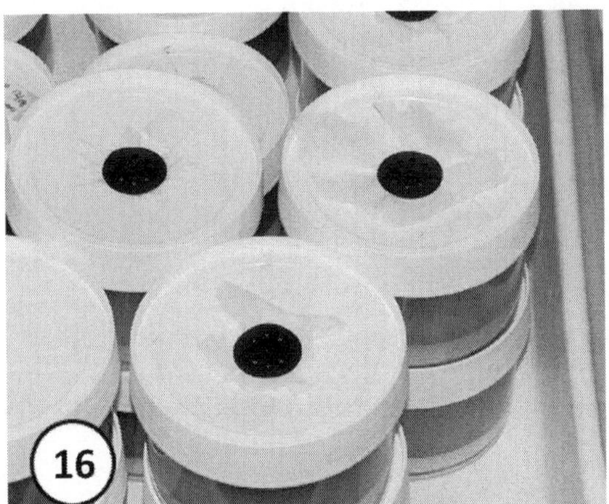

Fig. 16. Ovipostion container used for larval rearing after replacing the lids. Photographer, Mr. Tobin Rowland, Vector and Parasite Biology, Entomology Branch, WRAIR.

Fig. 17. Larval rearing container showing pupae (white arrow). Photographer Mr. Tobin Rowland, Vector and Parasite Biology, Entomology Branch, WRAIR.

The twenty-five mL tubes have a layer of plaster of Paris at the bottom and snap caps lids that are cut 2mm in the middle and a piece of organza is glued to cover the cut opening. Single gravid sand flies are isolated in the containers that are already soaked in water and dried with a paper towel. Sand flies are transferred using a mouth aspirator with a HEPA filter. Few drops of 20–30% sugar solution are placed on the organza to provide an energy source. The presence of sugar water during oviposition provides an essential energy source for females and results in increased survival thus affecting the total number of eggs deposited in the containers (Rangel et al. 1987, Schlein and Jacobson 1999). This method of oviposition is used for wild caught sand flies with unknown identification or in which infection with parasites has to be determined.

After isolation the tubes or oviposition containers are transferred to a plastic box containing a wet sponge. The density of sand flies in the ovipostion container directly affects the number of eggs per female (Killick-Kendrick and Killick-Kendrick 1991, Modi and Rowton 1999, Mann and Kaufman 2010). More eggs were obtained in ovipostion containers with increased density of flies thus for maintaining already established colonies is to add 20–30 females in a 300 ml oviposition container that is later used as a rearing container.

The second method is used to maintain colonies of sand flies where the oviposition container also serves as a larval rearing container. A 300–500 ml container with 8 holes in the bottom of the container and 1–2 cm of plaster of Paris, is soaked in water. After the plaster had fully soaked with water the surface and sides of the containers are wiped dry with paper towel before releasing blood fed females in the containers.

The top of the container is covered with a piece of organza cloth that is secured with a rubber band. A small entry hole is made in the organza to allow the entry of mechanical aspirator's collection tube. Gravid sand flies are collected using the mechanical aspirator or a mouth aspirator with HEPA filter. Twenty to 30 gravid females are isolated in the container to obtain large number of eggs per container for mass rearing sand flies. The oviposition containers with flies are placed in a large plastic container with a wet piece of sponge. The Sugar water soaked cotton balls are placed on top of the organza (Fig. 15). The lid of the plastic box is closed to maintain high humidity inside the box and the whole assembly is transferred to an environmental chamber (Fig. 12). The level of humidity in the oviposition containers directly affects the number of eggs laid and inversely affects the longevity of blood fed females. The moisture should be maintained in such a manner that no condensation of water droplet forms on the walls of the oviposition containers. Flies raised in constant saturated humid environment laid more eggs compared to dry environment and fewer eggs were retained in the dead females (Milleron et al. 2008). Oviposition cups are checked daily and after eight to ten days remaining live flies are transferred back to the adult cages (Fig. 14) for refeeding to obtain additional eggs. Most gravid females oviposit within 6–8 days at 26–28°C.

All the dead flies should be removed from the oviposition containers to avoid fungal growth that interferes with the growth of first instar larvae. Rearing containers should be checked frequently to determine the moisture level by feeling the plaster of Paris from bottom of the rearing containers from the drilled holes and containers placed in a shallow water bath containing water to replenish moisture in the containers. The oviposition container should be checked twice weekly for fungal growth, moisture replenishment and larval hatch.

In many sand fly species their eggs start hatching after 8 days at 26–28°C but staggered hatching has been reported from fourteen to forty days for *Lu. cruiciata* (Coq.) from Florida and *Lu. diabolica* (Hall) and *Lu. vexator* from Texas (Endris et al. 1982). In most cases after 8–10 days first instar larvae start hatching. At this time aged food that is sieved through an organza cloth is sprinkled on the larvae in a moist larval rearing container. The organza cloth used to cover the oviposition container is replaced with a proper lid with a hole in the center that is covered with a piece of organza cloth to allow for air exchange and later for placing Karo® drops or 20–30% sugar solution in cotton balls for emerging adults (Fig. 16). The plastic trays with larval rearing containers are transferred into the environmental chambers. The larvae undergo 4 larval molts. The time required for each molt is temperature dependent and differs for different species.

The duration of *Lu. anthophora* (Addis) larvae was determined in 20 groups of 50 first instar larvae each that were reared in 300 ml containers in an environmental chamber (Hot Pack model 434310) maintained at 26±2, 80% relative humidity, and 16:8 L:D photoperiod. Larvae were checked on alternate days and new food and moisture was replenished. The head capsule width and body length of 10 larvae was measured at 10X magnification daily to determine the duration of larval instars. Data was analyzed using one way-analysis of variance and Duncan's New Multiple range test (Sokal and Rholf 1981). A significant correlation was found between larvae body length head capsule width with time. The duration of first instar was 5 d, 2nd instar's duration was 3 d, 3rd instar's duration was 3 d and 4th instar's duration was 5 d. Thus overall larvae took 15–17 days for pupation after eclosion from the eggs. Adults emerged from pupae 8 d after pupation (Mahmood 1990). Fourth instar larvae go through a wandering stage and pupate at the surface of the larval growth medium (Fig. 17).

Emergence occurs in the same larval rearing containers (Fig. 18). The emerging adults are released on alternate days in the adult cage or for experimental use in other containers. Cotton balls soaked in a few drops of sugar solution are placed on top of the container as energy source before releasing the sand flies in adult maintenance cages (Fig. 14). From the first day of releasing adults in cages sand flies are provided with fresh slices of apple daily and 20–30% sucrose or Fructose solution in a cotton ball that is placed on top of the cage. The sugar water soaked cotton ball is replaced every third day to avoid fermentation. The same goal can be achieved by providing corn syrup or 1:1 diluted Karo® solution. The organza top should be cleaned with water soaked paper towel before adding fresh sugar source solution.

Fig. 18. Larval rearing container showing adult emerging sand flies. Photographer, Mr. Tobin Rowland, Vector and Parasite Biology, Entomology Branch, WRAIR.

Males emerge earlier than females and males of *Lu. anthophora* were capable of mating 13 h after emergence and their terminalia were fully rotated by 180° by this time. Mated females are ready to blood feed within 1–2 d after emergence. A more synchronous blood feeding can be achieved by blood feeding females 48 h after emergence and then refeeding 2 d later (Mahmood 1990).

Spent rearing diet should be either frozen for few days at –20°C orautoclaved to kill any remaining eggs, larvae and pupae before discarding in regular trash. Similarly, any adult sand flies that are not used in an experiment should be killed by freezing or by using CO_2 before discarding to avoid escape in the environment and establishment of sand flies in areas where they are not the native species.

Blood Feeding

Unlike mosquitoes, many species of sand flies do not require dusk and dawn period with gradually reduced light intensity in fact species like *Lu. anthophora* mate during the day, frequently on the host while females are blood feeding. Two methods of blood feeding have been frequently used for colony maintenance and include: 1) feeding on animals such as hamsters, chicken, rabbits and 2) *in vitro* blood feeding using either chicken or pig intestinal membranes (obtained from salted hog casing used to prepare sausages).

Feeding on Animals

Blood feeding on live anesthetized animals is the most common method used while establishing new colonies. Sand flies feed on many different types of hosts in nature depending on different species and their breeding habitats (Tesh et al. 1971). Initially, field collected sand flies may not blood feed on anesthetized hamsters and other hosts present in the environments such as lizards, birds, rabbits, and humans, etc. may be required (Killick-Kendrick and Killick-Kendrick 1991). Once a colony is established it can be slowly adapted to feed on a small host like hamster and mice thus reducing the need for large animal rooms (Modi and Rowton 1999).

Most colonized species are blood fed on an anesthetized adult Syrian hamsters or mice using (Ketamine 10mg* xylazine 2 mg/100 g body weight) (Modi and Rowton 1999) or by injecting 0.2 ml Ketamine hydrochloride (Ketaset®) intramuscularly using a 27 gauge needle and tuberculin syringe into a hamster. Many sand fly species can be fed on unshaved hamsters because sand flies readily feed on the ears and foot pads of the anesthetized animal (Endris et al. 1982) (Mahmood 1990). Beside the above two chemicals, a third combination is also used by intra peritoneal injection of a mixture of 10 mg/kg xylazine (Xyla-ject, Phoenix Pharmaceutical, St. Joseph, MO) and 30 mg/kg tiletamine-zolazepam (Telazol; Fort Dodge Laboratory, Fort Dodge, IA) (Rowton et al. 2008). The hamster is placed in the cage (31 cm^3) or a 1 gallon ice cream carton cage with its shaved belly facing up on a small sheet of plastic with an absorbent paper towel to avoid contamination of the cage because of involuntary discharge of urine or feces by the hamster inside the cage. The cage is covered with a black cloth for species feeding on lizards at night and females are allowed to feed for ≥ 1 h on an anesthetized hamster before it recovers from the anesthesia.

Blood fed females should not be disturbed just after blood feeding and any physical manipulation should be performed at least 1 h after blood feeding to allow time for diuresis only if experimental procedures require manipulation of females in this time period. Otherwise, it is better to leave fresh fed females inside the cages for the first 24–48 h after blood feeding. This allows the females to undergo diuresis, peritrophic membrane formation, and reduces potential injury to their internal organs during aspiration and transfer to a new cage.

Some sand flies, like *Lu. vexator,* prefer to blood feed on unrestrained live lizards. The mouth of the lizards can be taped leaving the nostrils open for breathing. While blood feeding hungry sand flies on larger animals, unfed flies are confined in 25 ml tubes with a tight organza screen on top and are directly pressed on the ears of the rabbit or body of the calves (Endris et al. 1982).

The anesthetized animal should be checked carefully before removal from the feeding cage and all sand flies attached to or mating on the animal should be carefully removed. Same precautions apply while removing the soiled paper towels and piece of plastic from the cage. If blood feeding during the day, the cage can be covered with dark cloth during feeding or the day and night should be switched inside an incubator so the flies can be fed during day by the workers. Sand flies can be blood fed twice a week to obtain maximum blood feeding. Adaptation time to blood feed on new hosts in new colonies varies and is dependent on the species. Sand flies are attracted to the hairless mice (nude mice) more than to BALB C mice (Volf and Volfova 2011).

In vitro Blood Feeding through a Bio-membrane

During maintenance of large number of colonies and experimentation for determining their vector competence, repellency to various repellent, or development of pesticides require mass rearing of sand flies for replication of statistically acceptable sample sizes for the results to be valid. Mass rearing requires production of large numbers of eggs that in turn is dependent on availability of large number of healthy hosts and a proper animal facility, thus creating a large budgetary burden on the facility. Many research laboratories are opting to blood feed sand flies through biomembranes due to higher maintenance costs of animal facilities and strict regulations for obtaining animal permits. This technique is also preferred for infecting the flies by feeding on blood mixed with cultured *Leishmania* parasites and readers interested in such studies are referred to (Galun 1979, Tesh and Modi 1984, Gosh 1994) and to an excellent chapter by Bates (1997).

For *in vitro* blood feeding two types of animal membranes, 1–3 d old chick's breast skin membranes and washed hog sausage casing, have been used successfully (Ward et al. 1978, Rowton et al. 2008, Mann and Kaufman (2010) (Figs. 19 and 20). Chicken skin is prepared by euthanizing 1–3 d old chicks (Ward et al. 1978) or 8 wk old Leghorn chicken, with CO_2 in a plastic bag. The feathers from the skin overlying breast of the chicken are removed and the skin is cut away carefully from the breast region and is carefully removed from the musculature of the chicken. Skin is saved unwashed or washed briefly with ethanol or left in 70% ethanol for 1 h (Rowton et al. 2008) and then washed with sterile saline before freezing at –20°C. Skin removed in this way from a large number of chicks is placed between layers of wax paper or layers of aluminum foil and stored in a freezer zip-lock bag and frozen at –12°C in a refrigerator freezer for later use.

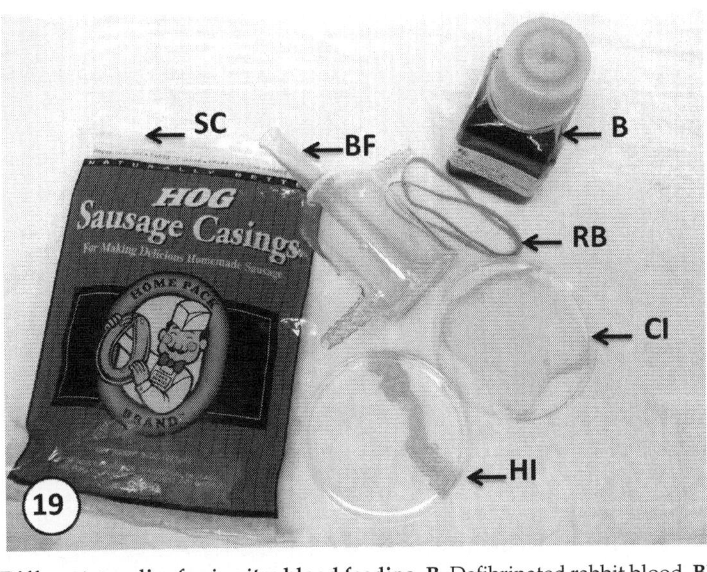

Fig. 19. Different supplies for *in vitro* blood feeding. B. Defibrinated rabbit blood, **BF.** Glass blood feeder, **CI.** Cut washed and spread piece of hog intestine, **HI.** Washed piece of Hog intestine, **RB.** Rubber band, **SC.** Bag of salted hog sausage casings. Photographer Dr. Farida Mahmood, PHCRS, JBSA Fort Sam Houston, TX.

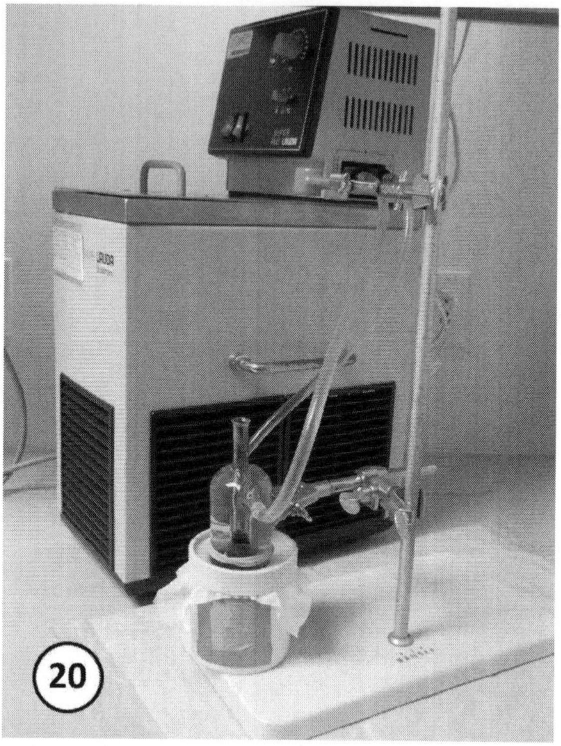

Fig. 20. *In vitro* blood feeding setup for feeding sand flies. Photographer Dr. Farida Mahmood, PHCRS, JBSA Fort Sam Houston, TX.

Any type of glass blood feeder (Perpatual Systems, Rockville, MD) used for mosquitoes blood feeding (Rutledge et al. 1976) can be used for blood feeding sand flies as long as it fits on the blood feeding container such that no space is left between the membrane and the stretched organza or the netting screen of the cage. The membrane to be used is taken out of the freezer, thawed and washed with sterile saline or distilled water before use. The thawed membrane is stretched on the open end of the jacketed glass blood feeder, taking care that the feathered side of the membrane faces outside and is secured tightly using rubber bands. This whole assemblage is attached to a hot water circulator at 37–39°C (model 2219 Multitemp II; LKB, Brommer, Sweden, or Hake Circulator, Series F, Brinkman Instrument Inc.) (Fig. 20). If a hot water circulator is unavailable, the rubber tube attached to the glass feeder can be directly attached to a hot water sink faucet after carefully adjusting the water in a beaker to the desired temperature using a thermometer. The other end of the feeder should be attached to rubber tubing that can be left in the sink for disposal of incoming water from the feeder. This allows a 37°C temperature for 30 min to 45 min. After adjusting the correct water temperature in the glass feeder's jacket the feeder is fitted on the organza screen of the sand fly container or cage. The funnel inside the water jacket is filled with blood using a plastic syringe without a needle. Blood feeding through a chicken membrane is comparably similar to the results obtained by direct feeding on the anesthetized Syrian hamsters (Rowton et al. 2008).

To make artificial feeding more economical, some hematophagous insects like mosquitoes, triatomine bugs and some species of sand flies can be fed through hog intestine membrane by using washed salted sausage casing (Mahmood 1999, 2004, 2006, Mann and Kaufmann 2011). The salted hog sausage casing is commercially available from many sources. A small piece of casing is cut and side is cut to open the intestine in a square piece (Fig. 19, CI and HI). The piece of intestine is thoroughly washed with distilled water to remove the entire salt and let stand in distilled water for about 15 minutes. The piece is then spread on a paper towel and dried with another paper towel. The rest of the procedure for attaching the membrane to the blood feeder and blood feeding sand flies is the same as above for chicken membrane feeding.

Defibrinated blood from different animal sources is available commercially and can be compared using the above two techniques. The main point of concern for blood meal selection is host preference by the sand flies and effect on their egg production for maintenance of colonies and mass production for research needs. Rowton et al. (2008) did not observe any difference in feeding success and production between flies fed on human blood through a chicken membrane and on hamsters suggesting this technique to be an economical alternative to live animals for colony maintenance. Mann and Kauffman (2011) were able to colonize *Lu. shanonni* from Florida using hog sausage casing and maintained the colony for 4 generations by *in vitro* blood feeding technique. Harre et al. (2001) blood fed *P. papatasi* through a chicken membrane on blood of 8 different mammal species and compared effect on the fecundity and survival rates of *P. papatasi*. Such studies should be conducted before using *in vitro* blood feeding technique to replace blood feeding on live animals. Blood of different mammalian species may be needed to obtain optimal fecundity and survival of different sand fly species. Also important is the long-term use of *in vitro* membrane feeding on other important traits such as selection of animal host in the laboratory studies (Rowton et al. 2008). After blood feeding the sand flies should not be disturbed for at least 24 h for completion of diuresis and formation of a peritrophic membrane in their midgut.

The most important concern in any colonized species is keeping the genetic variability of the colony and it can be achieved by adding progeny of field collected sand flies when possible. The genetic variability can also be maintained by establishing two separate starter colonies and whenever a bottle neck effect is observed males or females of the second colony can be added to the suffering colony.

The main concern during construction and later handling of healthy and infected flies is their containment inside an insectary and the health of individuals working in areas where transmission of various sandflyborne diseases is endemic. We refer the researchers involved in sandflyborne human and animal pathogen studies to Arthropod Containment Levels (ACLs) publication by Liebert (2003). The main concerns as related to construction of a sand fly insectary and colonization are discussed below.

Arthropod Containment Levels (ACLs) for Sand Flies

The following routine maintenance procedures should be followed in all insectaries where sand fly colonies are maintained.

1. A standard of procedures or a response procedure should be developed to respond to accidental release of healthy and infected sand flies.
2. The storage area, for supplies should be maintained outside the insectary to decrease the number of hiding areas for escaped sand flies.
3. All electric inlets should be covered.
4. A negative air pressure curtain should be maintained all the time and should be tested for the air pressure at different distances from the insectary door.
5. All containers in the insectary should be labeled with the name of the researcher, sand fly species' name, date, locality of colonization and the experiment title.
6. No infected vertebrates should be maintained in the insectary to blood feed the sand flies.
7. All live sand flies should be maintained in a screened off area in the main insectary room.
8. All sand flies should be handled with a battery powered aspirator. If use of mouth aspirator is necessary for a certain procedure than always use aspirators with a HEPA filter.
9. A mouth aspirator should never be used to transfer field collected or laboratory infected sand flies.
10. All field collected sand flies should be manipulated in a glove box during transfer from the collection net to a cage. A glove box can be prepared with strong transparent plastic or plexi glass sheets (see Chapter 1).
11. No live sand flies should be manipulated inside a biosafety hood. The sand flies are too small to be manipulated inside a biosafety level II

hood due to the circulating air pressure and would be sucked in the HEPA filter.

12. Rearing container cleaning and the disposal of rearing material should be done after freezing the container in a –20°C freezer for at least overnight.
13. All immature rearing containers should be of unbreakable plastic. The same is true for sand fly cages.

The above practices apply to all ACLs of insectaries. Below are some recommendations for handling native and exotic infected and uninfected sand flies.

Native Sand Flies

Sand fly species native to an area where sand fly colonies are maintained can be maintained at arthropod containment level-1 (ACL-1) if no experimentation is conducted using biosafety level II pathogens. The ACL-1 facilities can be used to maintain colonies of uninfected native sand fly species or those infected with a non-pathogen. Some arthropod species that are not viable in nature due to the geographic location of insectary facilities and harsh environment during part of the year thus eliminating any accidentally released insects during cold winters (Liebert 2003). If a biosafety level II pathogen is used to infect such local arthropods than the ACL containment level is automatically raised to the biosafety containment level of the pathogen used (Leibert 2003).

Exotic Species of Sand Flies

As general rule, exotic sand fly species require biosafety level II practices to avoid their escape in a nonnative area. The United States Department of Agriculture's (USAD) permit is required to bring in species that are not native to United States. Exotic sand fly species that have a potential of establishment in a new geographical area, or are infected with a pathogen are kept in ACL-2 and above depending on the pathogen species involved and the environmental conditions in an area. The ACL is selected in consultation to the organizations' proper IACUC (for use of infected animals), and IBC (for genetic manipulation experiment).

Pathogen Transmission Studies

For experiments involving handling of pathogens during sandfly pathogen transmission studies precaution should always be taken, including:

1. Use of gloves, while feeding the animals experiments.
2. Use of personal protection clothing and equipment such as respirators with HEPA filters for arthropod associated allergens, headgear or covers during working with the live specimens.
3. Arthropods should be located in incubators or double-screened small areas.
4. All containers should be autoclaved and laboratory surfaces should be cleaned or decontaminated with disinfectant that kills the parasites or pathogens but would not harm the insects
5. Infected arthropods should be kept in locked environmental chambers or incubators.
6. Initial serum samples are taken for individuals working in the ACL-2 or above insectary.
7. Access is restricted to individuals working in the insectary or individuals accompanied by them and are made aware of the pathogen being under investigation.
8. Institutional protocols should be followed during development of and while conducting experiments in the insectaries.

To conclude we have tried to provide the most important information to help a researcher for colonizing sand flies from the field and maintaining laboratory cultures at affordable budget.

Acknowledgements

This chapter is a tribute to Dr. David G. Young as a great teacher who provided many of us with an opportunity to work with sand flies and taught us sand fly biology and identifications. We also thank Walter Reed Army Institute of Research to let us use photographs of different steps in sand fly rearing.

References

Addis, C.J. 1945. *Phlebotomus (Dampfomyia) anthophorrus*, n. sp. and *Phlebotomus diabolicus* Hall from Texas (Diptera: Psychodidae). J. Parasitol. 32: 119–127.

Adler, S. 1964. Factors determining the behaviour of *Leishmania* sp. in sandflies. Harefuah. 14: 1–5.

Alexander, B. 1995. Bartonellosis in Colombia and Ecuador: A review. Am. J. Trop. Med. Hyg. 52: 354–359.

Alexander, B. 2000. Sampling methods for Phlebotomine sandflies. Med. Vet. Entomol. 14: 109–122.

Anonmyous. 2007. Leishmania in relation to Service in Iraq/Afghanistan, U.S. Armed Forces. Medical Surveillance Monthly Report.

Bates, P.A. 1997. Infection of phlebotomine sandflies with *Leishmania*. pp. 112–120. *In*: J.M. Crampton, C.B. Beard and C. Louis [eds.]. The Molecular Biology of Insect Disease Vectors. A Methods Manual. Chapman & Hall, NY, USA.

Barraud, P.J. 1929. A simple method for carriage of mosquitoes over long distances. Ind. J. Med. R. 28: 281–285.

Benedeck, T. 1940. American leishmaniasis, Report of the first autochthonous case in United States. J. Trop. Med. Hyg. 43: 147–155, 164–168.

Boggiatto, P.M., A.E. Ramer-Tait, K. Metz, E.E. Kramer, K. Gibson-Corley, K. Mullin, J.M. Hostetter, J.M. Gallup, D.E. Jones and C.A. Petersen. 2010. Immunologic indicators of clinical progression during canine *Leishmania infantum* infection. Clinical and Vaccine Immunology 17: 267–273.

Claborn, D., P. Masuoka, M. Morrow and L. Keep. 2008. Habitat analysis of North American sand flies near veterans returning from leishmania-endemic war zones. International Journal of Health Geographics 7: 1–65. http://www.IJ-Healthgeographics.com/content.7/1/65.

Ellis, S.B., G. Appenzeller, H. Lee, K. Mullen, R. Swenness, G. Pimentel, E. Mohareb and C. Warner. 2008. Outbreak of sandfly fever in central Iraq, September 2007. Military Medicine 173, 10: 949.

Endris, R.G., P.V. Perkins, D.G. Young and R.N. Johnson. 1982. Techniques for laboratory rearing of sand flies (Diptera: Psychodidae). Mosq. News 42: 400–407.

Endris, R.G., R.B. Tesh and D.G. Young. 1983. Transovarial transmission of Rio Grande virus (Bunyaviridae): phlebovirus by the sand fly, *Lutzomyia anthophora*. Am. J. Trop. Med. Hyg. 32: 862–864.

Endris, R.G., D.G. Young and J.F. Butler. 1984. The laboratory biology of the sand fly *Lutzomyia anthophora* (Diptera: Psychodidae). J. Med. Entomol. 21: 656–664.

Feliciangeli, M.D. 2004. Natural breeding places of Phlebotomine sand flies. J. Med. Vet. Entomol. 18: 71–80.

Galati, E.A.B., V.L.B. Nunes, M.E.C. Doral, G. Cristaldo, H.C. Rocha, R.M. Gonclalves-Andrade and G. Naufel. 2001. Attractiveness of Black Shannon trap for Phlebotomines. Mem. Inst. Oswaldo Cruz. Rio de Janeiro 96: 641–647.

Galun, R. 1979. Combination of *in vitro* feeding techniques and tissue culture for study of arthropod-born disease agents. pp. 399–406. *In*: K. Maramorosch and H. Hirumi [eds.]. Practical Tissue Culture Applications. Academic Press, NY.

Ghosh, K.N. 1994. A modified artificial membrane feeding method for the study of the transmission dynamics of leishmaniasis. Trans. R. Soc. Trop. Med. Hyg. 88: 488–489.

Harre, J.G., K.M. Dorsey, K.L. Armstrong, J.R. Burge and K.E. Kinnamon. 2001. Comparative fecundity and survival rates of *Phlebotomus papatasi* sandflies membrane fed on blood from eight mammal species. Med. Vet. Entomol. 15: 189–196.

Hertig, M. 1939. Phlebotomus and Carrion's disease. Proc. Sixth Pacific Sci. Cong. 5: 775–779.

Kern, F. and J.K. Pedersen. 1973. Leishmaniasis in the United States: a report of ten cases in military personnel. JAMA 226: 872–874.

Killick-Kendrick, R. 1978. Recent advances and outstanding problems in the biology of phlebotomine sandflies. Acta Trop. 35: 297–313.

Killick-Kendrick, R. and M. Killick-Kendrick. 1991. The initial establishment of sand fly colonies. Parasitologia 33: 315–320.

Lane, R.P. 1993. Sand flies (*Phlebotominae*). pp. 78–119. *In*: R.P. Lane and R.W. Crosskey [eds.]. Medical Insects and Arachinds. Chapman and Hall, London, England.

Libert, M.A. 2003. Arthropod Containement Levels (ACLs). Vector-Borne and Zoonotic Diseases 3: 75–90.

Lawyer, P.G. 1984. Biology and colonization of the sand fly *Lutzomyia diabolica* (Hall) (Diptera: Psychodidae) with notes on its potential relationship to human cutaneous leishmaniasis in Texas, USA. Ph.D. dissertation. University of Florida, Gainesville, FL.

Lawyer, P.G., E.D. Rowton, P.V. Perkins, R.N. Johnson and D.G. Young. 1991. Recent advances in laboratory mass rearing of phlebotomine sand flies. Parasitologia 33: 361–364.

Magill, A.J., M. Grogl, R.A. Gasser, W. Sun and C.N. Oster. 1993. Visceral infection caused by *Leishmania tropica* in veterans of Operation Desert Storm. N. Engl. J. Med. 328: 1383–1387.

Mahmood, F. 1990. Different aspects of the biology of phlebotomine sand flies (Diptera: Psychodidae) and their age structure in a focus of cutaneous Leishmaniasis in northeastern Colombia. Ph.D. dissertation. University of Florida, Gainesville. FL. 290 pp.

Mahmood, F. 1999. Laboratory Maintenance of mosquitoes. pp. 3–54. *In*: K. Maramorosch and F. Mahmood [eds.]. Maintenance of Human, Animal and Plant Pathogen Vectors. Science Publishers, Inc. Enfield, USA.

Mahmood, F. and D. Borovsky. 1992. Biosynthesis of trypsinlike and Chymotrypsinlike enzymes in immature *Lutzomyia anthophora* (Diptera: Psychodidae). J. Med. Entomol. 29: 489–495.

Mahmood, F. and D. Borovsky. 1993. Biosynthesis of serine proteases in *Lutzomyia anthophora* (Diptera: Psychodidae). J. Med. Entomol. 30: 683–688.

Mahmood, F.D.A. Carlson and D. Borovsky. 1992. Biosynthesis and metabolism of Juvenile hormone III from methyl farnesoate by exposed corpora allata of *Lutzomyia anthophora.* J. Med. Entomol. 29: 548–555.

Mahmood, F., R.E. Chiles, Y. Fang, E.N. Green and W.K. Reisen. 2006. Effect of time after infection, mosquito genotype, and infectious viral dose on the dynamics of *Culex tarsalis* vector competence for western equine encephalomyelitis virus. J. Am. Mosq. Control Assoc. 22: 272–281.

Mahmood, F., Y. Fang, R.E. Chiles and W.K. Reisen. 2004. Methods for studying the vector competence of *Culex tarsalis* for Western equine encephalomyelitis virus. J. Am. Mosq. Control. Assoc. 20: 277–282.

Mann, R.S. and P.E. Kaufman. 2010. Colonization of *Lutzomyia shannoni* (Diptera: Psychodidae) utilizing an artificial blood feeding technique. J. Vec. Ecology 35: 286–294.

Mann, R.S., P.E. Kaufman and J.F. Buttler. 2009. *Lutzomyia* spp. (Diptera: Psychodidae) response to olfactory attractant and light-emitting diode-modified Mosquito Magnet X (MM-X) traps. J. Med. Entomol. 46: 1052–1061.

Marchais, R., F. Bounchet and P. Bounchet. 1991. Laboratory rearing of *Phlebotomous perniciosus* (Diptera: Psychodidae) and fungal growth problem. Parasitologia 33: 393–397.

McHugh, C.P., M. Grogl and R.C. Kreutzer. 1993. Isolation of *Leishmania mexicana* (Kinetoplastida: Trypanosomatidae) from *Lutzomyia anthophora* (Diptera: Psychodidae) collected in Texas. J. Med. Entomol. 30: 631–633.

McHugh, C.P., P.C. Melby and S.G. LaFON. 1996. Leishmaniasis in Texas: Epidemiology and clinical aspects of human cases. J. Trop. Med. Hyg. 55: 547–555.

Milleron, R.S., C.R. Meneses, D.A. Elnaiem and G.C. Lanzaro. 2008. Effect of varying moisture on egg production and longevity of *Lutzomyia longipalpis* (Diptera: Psychodidae). J. Med. Entomol. 45: 160–165.

Modi, G.B. and R.B. Tesh. 1983. A simple technique for Mass rearing *Lutzomyia longipalpus* and *Phlebotomus papatasi* (Diptera: Psychodidae) in the laboratory. J. Med. Entomol. 20: 568–569.

Modi, G.B. and E.D. Rowton. 1999. Laboratory maintenance of phleobotomine sand flies. pp. 109–121. *In*: K. Maramorsch and F. Mahmood [eds.]. Maintenance of Human, Animal, and Plant Pathogen Vectors. Science Publishers, Enfield NH, USA.

Montoya-Lerma, J., H. Cadena-Pena and C. Jaramillo-Salazar. 1998. Rearing and colonization of *Lutzomyia evansi* (Diptera: Psychodidae), a vector of visceral leishmaniasis. Mem. Inst. Oswaldo Cruze, Rio de Janeiro 93: 263–268.

Montoya-Lerma, J. and R.P. Lane. 1996. Factors affecting host preferences of *Lutzomyia evansi* (Diptera: Psychodidae) a vector of visceral leishmaniasis in Colombia. Bull. Entomol. Res. 86: 43–50.

Pessoa, S.B. and M.P. Barretto. 1948. *Leishmaniose tegumentar americana*. Ministério de Educaco e Saudé, Imprensa Nacional, Rio de Janeriro.

Peterkova-Koci, M. Robels-M. Robles-Murguia, M. Ramalho-Ortigao and L. Zurek. 2012. Significance of bacteria in oviposition and larval development of the sand fly *Lutzomyia longipalpus*. Parasites and Vectors. pp. 1–8.

Rangel, E.F., N.A. Souza, E.D. Wermelinger and A.E. Barbosa. 1987. Biology of *Lutzomyia intermedia* Lutz & Neiva, 1912 and *Lutzomyia longipalpis* Lutz & Neiva, 1912 (Diptera: Psychodidae), under experimental conditions. Aspects related to oviposition. An. Soc. Entomol. Brasil 16: 289–296.

Reuss, S.M., M.D. Dunbar, B. Maron, C. Mays, J.L. Owen, M.F. Mallicote, L.L. Archer and J.F.X. Wellehan Jr. 2012. Autochthonous *Leishmania siamensis* in Horse, Florida, USA. Emerg. Infec. Disease 18: 1545–1547.

Rowton, E.D., K.M. Dorsey and K.L. Armstrong. 2008. Comparison of *in vitro* (Chicken-skin membrane) versus *in vivo* (live hamster) blood-feeding methods for maintenance of colonized *Phlebotomus papatasi* (Diptera: Pscyhodidae). J. Med. Entomol. 45: 9–13.

Rutledge, L.C., M.A. Moussa and C.J. Belletti. 1976. An *in vitro* blood-feeding system for quantitative testing of mosquito repellents. Mosquito News 36: 283–293.

Schlein, Y. and R.L. Jacobson. 1999. Sugar meals and longevity of the sandfly *Phlebotomus papatasi* in an arid focus of *Leishmania major* in the Jordon valley. Med. Vet. Entomol. 13: 65–71.

Schlein, Y., S. Borut and R.L. Jacobson. 1990. Oviposition diapause and other factors affecting the egg-laying of *Phlebotomus papatasi* in the laboratory. Med. Vet. Entomol. 4: 69–78.

Smith, R.O.A. 1925. A note on a simple method of breeding sandflies. Ind. J. Med. Res. 12: 741–742.

Sokal, R.R. and F.J. Rohlf. 1981. Biometry. W.H. Freeman and company, USA. 859 pp.

Souza, N.A., C.A. Andrade-Coelho, M.L. Vilela, A.F. Barbosa and E.F. Rangel. 1999. A new larval diet for colonization of Phlebotomine sand flies. Mem. Inst. Oswaldo Cruz, Rio de Janeiro 94: 845–847.

Sudia, W.D. and R.W. Chamberlain. 1962. Battery operated light traps, an improved model. Mosq. News 22: 126–129.

Tesh, R.B. 1988. The genus *Phlebovirus* and its vectors. Ann. Rev. Entomol. 33: 169–181.

Tesh, R.B., B.N. Chaniotis, M.D. Aronson and K.M. Johnson. 1971. Natural host preferences of Panamanian phlebotomione sand flies as determined by precipitin test. Am. J. Trop. Med. Hyg. 20: 150–156.

Tesh, R.B., B.N. Chaniotis, P.H. Peralta and K.M. Johnson. 1974. Ecology of viruses isolated from Panamanian phlebotomine sandflies. Am. J. Epidemiol. 93: 491–495.

Tesh, R.B. and G.B. Modi. 1983. Growth and transovarial transmission of Chandipura virus (Rhabodoviridae: Vesiculovirus) in *Phlebotomus papatasi*. Am. J. Trop. Med. Hyg. 32: 621–623.

Tesh, R.B. and G.B. Modi. 1984. A simple method for experimental infection of Phlebotomine sand flies with *Leishmania*. Am. J. Trop. Med. Hyg. 33: 41–46.

Trainor, K.E., B.F. Porter, K.S. Logan, R.J. Hoffman and K.E. Snowden. 2010. Eight cases of feline cutaneous leishmaniasis in Texas. Vet. Pathol. 47: 1076–1081.

Volf, P. and V. Volfova. 2011. Establishment and maintenance of sand fly colonies. J. Vec. Ecology 36: 51–59.

Ward, R.D., R. Lainson and J.J. Shaw. 1978. Some methods for membrane feeding of laboratory reared, neuotropical sandflies (Diptera: Psychodidae). Ann. Trop. Med. Parasitol. 72: 269–276.

Young, D.G. 1979. A review of the bloodsucking Psychodid flies of Colombia (Diptera: Phlebotominae and Sycrocinae). Bulletin 806 (Technical). Agricultural Experiment Stations, Institute of Food and Agricultural Sciences, University of Florida, Gainesville. 266 pp.

Young, D.G. and M.A. Duncan. 1994. A guide to the identification and geographic distribution of *Lutzomyia* sand flies in Mexico, the West Indies, Central and South America (Diptera: Psychodidae). Mem. Am. Entomol. Inst. 54: 881.

Young, D.G. and P.G. Lawyer. 1987. New World vectors of the leishmaniasis. *In:* Current Topics in Vector Research. Springer-Verlag. New York. 4: 29–71.
Young, D.G. and P.V. Perkins. 1984. Phlebotomine sand flies of North America (Diptera: Psychodidae). Mosq. News 44: 263–304.
Young, D.G., P.V. Perkins and R.G. Endris. 1981. A larval diet for rearing phlebotomine sand flies (Diptera: Psychodidae). J. Med. Entomol. 18: 446.

7

Establishment and Maintenance of Small Scale Tsetse Colonies

Serap Aksoy

ABSTRACT

Tsetse flies (Diptera: Glossinidae) are medically and agriculturally important vectors of the protozoan parasites African trypanosomes which cause devastating diseases in humans and animals in large regions of sub-Saharan Africa. Laboratory colonies of *Glossina* are established to develop an understanding of the basic biology of parasite transmission in the fly as well as to develop and implement vector-based disease control strategies. Here, some unique aspects of tsetse biology are described that have an impact on laboratory rearing methods and the maintenance and feeding conditions needed to establish small research colonies are reviewed.

Introduction

The genus *Glossina* contains 31 taxa (21 species and eight subspecies), assigned to four species groups: *Austenina* (=*fusca* group), *Nemorhina* (=*palpalis* group), *Glossina* s. str. (=*morsitans* group) and *Machadomyia* (=*austeni* group). It has been suggested that the superior claspers of the male

Yale University School of Medicine, Department of Epidemilogy and Public Health, 60 College St., 606 LEPH, New Haven, CT 06510, USA.

Addendum on page 327

become progressively more complicated through the groups *fusca-palpalis-morsitans* indicating the ancestral nature of the *fusca* group (Newstead et al. 1924). The analysis of 14 enzyme loci has also indicated that members of the *fusca* group appear to be the most divergent (Gooding et al. 1991). The recent phylogenetic analysis of different *Glossina* species based on their nuclear gene sequence analysis, further supports their taxonomic placement into four subgenera, where the *fusca* group forms the ancestral subgenus, followed by the closely related *morsitans, palpalis,* and *austeni* species groups (Chen et al. 1999). The genus is restricted to sub-Saharan Africa where its different species occupy diverse habitats. The *palpalis* and *fusca* groups require high humidity and live mainly in riverine gallery and primary forests of Central and West Africa or in relict forests of eastern Africa while the *morsitans* group species have a wide distribution in drier savanna and thicket habitats.

Tsetse transmits the protozoan African trypanosomes which cause diseases in humans as well as in domestic animals. At the beginning of this century, human sleeping sickness was one of the most important public health problems of Africa. By the mid 1950's, the incidence of disease had been reduced in endemic countries following massive campaigns that combined detection and treatment of symptom-free cases and chemoprophylaxis. Today, the breakdown of health facilities in many countries as a consequence of war or civil strife has led to a huge resurgence of sleeping sickness disease, with highest incidences being seen in Zaire (Ekwanzala et al. 1996), Uganda (Lancien 1991), Sudan (McKinley, July 18, 1997) and Angola while increasing numbers are being reported from neighboring Congo and Central African Republic. *Trypanosoma brucei* complex parasites cause the devastating human epidemics; *T. b. gambiense* and *T. b. rhodesiense* cause chronic Gambian sleeping sickness in West Africa and acute Rhodesian sleeping sickness in East Africa, respectively. The *T. b. brucei* and related parasites also cause various diseases in domestic animals, collectively called nagana, which are often cited as the major hindrance for the agricultural and economic development of large areas of sub-Saharan Africa. While all species of flies can transmit trypanosomes, the *palpalis* group species can easily adapt to life near or within human villages and exhibit a preference for human bloodmeals and hence are efficient vectors in West Africa while the *morsitans* group is primarily responsible for disease transmission in East-Africa. Both sexes of adult tsetse are strict blood feeders, and hence can contribute to parasite transmission. Parasite transmission in the fly is influenced by many factors including the fly species, age and sex, source of the bloodmeal, and the strain of parasite. However, once the flies are infected, they remain infected the rest of their lives and since they require a bloodmeal every 2–3 d in nature and can live upto 5–6 mon in the field, they are very efficient vectors of trypanosomiasis.

Because of the extensive antigenic variation phenomenon the parasite displays in its mammalian host, efforts to develop a mammalian vaccine have not been successful. Disease management is reliant on the surveillance of active cases followed by use of chemotherapy and on the reduction/ elimination of tsetse populations by targets and traps. Odor-baited (acetone, octenol, phenols or cattle urine) box traps are used for *morsitans* and biconical or pyramidal traps are used for the *palpalis* group. *Glossina* are active during the day with most activity being confined to the periods of dawn and dusk. In addition, the genetic control strategy, Sterile Insect Technique (SIT) has been successfully used for tsetse control. In SIT, large numbers of male flies are sterilized by irradiation and released into the field where they mate with virgin wild females. The females which receive irradiated sperm can not produce offspring and hence this approach leads to population suppression and eventual eradication. Since young tsetse females mate once and become refractory for further matings and since tsetses have a low reproductive rate, they are ideal targets for the implementation of SIT which in fact has been recently used to successfully eradicate the *G. austeni* population from the island of Zanzibar (Msangi et al. 1999). One of the criticisms against SIT has been its dependence on the maintenance of large breeding colonies which can be expensive and technically challenging. However, the tsetse breeding facility recently established in Tanga on the Tanzanian mainland has been able to successfully maintain almost 1,000,000 reproducing females which were used for the male releases in Zanzibar indicating that the technical aspects of the rearing process is no longer a handicap.

Maintenance Methods

Establishment of colonies involves collecting puparia, sexing and sorting the emerging adults, setting up matings between males and females, separating the mated females into new cages and maintaining the breeding females by providing a bloodmeal every 30–36 hr. These females will deposit their first fully developed larva when they are about 18 d old and will continue to deposit a single larva every 9 d over their 3–4 mon lifespan in the insectary. Below, are described the conditions required for maintaining a small size colony that can be used for research purposes.

Life-history of Tsetse

Female tsetse are viviparous, ovulation proceeds alternately from the two polytrophic ovarioles and the egg is fertilized by sperm from the paired spermatheca. Each female produces a mature larva every 9–10 d and in its 3–4 mon lifespan, each female can deposit 10–12 offspring. At 25°C, the egg stage requires 4 d, and 1, 1.5 and 2.5 d are required for the 1st, 2nd and 3rd larval instars, respectively. During its intrauterine life, the larva receives nutrients from its mother by secretions from the uterine gland via an extensive network of ducts. The deposited larva rapidly burrows into the moist soil to a depth of 1–2 inches and within 1–2 hr hardens to form a black barrel-shaped puparia with two characteristic polypneustic lobes. The puparial stage takes about 30–34 d, females emerge first with the males requiring 1–2 additional d. Males become sexually active in about 1 wk after eclosion, whereas females require 2–8 d depending on the species. The copulatory response is mediated through a contact pheromone which is present in the cuticular wax of the female. Copulation can take several hr during which time the male transfers the components of the spermatophore followed by the spermatozoa. Young females only mate once and utilize the sperm stored in the spermathecae and become refractory for mating thereafter.

Source of Material for the Colony

Tsetse colonies can be initiated either by field collected puparia or alternatively, from puparia deposited by field collected adult females. The field-caught females can be brought into the insectary and maintained long enough to deposit puparia to initiate the colony. The field collected puparia have the advantage in that the emerging adults are free pathogens since the field collected flies may already be infected with trypanosomes and hence would require special handling procedures. However, field collected pupae can carry parasitoids and hence some quarantine care is necessary. The difficulty with obtaining puparia from the field is that it may be a challenge to locate puparial deposition sites for tsetses when large numbers of flies are needed to begin the colony. Since the deposited larvae quickly bury their way into the soil and pupate, soil from suitable habitats would have to be collected and sifted to obtain the puparia. In either case, however, it is important that sufficient material is obtained to ensure ample sampling of the gene pool for success of colony. It is of course also possible that puparia be supplied from existing colonies for establishment at a new site. When transporting puparia, it is important to use at least one week old puparia and to provide adequate amount of air in the shipping container and adequate insulation to prevent exposure to temperature fluctuations. In some countries, however, maintaining tsetse fly colonies requires special permits, so that the appropriate agencies should be consulted before any effort is made.

Puparia Collection and Emergence

In colonies where less than 1000 breeding females are maintained, the deposited puparia can be collected weekly. The puparia are stored in tubes or petri dishes with the container open to air, paying attention that the thickness of the puparia in each container is less than an inch. Cages are directly placed over puparia storage tubes to allow the adults to emerge into the cages. Once the adults begin to emerge the cages may be replaced daily by placing a cork or rubber stopper at the opening and the emerged adults can then be stored, fed and mated as described below. Typically, at least 85% of the puparia should emerge if the general health of the colony is satisfactory.

Cages and Trays

The size of cages used for emergence and maintenance may vary according to the colony size. Cages can either be made from metal frames or from PVC tubing. Rectangular stainless steel wire frames of 18x8x4 cm having a closed end fitted with a corked hole can typically hold 20–30 adults (*Glossina morsitans morsitans*) while circular cages of 20 cm in diameter made from PVC tubing (4 mm thick) cut into 5 cm lengths are used to hold about 100 females. All cages must be covered with a black Terylene netting through which the flies may feed and through which the larvae, but not the adults may pass. For most species 250 denier black Terylene netting with about 16 holes/cm^2 is suitable. The netting can either be sawn into sleeves into which the metal frames can be tightly inserted and the netting can be stretched and attached with tape to the end plates or it may be glued to the PVC frames using PVC cement. One of the end plates has an opening into which the puparia containing tube can be inserted as shown in Fig. 1.

When the adults emerge into cages, the opening placed over the puparia storage tubes can be closed with a cork or rubber stopper and adults can be maintained in the same cages. The cages are kept on trays which can be made from metal, plastic or plexiglass. It is important that the trays are constructed in a way that only a small portion of the bottom of each cage is in contact with the tray. This can be accomplished by having two appropriately placed ridges or rods in the tray on which the cages could rest. This is especially needed so that larva deposited by the breeding females can escape from the cage, squeeze through the netting and fall onto the tray (Fig. 2). The puparia can be collected from the trays and placement of some coarse grain sand in the trays makes collection of puparia easier. One side of the cage should be reserved for feeding and the opposite side should be placed face-down on the tray. The cages should be dated and the number of flies housed in each cage should be marked and as flies die

Fig. 1. Maintenance of puparia.

Racks of rectangular metal emergence cages placed over puparia holding tubes are shown, picture taken from Dr. Ron Gooding's insectary at University of Alberta, Edmonton, Canada.

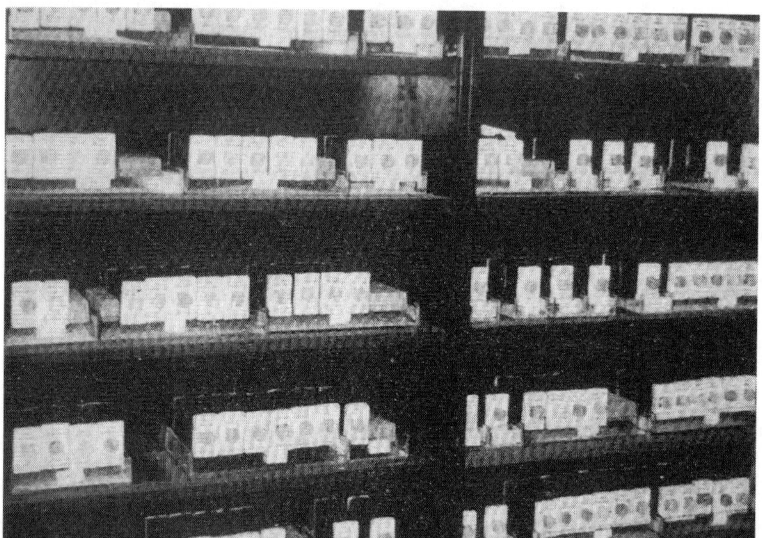

Fig. 2. Maintenance of adult tsetse flies.

Shelves containing adult maintenance cages placed on plastic holding trays can be seen. The puparia are collected weekly from the trays and maintained as shown in Fig. 1 above, picture taken from Dr. Ron Gooding's insectary at the University of Alberta, Edmonton, Canada.

they should be removed and recorded. The cages should be handled with care, especially cages housing the breeding females should not be disturbed as often this would result in aborted larvae or eggs. In large insectaries, to assess the quality of the colony, either the size and/or the weight of the collected puparia is determined. The weight of the puparia collected over a 24 h period can either be weighed in batches or as individuals.

The mean puparial weights (mg) for different *Glossina* species are: *G. tachinoides* (17.3 ± 1.9), *G. palpalis* (30.9 ± 3.8), *G. f. fuscipes* (32.8 ± 3.7), *G. austeni* (23.1 ± 3.6), *G. pallidipes* (35.9 ± 3.9), and *G. brevipalpis* (64.4 ± 5.7) (Gooding et. al. 1997). It is recommended that in a healthy colony no more than 10% of the puparia should be below the lower value of the mean. Alternatively, a size-sorting machine can be used to separate different sized puparia by passing the collected puparia through a sieve system (Fig. 3, Zelger 1976).

A significant increases in the number of smaller puparia indicates that the health of the colony is being compromised and it would be important at this point to evaluate the nutritional value of the bloodmeal source and the environmental conditions in the facility to prevent significant loss of fecundity. In addition to suboptimal environmental conditions, excessive crowding in cages or bacterial or chemical contamination may also contribute to reduced fecundity. Alternatively, a chart can be maintained

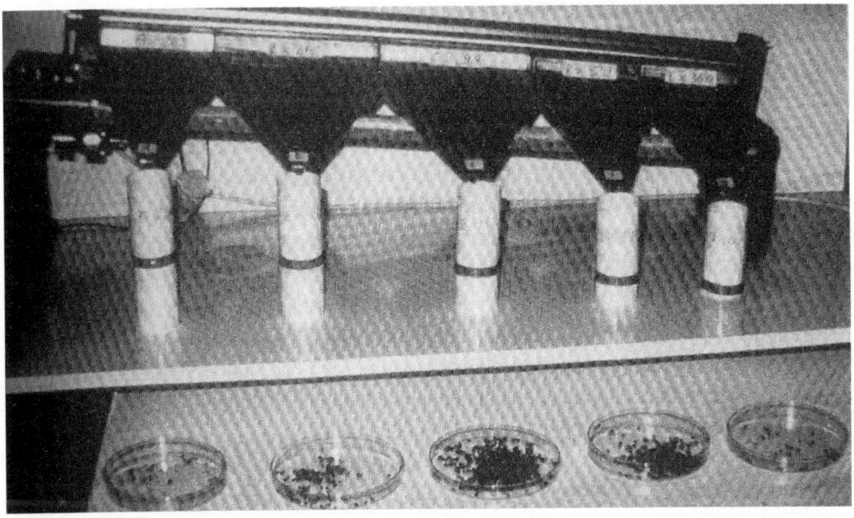

Fig. 3. Quality Control.

The puparia sizing apparatus which is used to assess the quality of a tsetse colony is shown. The collected puparia are passed down the trough which contains different sized openings. The range of different sized-puparia collected can be seen on the petri dishes. Picture taken at the International Atomic Energy Agency's Laboratories at Seibersdorf, Austria.

where the numbers of emerging adults and the collected puparia are noted weekly. The trays should also be examined for spontaneous abortions which often indicate that the general well-being of the breeding females has been compromised.

Temperature, Relative Humidity and Lighting Requirements

Both puparia and adults of most species are maintained at 24°C, and under fairly humid conditions, about 70% humidity for the *morsitans* group and 80–85% humidity for the *palpalis* group flies is recommended. The flies are maintained under subdued, indirect light and while tungsten lamps are preferred, fluorescent lighting in rooms in which human activity is minimal is generally sufficient. When newly emerged adults are being sorted into different sexes to initiate matings, the cages can be immersed in crushed ice to immobilize the flies, however this exposure time to low temperatures should be minimized and flies should be fed several hours (6–8 hr) after recovery. Early mortality rates when the flies are about 10–14 d are generally due to excessive chilling or rough handling conditions. When handling flies, they should also not be chilled directly after feeding and in general, if there are fluctuations in temperature and humidity conditions in the facility, flies should be fed several hours after the appropriate adjustments are made.

Feeding and Source of Bloodmeal

Flies should be fed under environmental conditions that are as close to those used for maintenance as described above. Originally, most tsetse colonies were maintained on lives hosts such as goats, guinea pigs or rabbits. Given its conveniences however, in recent years, most insectaries have switched to using an *in vitro* feeding system where fresh or freshly frozen defibrinated blood is provided to the flies through a silicone membrane (Bauer and Wetzel 1976, Moloo 1971). The details of the preparation of the membrane and the feeding trays have been described (Gooding et al. 1997). While large insectaries have used corrugated aluminum trays, etched glass plates can also be used for feeding if the colony size is relatively small. It is important to maintain the trays and the membranes sterile, hence the glass plates are washed and autoclaved after each use, and the aluminum trays and membranes are washed and kept in an oven at 120°C overnight before use. The blood is maintained at body temperature by placing the trays on a suitable heating apparatus that has a rheostat for controlling its temperature. A thin layer of blood is poured onto the tray and the membrane is placed on the blood gently so as not to trap air bubbles (Fig. 4). Blood is uniformly distributed under the membrane by use of a sterile plastic rod. If flies are

Fig. 4. Blood-feeding apparatus: the aluminum tray and the silicone membrane placed over the poured blood can be seen. The trays are placed on a heated surface to maintain an ambient temperature of blood. Picture taken at the International Atomic Energy Agency's Laboratory, Seibersdorf, Austria.

found dead with indigested blood in their guts shortly after feeding, it may be due to either contaminated blood or to blood that was too hot above the ambient temperature.

Cages are placed with the feeding surface of the cage next to the membrane and flies are allowed to feed for about 10 minutes. Several batches of cages can be fed on the same membrane, successively. This *in vitro* feeding approach can also be used to establish parasite infections in tsetses using frozen infected blood after growing up the trypanosomes in rats or rabbits (Welburn and Maudlin 1987). Various sources of blood can be used for feeding the colony. In the IAEA insectary at Seibersdorf, large volumes of blood obtained from a variety of animals from local slaughter houses are mixed, defibrinated, microbially decontaminated by exposure to 1.0–1.5 KGy gamma-irradiation and then frozen to –20°C in large batches. As needed, one batch is defrosted, a microbiological test and a bioassay are run to evaluate its level of microbial contamination and to assess its nutritional suitability. A group of fertile females are fed on the batch and their fecundity is evaluated over time before the batch is used to feed the entire colony. Alternatively, for smaller colonies, fresh defibrinated rabbit or bovine blood can be purchased weekly to feed flies. Freshly frozen blood can be stored at –20°C in small aliquots and can be defrosted and used as needed.

If finding a reliable source of blood is difficult, however, and sterile feeding conditions can not be easily achieved, an *in vivo* feeding system is recommended. Often, the ears and shaved backs of 5–6 mon old rabbits that are trained to sit in restraining boxes are used by holding the cages in place by elastic belts (Fig. 5). It is important that the animals must receive a diet free of antibiotics and insecticides as both will reduce fecundity.

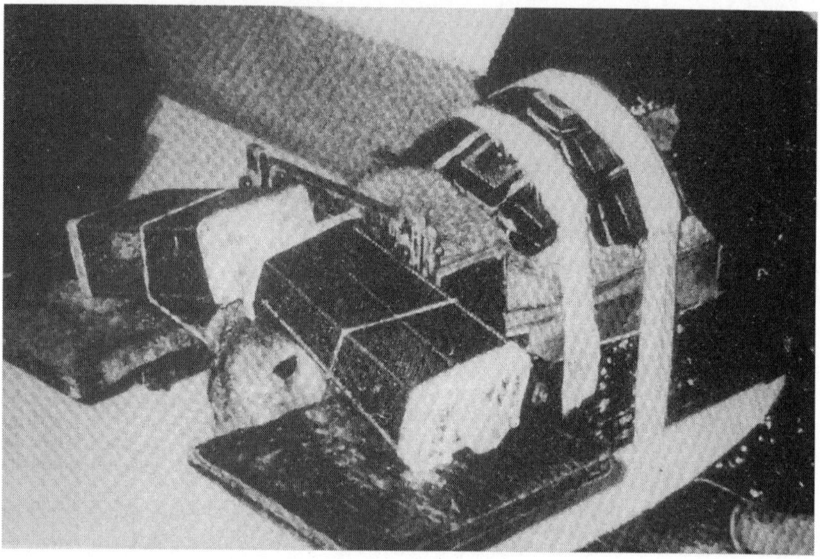

Fig. 5. A rabbit is shown immobilized in a restraining cage with tsetse cages attached to its ears and back. Picture taken from Dr. Ron Gooding's insectary at the University of Alberta, Edmonton, Canada.

Breeding Tsetse

In order to obtain the maximum insemination rates in females, it is important to separate newly emerged females and males. This is done by chilling the emerged new adults and sexing flies manually. In males, the posterior end of the abdomen is pointed, the superior claspers or external genitalia are heavily sclerotized and the fifth abdominal sternum is largely sclerotized and carries many setae. In contrast, the females's abdomen is truncated and its midline region is not sclerotized and has a pale appearance. Although some important variations are noted between species, males (*G. morsitans morsitans*) reach full sexual capability within a week and females are sexually receptive when they are 2–4 d old. The females of *G. brevipalpis* are receptive 1 d after eclosion (Moloo and Kutuza 1988) whereas females of *G. pallidipes* are not receptive for at least a week. At the appropriate times, a 1:2 ratio of males to females are used to establish cage matings. Males are separated from the females after 24–48 hr, and they can be used for two or three additional matings. The pregnant females are maintained for 12–16 weeks during which time each female deposits one pupa every 9–10 d. The source of bloodmeal, environmental conditions of the insectary facility and crowding in cages can all affect the fecundity of colonized flies as well as the pupal emergence rates. Alternatively, failure of males flies to mate or insufficient insemination can also reduce fecundity. In order to ensure that copulation occurs, visual inspection of cages after males and females have been placed together is recommended. In general between 45–65% of the puparia must be retained to sustain the colony and about 10% of the entire colony size needs replacements on a weekly basis (Gooding et al. 1997).

References

Bauer, B. and H. Wetzel. 1976. A new membrane for feeding *Glossina moristans* (Westwood) (Diptera: Glossinidae). Bull. Entomol. Res. 65: 319–326.

Chen, X., L. Song and S. Aksoy. 1999. Concordant evolution of a symbiont with its host insects species: Molecular phylogeny of genus *Glossina* and its bacteriome-associated endosymbionts. *Wigglesworthia glossinidia*. J. Mol. Evol. 48: 49–58.

Ekwanzala, M. J. Pepin, N. Khonde, S. Moliho, H. Bruneel and P. De. Wals. 1996. In the heart of darkness sleeping sickness in Zaire. Lancet 348: 1427–1430.

Gooding, R.H., U. Feldman and A. Robinson. 1997. Care and Maintenance of tsetse colonies. pp. 578. *In*: J.M. Crampton, C.B. Beard and C. Louise [eds.]. The Molecular Biology of Insect Disease Vectors. Chapman and Hall, Cambridge.

Gooding, R.H., S.K. Moloo and B.M. Rolseth. 1991. Genetic variation in *Glossina brevipalpis*, *G. longipennis*, and *G. pallidipes*, and the phenetic relationships of *Glossina* species. Med. Vet. Entomol. 5: 165–173.

Lancien, J. 1991. Campaign against sleeping sickness in South-West-Uganda by trapping tsetse flies. Ann. Soc. Belg. Med. Trop. 71 Suppl. 11: 35–47.

McKinley, J.C. July 18, 1997. Deadly Epidemic Emerges in Sudan. In New York Times. New York City. 1 pp.

Moloo, S.K. 1971. An artificial feeding technique for *Glossina*. Parasitology 63: 507–512.

Moloo, S.K. and S.B. Kutuza. 1988. Large-scale rearing of *Glossina brevipalpis* in the laboratory. Med. Vet. Entomol. 2: 201–202.

Msangi, A., N.E. Kiwia, F. Mramba, W.A.M. Kitwita, I. Malele, M.B. Byamungu, G. Kasilagila, V.A. Dyck and A.G. Parker. 1999. Eradication of *Glossina austeni* (Newstead) on Ungujaisland (Zanzibar) by the sterile insect technique. 2. Mass-production and quality assessment of sterile flies. pp. 219–230. *In*: Animal Trypanosomosis: Vector and Disease Control using Nuclear Techniques. Proceedings of the second FAO/IAEA Seminar for Africa. 27 November–1 December 1995. Zanzibar, United Republic of Tanzania. Backuys Publishers, Leiden. 303 pp.

Newstead, R., A.M. Evans and W.H. Potts. 1924. Guide to the study of tsetse flies. Mem. Liverpool School Trop. Med. 1: 332.

Wellburn, S.C. and I. Maudlin. 1987. A simple *in vitro* method for infecting tsetse with trypanosomes. Ann. Trop. Med. Parasitol. 81: 453–455.

Zelger, R. and K. Russ. 1976. Puppentrennmaschine mit zwe Qalzepaaren. Z. Angew. Zool. 63: 257.

8

Rearing Aphids to use in Virus-Vectors Studies

Catherine Katsar[1] and *Stewart Gray*[1,2,*]

ABSTRACT

Numerous systems exist to rear aphids for use in various experimental procedures. Aphids are among the easiest of insects to maintain in the laboratory, however certain rules need to be strictly adhered to in order to generate physiologically normal insects that will behave in a normal manner. In this chapter we outline the specifics of one system that has been in use for over 50 years to maintain multiple cereal aphids for use in virus transmission studies. Details on facilities, maintenance schedules, handling, and potential pitfalls are provided. In addition, general information on starting new colonies and selecting appropriate rearing conditions are provided. The system outlined here is easily adaptable for use with many other aphid and host plants.

Key Words: work room, cages, contamination, viruses.

Introduction

Aphids are the most important vectors of plant viruses and the most widely studied insect vectors. Numerous laboratories maintain aphid colonies for use in various experimental and commercial applications. It is likely that there are as many methods and systems for rearing aphids as there are

[1] Plant Protection Research Unit, USDA, ARS.
[2] Department of Plant Pathology, Cornell University, Ithaca, NY 14853, USA.
* Corresponding author

colonies being maintained. Therefore, this chapter is not intended to present an overview of all the possibilities one may employ to rear aphids for use in virus transmission studies, but rather to present an overview of how our laboratory maintains aphid colonies. Many of the techniques, procedures and equipment used in our system can be adapted for use with other aphid species—plant host combinations. The reader is also directed to several other reviews that describe alternative aphid rearing systems (Austin et al. 1991, Blackman 1987, Campbell Christopher and Hutchison William 1995, Forbes, Frazer and Chan 1985, Girma et al. 1990, Nault 1969, Ridland et al. 1988, Sylvester 1987, Van Rensburg 1981, Webb Susan et al. 1994).

Currently we maintain *Myzus persicae* and six species of grain aphids, *Rhopalosiphum padi, R. maidis, R. rufiabdominalis, Sitobion avenae, Schizaphis graminum,* and *Metapolophium dirhodum. Myzus persicae* is maintained on radish or turnip while all of the grain aphids are maintained on winter barley. In addition, to these species, multiple clones (biotypes) are maintained of *S. graminum, R. padi* and *R. maidis.* As you can imagine the continued maintenance of all these species and clones can utilize considerable resources and the utmost care must be taken to insure the colonies (especially the colonal populations) remain pure. The system that has been designed and implemented to maintain all the aphids and guarantee a steady supply of experimental material has been evolving over several decades (Power and Gray 1995, Rochow 1969) and is likely to continue to evolve as our needs change. Nevertheless, the current system is a highly effective and reliable way to generate aphids for use in virus transmission studies and genetic studies.

Starting New Colonies

Aphid Selection

Generally one chooses the best vector for a given virus, i.e., one that transmits the virus efficiently among the virus host plants. This information may be available from the literature (Blackman and Eastop 1984) or may need to be determined empirically. However, it must be recognized that there may be considerable variability in the transmission efficiency among clones within an aphid species and/or within isolates of a particular virus (Gray 1999). Regardless of what the selected aphid is, it is of utmost importance to have the aphid vector identified by a trained taxonomist and to keep type specimens for future reference.

Colonies can be started from field collected aphids or from aphids received from colleagues that maintain colonies. Although less likely from the latter source, newly acquired aphids may harbor unseen parasites or pathogens. For this reason newly established colonies should be quarantined away from any other established colonies until it is known for certain they are contaminant free.

One method to minimize the chance of parasite/pathogen contamination is to start all colonies with first instar nymphs. Apterous, parthenogenic, female adults are placed in small containers with tightly fitting lids containing moist filter paper or leaves from healthy host plants. The aphids are allowed to produce progeny over a 12–24 hour period. The newly emerged nymphs are then transferred to healthy plants to initiate the colony. In general, pathogens and parasites are not transovarially transmitted to the progeny, although this rule does not apply to viruses that replicate in aphids (Ammar 1994). The use of first instar nymphs will eliminate any plant viruses that are transmitted in a noncirculative or circulative, nonpropagative manner (Gray and Banerjee 1999). Contamination of an aphid colony by pathogens (see also section on Colony Contaminants) may not be immediately obvious, but over the long term the effects of the pathogen will be seen as a reduction in vigor of the aphid colony, e.g., higher morality, lower fecundity and reproductive rates. Parasite contamination is generally obvious in that the parasite can be seen or mummified aphids are found.

Assuming the initial aphids used to start the colonies are pathogen-free, the domestication of field collected aphids may still be unsuccessful. The changes in environment will immediately create a sorting of suitable genotypes and may drastically alter the phenotypic characteristics and qualities for which the aphid species or clone was initially chosen. In many instances, the fitness of a field clone will be altered so the progeny have a high mortality rate and/or lower fecundity and will not survive the transfer to a new host and new environment. Starting colonies using newly emerged nymphs from several adults (not siblings) should help to increase the degree of genetic variability within the colony. However, as previously mentioned, the individuals within a genetically diverse population of aphids may differ in their ability to transmit viruses. We have found it more helpful to attempt to start several colonies; each initiated using the progeny of one adult. If the aphid clone survives the transition to domestication we then assess the ability of the clone to transmit virus and make a decision as to whether to continue or discard the colony. Several of the clonal populations of aphids maintained in our laboratory were initially collected in the late 1950's and have been continuously reared under similar conditions (Rochow 1969). The specificity of virus transmission and the efficiency of transmission of several viruses has remained similar throughout their domesticated history (Power and Gray 1995) suggesting that the virus transmission genotype is a stable characteristic and not subject to noticeable changes due to genetic mutation events. Aphid-host preference has also been shown to be a stable genetic characteristic within clonal lines (Di Pietro and Caillaud 1998). However, aphid colonies maintained continuously by parthenogenetic reproduction may accumulate genetic changes (Lushai et al. 1998). Based on the information that clonal variation in virus transmission efficiency does exist within aphid species, the genetic changes accumulating in parthenogenetic colonies may ultimately alter the virus transmission phenotype. To minimize the chances of an alteration in transmission phenotype due to the accumulation of genetic mutations we begin new colonies every 5–7 years using the progeny of one adult. The new colonies are maintained separately from the old and the transmission phenotype assessed for one year to establish it is similar to the historical records.

The establishment of aphid colonies using aphids sent by colleagues should duplicate the rearing conditions the aphids are adapted to. Acclimation to new hosts or to new environmental conditions may be accomplished over time, but recognize that any changes may also impose some genetic selection on the aphid population. This may not alter the virus transmission phenotype, but such occurrences should be tested for.

Host Plants

It is quite simple to rear aphids once they have been established on a suitable host plant. However the choice of host plant should be considered carefully. Information is available in the literature that describes or lists common hosts of many aphid species (Blackman and Eastop 1984). When choosing a host plant for an experimental system it is important to find a suitable one. The host plant needs to possess the proper chemical clues to induce feeding (Bernays and Chapman 1994, Klingauf 1987, Srivastava 1987). These clues included nutritive and non-nutritive stimuli. Non-nutritive stimuli important in host recognition and acceptance by an aphid include plant odor, color and phagostimulatory compounds such as phenols and flavanoids.

Nutritional levels affect host suitability and aphid development time. Aphid diet can also affect morph production. Alate (winged) individuals are more often produced in higher numbers on unsuitable host plants. In general, feeding on mature plant parts increases the proportion of alate individuals produced whereas feeding on seedlings or young plant tissue increases the proportion of apterous (wingless) individuals.

It is also important to consider what plant species will be used for virus transmission studies. For example, what diagnostic host(s) will be use to assess virus infection, and perhaps what virus maintenance and propagation hosts will be used especially if the virus is not mechanically transmissible and must be maintained in aphid inoculated plants. It is often easier if the aphids can be maintained on the same or closely related plant host as will be used as a host for the virus. Aphids may have trouble adapting to new hosts and this adaptation process may effect how the aphid feeds on a plant and profoundly alter the ability or efficiency by which an aphid acquires virus from or inoculates virus into a plant. We rear all the cereal aphids on a common host, barley cv Lud. This simplifies planting schedules, but foremost all the aphids are able to reproduce and survive well on this host. Furthermore, the transition to oats, which is preferred as a diagnostic host for virus transmission assays is without difficulty. Oats would not be an optimal host for rearing one or more of the aphid species due to a nonpreference for the host and an inability to maintain quality colonies for an extended period of time. For example, oats are a non-preferred hosts for *R. maidis* and cannot be used to rear the aphids, however, the aphid will feed well enough on young oat plants to transmit barely yellow dwarf virus which requires long access and inoculation feeding times (>12 hr) and phloem contact (Gray et al. 1991, Power et al. 1991).

It may be difficult to use common plants for all phases of virus transmission studies. For example, we use *M. persicae* to study the transmission of potato leafroll luteovirus. The important plants from the virus perspective are various tobacco species and *Physallis floridana*. The aphids do not feed well on these virus hosts and none are ideal for rearing the aphid. We use radish or turnip. If the plant hosts used in virus transmission assays are not preferred hosts of the aphid, care must be taken in interpreting transmission data. A lack of, or a low level of, transmission may not be due to aphid-virus or virus-plant interactions, but rather to aphid-plant interactions, i.e., feeding problems.

Another point to consider when choosing a plant host for rearing aphids is that not all genotypes within a plant species will be equally suitable. Numerous properties of plants have been shown to affect aphid feeding and reproduction including physical properties (e.g., leaf waxes) and chemical properties (Bergman et al. 1991, Dillwith et al. 1991, Dreyer and Campbell 1987, Goffreda et al. 1990, Leszczynski and Dixon 1992, Rustamani, Kanehisa and Tsumuki 1992, Stoner 1990, Weibull and Melin 1990, Zhang and Cao 1991). Any number of properties of a plant genotype can be manifested as a resistance to aphids and quantified as a change in fecundity, reproductive rate and/or longevity. It is wise to examine a number of plant genotypes to determine which will work best as a maintenance host for a particular aphid.

Facilities

If you are fortunate enough to be able to design aphid rearing and working facilities there are several features to consider. The physical layout should minimize the possibility of contamination, both viral and insect. If possible, rearing facilities should be planned so that there are separate areas for working with virus and virus-free materials, and a third space for a plant nursery. All work areas should be free of other insect pests, pathogens, or insecticide contamination. The specifics of our facilities are outlined below.

Healthy Plants

All aphids are reared on healthy plants. The plants are grown in an isolated greenhouse or growth chamber that never is treated with any type of pesticides. If possible, plant material is not maintained more than 1–2 weeks prior to being used for aphid colonies or transmission tests. Any unused plants are promptly discarded to prevent any unwanted insect or pathogen infestation. Plants removed from the facility are promptly caged prior to being moved into any aphid rearing or work rooms to prevent their infestation by any stray aphids.

Work Rooms

Colony maintenance activities or collection of aphids for use in virus transmission tests are conducted in a "No virus" work area that consists of a work room connected to rearing facilities (Fig. 1). Caged colonies are removed from the rearing chambers, aphids collected and plants (and extra aphids) disposed of within the work room without contaminating the rearing chambers or outside areas with any aphids. The description of handling events is described below in Section IV. Separate work areas within the work room minimize movement of aphids between areas if multiple persons work concurrently. A large sink is handy for rinsing cages when they are removed from plants to minimize the number of aphids moving about the room. We currently use a commercial freezer to place the colony plants after the aphids are collected from the plants. It is an effective and environmentally friendly way to dispose of the unwanted aphids remaining on the plants and in the pots.

A separate facility is used to work with aphids that have been exposed to virus-infected plants or fed on virus-containing solutions, and to maintain viruliferous aphids during inoculation access periods. Our "viruliferous" work area consists of an isolated work room directly connected to plant growth rooms (Fig. 2). This arrangement allows viruliferous aphids to be transferred from virus source plants/tissues to recipient plants and held in isolation during the inoculation access period. Again all plants in the work room and growth rooms are caged at all times except during the brief periods that aphids are placed on the plants. Following an inoculation access

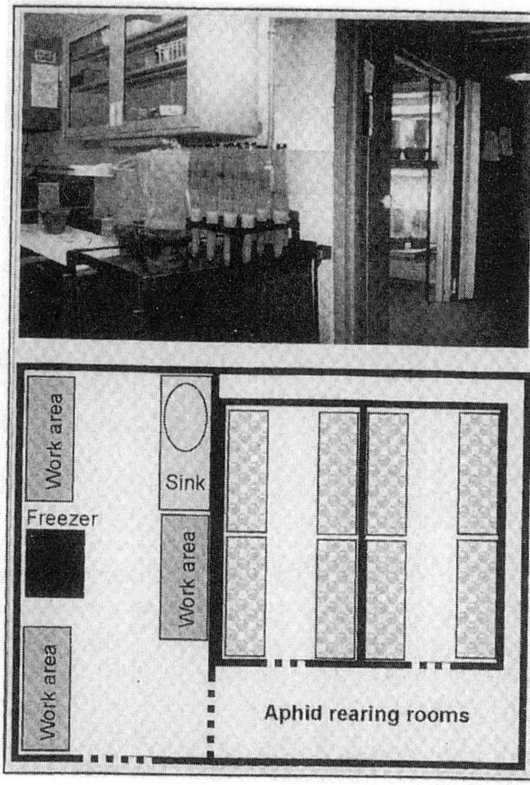

Fig. 1. The arrangement of work space and rearing facilities for working with healthy aphids. The area is separated from other work areas in which virus infected plant material may be used or kept. Aphid rearing chambers are modified cold rooms equipped with shelving and florescent lights. Plants are placed in trays are watered from the bottom to avoid disturbing the cages. The work area consists of multiple work stations, a freezer to dispose of colony plants following the collection of aphids, and a sink for rinsing cages, dishes and other equipment. Notice that all plants are caged while they are in the room to prevent the infestation by aphids that may get loose.

period the caged plants are transferred to a fumigation chamber located in an other location, uncaged and fumigated. Plants are then moved to an insect-free greenhouse. This is a separate greenhouse from where the healthy plants are raised. We minimize the use of insecticides in either greenhouse area to avoid problems of using insecticide-treated plants in any virus transmission test.

We have utilized modular constructed cold rooms as aphid rearing and plant growth rooms. These units are considerably less expensive than plant growth chambers which are not necessary since the plants are only maintained for a short period of time under these conditions and optimal plant growth conditions are not warranted. The cold rooms are outfitted

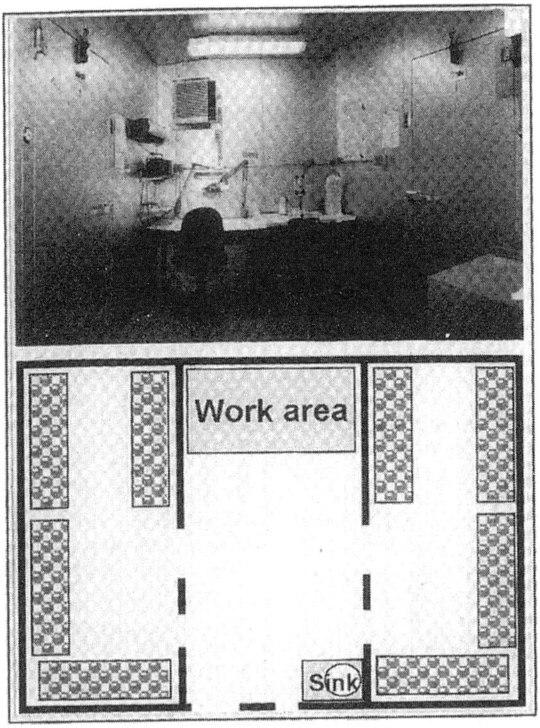

Fig. 2. Isolated work area plant growth rooms for working with viruliferous aphids and virus-infected plants. All plants would be caged prior to entering the space. Aphids that have been feeding on virus-infected leaves or plants are transferred to test plants in this area. All plants are then moved directly into one of the connecting growth rooms for an appropriate time required for virus inoculation. Dashed lines indicate doorwavs.

with small refrigeration units and simple temperature control mechanisms to maintain constant temperatures in the range of 15–25C. Simple shelving outfitted with fluorescent lights is adequate for most situations (Figs. 1 and 3). In general aphid colonies are kept under constant light at a temperature of 20±1C. Relative humidity in the room varies between 40 and 80%. High levels of humidity will exacerbate problems with fungal pathogens of aphids and plants.

Cages

Several things should be considered before designing cages for aphid studies. Cages may be used to enclose an entire plant or group of plants. The durability of the cage may also be an issue, depending on how often they will be used, how they will be handled, and how they are to be stored when not in use.

All cages should have adequate ventilation so that airflow into and out of the cage is not restricted. This will help reduce the relative humidity and free moisture within the cage and thus prevent disease development and problems of aphid entrapment in water droplets. Ventilation requirements will vary according to the size of the cage being used and the conditions of the room in which the aphids are being reared.

While cages need adequate ventilation, they must also be able to prevent escape of experimental organisms from the cage or intrusion of foreign pests into it. Any type of screening used for ventilation should be a good quality with a mesh aperture size of 210 um or less to prevent ingress or egress of the 1st instars. It is also important to pay attention to all the materials being used for construction of the cages. Certain plastics, glues, and sealers can have toxic effects on the aphids and/or the host plant. Thus, any new material should be tested for toxic/adverse effects both plant and insect prior to use.

Doors in cages are the most difficult areas to seal and invariable aphids can escape most frequently when cages are opened. We developed our system so that cages are not opened until the colony is harvested for use. Cages enclose single plants or clusters of plants and fit into the pots in which the plants are grown (Figs. 3 and 4). We do not recommend recaging colonies

Fig. 3. The aphid colony arrangement and cage design that is most commonly used in our rearing system for cereal aphids. Each pot contains approximately 15 barley seedlings that were sown into the middle of a 6′ pot. The cage is comprised of a plastic cylinder approximately 10 cm in height and of a diameter that fits within the pot. A cylinder of mesh netting, closed at the top, is glued to the plastic cylinder to form an enclosed arena. The plastic ring is pushed into the soil to prevent aphid escape. A layer of white sand on top of the soil helps to hold the cage more securely and allows aphids that drop off the plant to be easily found. Pots are placed in pans and watered from the bottom.

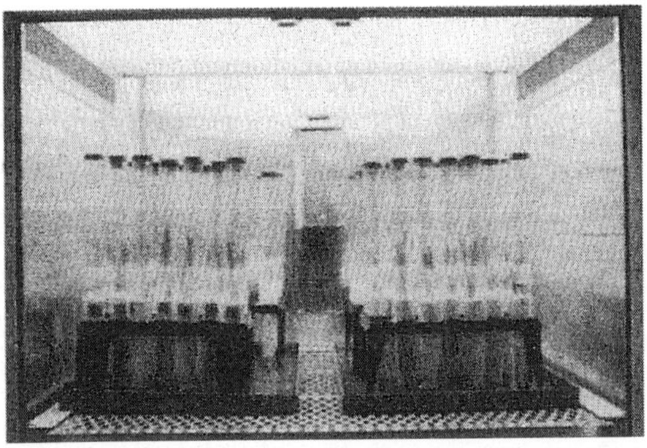

Fig. 4. An alternative cage and colony design that allows numerous colonies of aphids to be reared in close proximity. The plants grown in plastic cones. Approximately five seeds are sown into the center of the soil. The cage is a thin walled plastic cylinder that fits tightly into the cone. Holes are drilled approximately 3–4 cm from the bottom of the cage and covered with mesh. The top of the cylinder is fitted with a mesh covered cap. The top and bottom ventilation is required to prevent the buildup of condensation and the growth of fungi. Using this system we can maintain over 200 individual colonies within a small reach-in growth chamber. These colonies can be maintained for 2–3 weeks and each with yield 200–300 aphids.

once the cage is removed. This often leads to the escape of individuals that can contaminate other colonies. We have found that a "one use only" colony is the best insurance against contamination problems.

Colony Maintenance

Schedules

It is best to standardize and schedule all colony maintenance and sanitation practices. This ensures consistency of aphid quality within and among experiments, lessens the chance of pathogen, parasite and pest contamination, and eliminates the negative effects of overcrowding on colony development and vigor. Furthermore, a maintenance schedule will optimize resources to provide a constant supply of aphids for experiments when necessary or alternatively to just maintain a healthy colony for future needs. Colony maintenance activities can be a very time consuming process. Although well designed protocols and meticulous execution will save countless time and expense in salvaging contaminated colonies (we speak from experience), a balance must be achieved between colony maintenance and experimental needs.

The period of time between transfers to new hosts will vary depending on the aphid, host plant and the environment. We have found that 3 wk old colonies are optimal for providing an adequate number of quality aphids for all types of work. The aphids are good size, uncrowded and the plants are still vigorously growing. We can routinely collect several hundred aphids from one 3 wk old colony. One regular schedule is to begin new colonies each week in a two-day procedure. On the first day adults are collected and allowed to produce progeny which are transferred to new colony plants on day two. If the aphid is not being used for experiments and only being maintained, then new colonies are started every 3 wk. Maintaining the colonies at lower temperatures (<10°C) can slow the development and reproduction of aphids and extend the shelf life of colonies. However, low temperatures may induce the formation of sexual forms of the aphid and the parthenogenetic colony could be lost.

Transfer

All aphid cultures should be transferred in a different area from where they are reared. The transfer area should be chosen that could be easily washed down and sterilized. Our two-day transfer protocol is as follows. On day one a 3 wk old colony is harvested simply by removing the cage and shaking the leaves over a 11″x17″ sheet of white copier paper. The plant and remaining aphids are promptly placed in a –20°C freezer. Adults are collected from the paper using a fine camelhair brush and placed in a dish with a tight-fitting lid and containing moist filter paper and several healthy barely leaves. The following day new colonies are started using 15 1st instar nymphs. We have found that it is easier to transfer the nymphs from the leaf pieces in the dish to small pieces of moist filter paper and then place the filter paper on the leaves of the new colony plants. The nymphs will promptly crawl onto the plant as the paper dries. This eliminates excessive handling of the nymphs and the problem of dropping the nymphs into the soil while trying to place them on the plant. It is important to be very gentle when handling aphids to avoid damage to their stylets. A damaged stylet may prevent an aphid from properly feeding and in this way reduce transmission. Many negative virus transmission tests may be due to damaged aphids rather than an inability of the aphid to transmit the virus. Touching the underside of the aphid with the paintbrush is usually enough to stimulate stylet withdrawal from the leaf.

The easiest way to start new colonies is to transfer a leaf containing aphids from an old colony to a new plant. We caution against this technique since it allows the movement of any pathogens, parasites of the aphids and also the movement of unwanted plant pathogens or insect pests to the new colony. Colonies can be started with adults or older instars, but it is advisable to remove the aphids from the old colony into some collection container and move individuals to the new plant.

Transferring Multiple Species/Clones

If multiple species, or even more importantly, multiple clones of the same species, are being maintained great care must be taken during the collection and set up of new colonies. Strict adherence to sanitation practices during and between the transfer steps when aphids are not caged are critical. As previously mentioned, colony plants with unused aphids are placed into the freezer immediately after aphids are collected. The cage is placed in the sink and adults quickly collected into a dish. Alternatively, if numerous aphids must be collected for transmission experiments the aphids collected on the paper are transferred to a container. The cages are rinsed to remove all aphids and the counter tops and sink surfaces are thoroughly cleaned.

Individuals can be collected from the container into smaller dishes. If the aphids are excessively agitated or a fast moving species, e.g., *S. avenae*, the aphids can be placed in a refrigerator or on ice for a brief time to lower their metabolism. This will facilitate collection, confinement in a smaller area, and cleanup.

It is helpful to have flexible table lamp that can be positioned directly over the aphids being handled. Alates that will lift off from the table will for the most part alight on the light and fall back to the table. We never collect alates for routine colony setup because their origin may not be known. It is also very important to cage any plants that will be used to set up new colonies prior to bringing them into the transfer room. A stray alate alighting on an uncaged plant can be the start of problems that may not be recognized for weeks or in the case of multiple clones, the wayward alates progeny may never be recognized as different. It is best if the transfer of multiple species and clones can be separated in time or space. We routinely restrict transfer of clonal populations to different days of the week. A system to distinguish aphid biotypes/clones should also be developed and may include morphological or genetic markers, plant response, or insecticide resistance.

Colony Contaminants

Pathogen Contamination

Routine inspection of aphid colonies is important to detect pathogen contaminant early on. Entomopathogenic fungi are the most common problem associated with insect rearing (Latge and Papierok 1987). Fungi can often be visually detected by looking for the presence of fungal mycelia on aphids or aphid carcasses. The mummies are likely to remain attached to the plant and may be missed if aphids are shaken from the plant. A common symptom of fungal infection is a reduction in vigor of the colony.

If a fungal infection is suspected, but mycelia growth is not observed on the colony aphids can be collected into a dish containing moist filter paper and kept at 20–25°C. This may promote the fungal growth. There is no evidence that the fungal mycelia infection is transmitted to progeny therefore, fungal contamination can be controlled by quickly separating 1st instar nymphs from adults before they are contaminated with fungal spores. Maintaining humidity below 80% in the rearing chambers and cages, coupled with good sanitation practices, has controlled fungal infections in our hands.

Predators/Parasites

A variety of hymenopteran, including braconidae, have the ability to parasitize aphids. The presence of aphid mummies is indicative of parasites. Several insects also will prey upon aphids, e.g., coccinellids or lacewings (Latge and Papierok 1987). These are easily detected and careful sanitation practices will easily correct these problems.

Viruses

These are several aphid viruses, some of which can be detrimental to aphid colony survival (Laubscher and Von Wechmar 1992). The viruses are difficult to diagnose in many instances. Antiserum to some of the more common viruses has been prepared, but may not be widely available. The easiest way to determine if the colony if infected with a virus is electron microscopy (Gildow and D'arcy 1990, Laubscheret al. 1992). If viruses are detected in the aphid colony they may not have a severe pathogenic effect, however, there is very little information as to the effects of the aphid viruses on the transmission of plants viruses. The viruses are likely to be transovarily transmitted to progeny, but transmission efficiency is unlikely to be 100% (Laubscher and Von Wechmar 1992). It may be possible to obtain a virus-free colony by starting colonies with single 1st instar nymphs. Progeny can then be checked for the absence of virus.

Pathogen problems can be avoided for the most part by following a strict routine of sanitation and colony maintenance. In addition remember to quarantine any aphids that may be collected from the field or obtained from colleagues until their health can be established. Also remember to follow strict rules not to enter rearing facilities after being outside or working with aphids. You may be carrying contaminants or contaminated aphids on your clothes.

References

Ammar, E.D. 1994. Propagative transmission of plant and animal viruses by insects. Factors affecting vector specificity and competence. *In*: K.F. Harris (ed.). Advances in Disease Vector Research. Springer Verlag, New York. 10: 289–331.

Austin, A.B.M., G.M. Tatchell, R. Harrington and J.S. Bale. 1991. A method for rearing cereal aphids in a small space. Entomol. Expl. 61: 91–94.

Bergman, D.K., J.W. Dilwith, A.A. Zarabi, J.L. Caddel and R.C. Berberet. 1991. Epicuticular lipids of alphalpha relative to its susceptibility to spotted alfalfa aphids (Homoptera: Aphididae). Environ. Entomol. 20: 781–786.

Bernays, E.A. and R.F. Chapman. 1994. Contemporary topics in entomology. 2. Host plant selection by phytophagous insects. Chapman and Hall, New York.

Blackman, R.L. 1987. Rearing and handling aphids. *In*: A.K.M.a.P. Harrewijn (ed.). Aphids: Their Biology, Natural Enemies and Control. Elsevier, Amsterdam. 2b: 59–68.

Blackman, R.L. and V.F. Eastop. 1984. Aphids on the world's crops: an identification guide. John Wiley and Sons, New York.

Campbell, C.D. and W.D. Hutchison. 1995. Rearing methods and demographic statistics for a subterranean morph of the sugarbeet root aphid (Homoptera: Aphididae). Canadian Entomologist 127: 65–77.

Di Pietro, J.P. and C.M. Caillaud. 1998. Response to intraclonal selection for adaptation to resistant wheat in the English grain aphid (Homoptera: Aphididae). Environmental Entomology 27: 80–85.

Dillwith, J.W., R.C. Berberet, D.K. Bergman, P.A. Neese, R.M. Edwards and R.W. McNew. 1991. Plant biochemistry and aphid populations studies on the spotted alfalfa aphid *Therioaphis maculata*. Arch. Insect Biochem. Physiol. 17: 235–252.

Dreyer, D.L. and B.C. Campbell. 1987. Chemical basis of host plant resistance to aphids. Plant Cell Environ. 10: 353–362.

Forbes, A.R., B.D. Frazer and C.K. Chan. 1985. Aphids. *In*: P. Singh and R.F. Moore (eds.). Handbook of Insect Rearing. Elsevier, New York. 1: 353–359.

Gildow, F.E. and C.J. D'Arcy. 1990. Cytopathology and experimental host range of *Rhopalosiphum padi* virus, a small isometric RNA virus infecting cereal grain aphids. Invertebrate Pathology 55: 245–257.

Girma, M., G. Wilde and J.C. Reese. 1990. Influence of temperature and plant growth stage on development, reproduction, life span and intrinsic rate of increase of the Russian wheat aphid (Homoptera, Aphididae). Environ. Entomol. 19: 1438–1442.

Goffreda, J.C., E.J. Symkoviak, I.M. Sussex and M.A. Mutschler. 1990. Chimeric tomato plants show that aphid resistance and triacylglucose production are epidermal autonomous characters. Plant Cell 2: 643–650.

Gray, S.M. 1999. Intraspecific variability of luteovirus transmission within aphid vector populations. *In*: H.G. Smith and H. Barker (eds.). The Luteoviridae. CAB International Wallinford.

Gray, S.M. and N. Banerjee. 1999. Mechanisms of arthropod transmission of plant and animal viruses. Microbiol. Med. Biol. Rev. 63: 128–148.

Gray, S.M., A.G. Power, D.M. Smith, A.J. Seaman and N.S. Altman. 1991. Aphid transmission of barley yellow dwarf virus acquisition access period and virus concentration requirements. Phytopathology 81: 539–545.

Klingauf, F.A. 1987. Host plant finding and acceptance. *In*: A.K. Minks and P. Harrewijn (eds.). Aphids: Their Biology, Natural Enemies and Control. Elsevier, Amsterdam. 2a: 225–253.

Latge, J.P. and B. Papierok. 1987. Aphid pathogens. *In*: A.K.M.a.P. Harrewijn (ed.). Aphids: Their Biology, Natural Enemies and Control. Elsevier, Amsterdam 2b: 59–68.

Laubscher, J.M., M.A. Jaffer and M.B. Von Wechmar. 1992. Detection by immunogold cytochemical labelling of aphid lethal paralysis virus in the aphid *Rhopalosiphum padi* (Hemiptera: Aphididae). J. Invertebr. Pathol. 60: 40–46.

Laubscher, J.M. and M.B. Von Wechmar. 1992. Influence of aphid lethal paralysis virus and *Ropalosiphum padi* virus on aphid biology at different temperatures. J. Invertebr. Pathol. 60: 134–140.

Leszczynski, B. and A.F.G. Dixon. 1992. Resistance of cereals to aphids. The interaction between hydroxamic acids and glutathione S-transferases in the grain aphid Sitobion avenae F. Hom. Aphididae. J. Appl. Entomol. 113: 61–67.

Lushai, G., P.J. DeBarro, O. David, T.N. Sherratt and N. Maclean. 1998. Genetic variation within a parthenogenetic lineage. Insect Mol. Biol. 7: 337–344.

Nault, L.R. 1969. Laboratory rearing of aphids. J. Econ. Ent. 62: 261–262.

Power, A.G. and S.M. Gray. 1995. Aphid transmission of barley yellow dwarf viruses: Interactions between viruses and host plants. *In*: C.J. D'Arcy and P.A. Burnett (eds.). Barley Yellow Dwarf: 40 Years of Progress. APS Press, St Paul, MN. 259–291.

Power, A.G. and S.M. Gray. 1991. Aphid transmission of barley yellow dwarf virus inoculation access periods and epidemiological implications. Phytopathology 81: 545–548.

Ridland, P.M., R.J. Sward and R.B. Tompkins. 1988. A simple rearing system for cereal aphids especially suited to transmission studies with barley yellow dwarf viruses. Australasian Plant Pathology 17: 17–19.

Rochow, W.F. 1969. Biological properties of four isolates of barley yellow dwarf virus. Phytopathology 59: 1580–1589.

Rustamani, M.A., K. Kanehisa and H. Tsumuki. 1992. Aconitic acid content of some cereals and its effect on aphids. Appl. Entomol. Zool. 27: 79–87.

Shrivastava, P.N. 1987. Nutritional physiology. *In*: A.K.M.a.P. Harrewijn (ed.). Aphids: Their Biology, Natural Enemies and Control. Elsevier, Amsterdam. 2a: 99–121.

Stoner, K.A. 1990. Glossy leaf wax and plant resistance to insects in *Brassica oleracea* under natural infestation. Environ. Entomol. 19: 730–739.

Sylvester, E.S. 1987. Virus transmission studies. *In*: A.K.M.a.P. Harrewijn (ed.). Aphids: Their Biology, Natural Enemies and Control. Elsevier, Amsterdam. 2b: 59–68.

Van Rensburg, N.J. 1981. A technique for rearing the black pine aphid *Cinara cronartii* and some features of its biology. Homoptera: Aphididae. J. Entomol. Soc. South Afr. 44: 367–380.

Webb, S.E., L. Kok Yokomi Moh, D.J. Gray and C.M. Benton. 1994. *In vitro* rearing of grapevine aphid (Homoptera: Aphididae) on micropropagated shoot cultures of bunch grape and muscadine plants. Annals Entomol. Soc. America 87: 879–885.

Weibull, J. and G. Melin. 1990. Free amino acid content of phloem sap from Brassica plants in relation to performance of *Lipaphis erysimi* (Hemiptera: Aphididae). Ann. Appl. Biol. 116: 417–424.

Zhang, S. and R. Cao. 1991. A study on the interactions among several plant secondary compounds and aphids. J. Environ. Sci. 3: 89–94.

9

Leafhopper and Planthopper Rearing

Karl Maramorosch

ABSTRACT

Since the turn of the last century, rearing of leafhopper and planthopper vectors of bacteria, viruses, phytoplasmas and spiroplasms has been carried out in many laboratories. Certain species have been maintained easily for years, while related species have been difficult to rear and still others could not yet be kept in captivity. Mass production in the laboratory on host plants has been perfected for a few species and used for tests or resistance to vector-borne pathogens or plant resistance to the vectors. Rearing on artificial diets has been used to a limited extent.

Key Words: Plant hopper, aseptic feeding, cellulose nitrate cages, Leaf hoppers, Phytoplasmas

Introduction

Certain species of leafhopper and planthopper vectors of plant pathogens have been maintained in many laboratories and a variety of methods used by different workers. My own experience has been derived from working with L.M. Black at the Brooklyn Botanic Garden, New York in 1947–49, then during the following 12 years with L.O. Kunkel at Rockefeller University,

Department of Entomology, Cook College, Rutgers-The State University of New Jersey, New Brunwick, NJ 08901-8524, USA.

followed by 12 years at Boyce Thompson Institute in Yonkers, New York. I shall present in this chapter an overview of methods used by me and my associates, as well as some developed by others in Asia, Africa and Australia.

Leafhopper (*Cicadellidae*) systematics has been revised several times and names of vectors of plant pathogens changed very often. The chapter on leafhopper systematics by Nielson (1985) provided an excellent review of the leafhopper families. Among the 20 families of planthoppers (*Fulgoroidea*) two, the *Cixiidae* and the *Delphacidae*, are of great economic significance as plant pathogen vectors. More than 150 leafhopper and planthopper species are now known as vectors, but less than a dozen have been maintained in the laboratory (Brooks 1985). The subfamilies *Cicadellidae* and *Deltocephalinae* contain more than 60% of leafhopper vector genera that transmit 70% of plant pathogens—viruses, bacteria, phytoplasmas and spiroplasmas (Nielsen 1985).

Establishment of Colonies

Colonies of leafhoppers and planthoppers can be started from field collected insects or from "domesticated" insects obtained from other laboratories. Paul W. Oman, one of the foremost leafhopper taxonomists of the insects used a sturdy insect net to collect leafhoppers in the field by rapid sweeping over the tops of plants. Individual insects can be caught on leaves or stems of plants using insects sucking devices. Collected insects can be maintained on a variety of host plants and various types of cages have been devised for the maintenance of colonies and of individual insects. In 1924 Kunkel discovered that the causative agent of aster yellows disease was transmitted by the sixspotted aster leafhopper, *Macrosteles fascifrons* (Kunkel 1926). The vectors, first classified by China at the British Museum as *Cicadula sexnotata*, were found to act not only as mere vectors but also as alternative hosts of the presumptive virus, now known as phytoplasma. Kunkel found that when the phytoplasmas were acquired by leafhoppers feeding on diseased plants, they had to undergo an incubation period of at least 9 days before the insects were able to transmit these pathogens to susceptible plants. Kunkel postulated that during the incubation period the phytoplasmas multiply and circulate in their vectors.

The question of multiplication of plant pathogens in leafhopper vectors was a burning question for several years and the idea was vigorously disputed until Fukushi in Japan demonstrated that the rice stunt virus was able to multiply both in rice plants and in leafhopper vectors (Fukushi 1933, 1940). Fukushi's work was followed by Black's demonstration that a bacterial agent, causing clover club leaf disease, also multiplied in leafhopper vectors and clover plants (Black 1950). The same was proven for the wound tumor virus by Black and Brakke (1952) and for the aster yellows phytoplamas in aster leafhoppers by Maramorosch (1952a). Viruses, phytoplasmas and spiroplasmas that multiply in plants as well as in their insect vectors are therefore both insect and plant pathogens.

Colonies of leafhoppers and/or planthoppers are maintained on proper host plants. Certain vector species are nearly omnivorous and can breed on many different plant species. The sixspotted aster leafhopper *Macrosteles fascifrons* can be maintained on monocots and dicots, and permanent colonies have been kept for many years on rye, *Secare cereale*, which cannot be infected by the aster yellows phytoplasma and is therefore suitable for the breeding of stock colonies free of this phytoplasma. Barley, *Hordeum sativum*, also provides a suitable feeding source but though it remains symptomless when infected by the aster yellows phytoplasma, it can carry the pathogen and is, therefore, unsuitable for maintaining pathogen-free colonies. The leafhopper vector of *Spiroplasma kunkeli*, *Dalbulus maidis*, cause of corn stunt, as well as the related vector *D. elimatus*, can be maintained very easily on maize, Zea mays, as well as on teosinte, and not as well on a few species of *Gramineae*, but it will not breed on dicotyledenous plants. It is difficult to maintain leafhoppers and plant hoppers that under natural conditions feed only on deciduous trees. *Scaphytopius magdalensis*, a vector of blueberry stunt phytoplasma (Maramorosch 1955) survived on seedlings of a cross of *Vaccinium* sp. long enough to infect the plants, but the adults and their progeny died. No colony could be established because due to blueberry dormancy the insects were deprived of food. Attempts made many years ago at the Boyce Thompson Institute by Hartzell (personal communication) to grow a colony of *Macropsis trimaculata*, the vector of peach yellows phytoplasma, were unsuccessful. Not only the lack of a constant and proper host plant, but the slow development of these insects that, in nature, have but a single generation per year, were responsible for the encountered difficulties.

Tropical and subtropical species of Cicadellidae that produce several generations under field conditions are usually easy to breed and to mass produce for laboratory tests. In his classical experiments with aster yellows, Kunkel used glass lantern globes for individual as well as for groups of aster leafhoppers (Fig. 1). When Lindsay M. Black joined Kunkel's group at the Rockefeller laboratories in Princeton, New Jersey, he introduced cages made from sheets of transparent cellulose nitrate. The cages were light and could be made to fit pots 10 cm or 5 cm in diameter (Fig. 2). The latter were used by Black and his associates to cage agalian leafhoppers on crimson clover (*Trifolium incarnatum*) seedlings. Working in Black's laboratory at the Brooklyn Botanic Garden, I learned how to make such cages, using acetone to glue plastic saran monofilament screen into them (Fig. 3). In addition

Fig. 1. L.O. Kunkel (1884–1960) with lantern globe leafhopper cage, used by him at the Boyce Thompson Institute and at Rockefeller University. Picture taken by the author in 1959.

Fig. 2. Plastic cages from sheets of cellulose nitrate, introduced by L.M. Black (1907–1997) for leafhopper vector work.

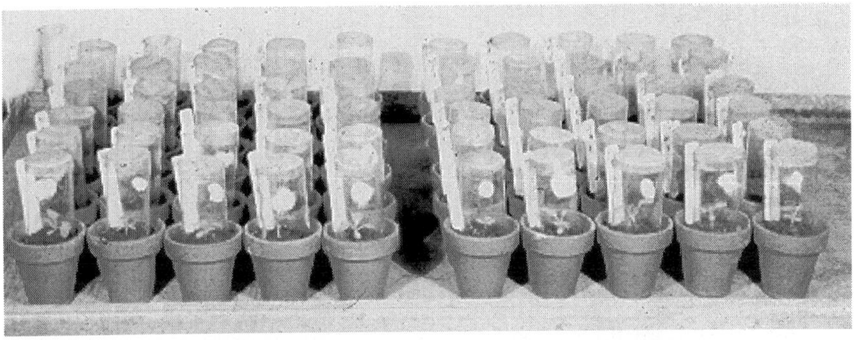

Fig. 3. China aster (*Callistephus chinensis*) seedlings under plastic cages.

Fig. 4. Leaf cage for single insect.

to cellulose nitrate, which presents a fire hazard, butyrate and other less flammable transparent plastics are now being used in various laboratories. Cages can also be constructed from saran monofilament screens and wire screens. Early experiments in Japan were conducted using bamboo cages.

When collected insects are released on caged plants, the plants have to be grown under conditions that prevent earlier exposure to insecticides or fungicides. Accidental contamination of plants can result in disastrous development (see PITFALLS). For confining single insects to plants and for rapid transferring of single insects to fresh plants, handy insect vector cages have been used in several laboratories. Such a device, developed by me and submitted to the Journal of Economic Entomology in 1951, was rejected outright by the editor-in-chief, who wrote a sarcastic letter, explaining that readers of the journal are not interested in keeping leafhoppers alive, but in killing them with insecticides. Undisturbed by this rejection, I submitted the description to the Journal of the New York Entomological Society and it was published promptly (Maramorosch 1951). The cage was made from pieces of translucent centrifuge tubes made from "lusteroid", in fact, cellulose nitrate. Holes in the tubes were cut with a paper punch and plastic or nylon screen used for ventilation at the top.

Early insect vector cages were fixed to two strips of cellulose nitrate connected by a shoe eyelet, permitting the scissor-like opening of the two parts, between which the leaf to be exposed was inserted. At the opposite end the two celluloid strips were clamped together by a binder clip to hold the cage in place on a leaf. This device permitted rapid transfer of insect from plant to plant without harming the insects in any way. The smallest cages that weighed 1.5 g were 1.2 cm in diameter and 1.8 cm in height. An improvement was later made by using small permanent magnets (Fig. 5). Stock colonies of leafhoppers were maintained in large wooden screen cages (Fig. 6). Screens of steel or lumite saran monofilament provided ventilation and glass was used on the back so as to be able to place a light source behind the glass and thus attract leafhoppers to the glass wall when the opposite door had to be opened. These large cages were on wheels, to move them easily from greenhouse insectary to a darkened room whenever adults or nymphs were to be removed.

For mass propagation much larger cages were built. At the International Institute for Tropical Agriculture (IITA) in Ibadan, Nigeria, cages that could hold more than 20,000 leafhoppers were routinely used by Dabrowski (1987) for breeding *Cicadulina* spp. Very large cages were also necessary to screen coconut palms exposed to *Myndus crudus* in Florida (Tsai and Maramorosch 1988) (Fig. 7).

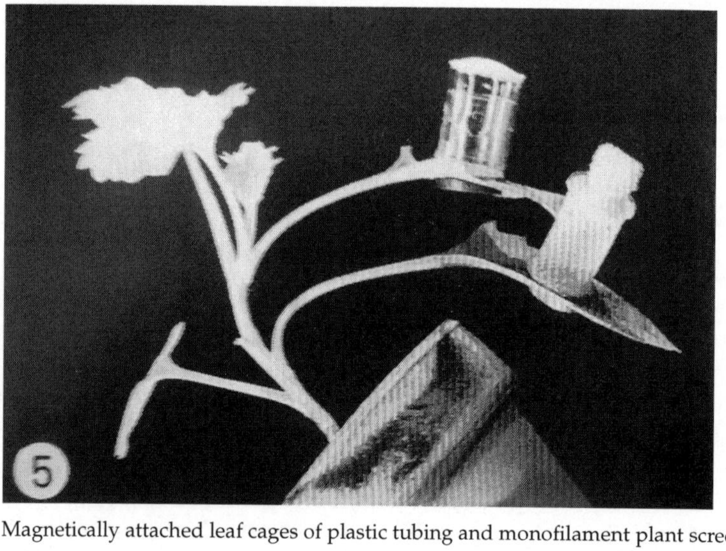

Fig. 5. Magnetically attached leaf cages of plastic tubing and monofilament plant screens.

Fig. 6. Wooden cages on wheels for leafhopper stock colonies.

Fig. 7. Large outdoor cages on palm seedlings.

Artificial Feeding

Rearing leaf- and planthoppers on artificial diets has met with considerable difficulties. Following the introduction and large scale use of such diets for other insects, attempts have been made to devise a diet as well as a means of feeding that would be suitable for the insects. In 1927 Carter kept beet leafhoppers, *Circulifer tenellus*, on a dilute cane sugar solution. He also experimented with the brown leafhopper *Laodelphax striatellus*, maintaining it on distilled water and feeding the insects through animal membranes.

Mosquito workers have used successfully such membranes before parafilm membranes became available.

The growth of *M. fascifrons* leafhoppers on a complex synthetic diet was first achieved by Mitsuhashi and Maramorosch in 1963. Sterile cotton wicks were used and though the leafhoppers were kept alive they did not complete their life cycle. It was not necessary to feed leafhoppers through membranes but membrane feeding could prevent fungal and bacterial contamination. In 1971 Mitsuhashi and Koyama were the first to rear *L. striatellus* planthoppers continuously on an artificial diet (Mitsuhashi 1979). Only three leafhopper species have been reared continuously on artificial diets: *Marcosteles orientalis*, *Inazuma dorsalis*, and *M. fascifrons*. The planthopper *L. striatellus* has also been grown continuously on an artificial diet (Mitsuhashi 1974). Koyama (1973) was able to rear continuously on artificial diets the brown planthopper *Nilaparvata lugens*, as well as *I. dorsalis*.

Koyama (1973) improved the artificial feeding method further to rear the brown planthoppers *N. lugens*, *I. dorsalis*, and *Nephotettix cinsticeps* on synthetic diets. Brooks (1985) pointed out that a feeding stimulus for leafhoppers or planthoppers must be provided before artificial synthetic media can be evaluated successfully. Certain physical factors, such as the height of the rearing vessel might be critical to rearing first instar nymphs of *L. striatellus*. The color of the light is also important and yellow was found most suitable. The same author found that while *M. fascifrons* could feed satisfactory under conditions provided by her, the planthopper *Peregrinus maidis* could not survive. Obviously several physical factors have not yet been identified. When oviposition stimuli were investigated, it was found that salicylic acid stimulated the probing of *N. lugens*, but not of *L. striatellus*.

The listed 3 species of leafhoppers and the planthopper *L. striatellus* can be reared continuously on artificial diets (Mitsuhashi and Koyama 1975) but attempts to maintain other species have met with considerable difficulties (Brooks 1985, Mitsuhashi 1979). The oviposition and the completion of the life cycle apparently requires as yet undetermined factors.

Aseptic Rearing

In 1963 Mitsuhashi and Maramorosch described a simple method for aseptic rearing of *M. fascifrons*. Leafhopper eggs were surface sterilized and hatching first instar nymphs from these eggs placed on a wick saturated with an aseptic synthetic diet. Third instar nymphs completed one molt but they were unable to complete the life cycle because the wick proved too sticky.

Consequently Mitsuhashi (1979) devised a method of rearing leafhoppers aseptically on aseptic plants, grown from surface sterilized seed. Using sterile plant seedlings, several leafhoppers: *Nephotettix apicalis*, *N. cincticeps*, *D. maidis*, *I. dorsalis*, *Agalliopsis novella*, *Agalia constricta*, *L. striatellus*, *M. fascifrons*, and *N. lugens*, have been maintained.

Manipulation of Leafhoppers and Planthoppers

Several methods have been used to manipulate adults or nymphs for experimental purposes. A simple method for anesthetizing the insects was by using carbon dioxide (Maramorosch 1956). If a cold room is available, manipulation at 0°C can be used and this temperature is well tolerated by immobilized insects.

For the injection of pathogens into abdomens of adults or first instar nymphs, a device based on the original stage used in Poland by Rudolf Weigl for body lice (Weigl 1947) has been modified (Fig. 8). This operating stage facilitated rapid inoculation of small insects (Maramorosch and Jernberg 1970). For several years this multiple-insect holder has been used with leafhoppers, mosquitoes, beetles and fruit flies. With leafhoppers, groups of 20 insects were placed in test tubes, anesthetized with carbon dioxide for 1–2 minutes and arranged individually on the stage. While the insect soon recovered from anesthesia, they could not escape the firm grip of the holders (Fig. 8).

A much simpler method has been used in recent years, which does not require the use of the adjustable insect holder. Insects anesthetized in a test tube by carbon dioxide were placed on a parafilm membrane stretched

Fig. 8. Multiple insect holder, adapted from Rudolf Weigl's original holder for body lice, used for leafhoppers, aphids, mosquitoes, etc.

over a glass beaker, then covered by a second parafilm membrane placed gently on top, and fastened by a rubber band. The recovering insects can be injected, puncturing through the upper membrane, but they cannot escape until the top membrane is removed.

Pitfalls

Numerous difficulties can be encountered culturing leafhoppers and planthoppers. A few will be mentioned here. Proper cages, aeration, host plants, light, humidity and temperature (Decker and Cunningham 1967) can be critical factors when a species collected in the field is brought to an insectary. Despite the expertise of entomologists and plant pathologists who have been able to maintain certain species for many years, other species have not yet been maintained in captivity.

The presumptive vector of the lethal yellowing phytoplasma in Florida and the Caribbean, *Myndus crudus*, has been discovered by James H. Tsai, who repeatedly attempted to grow a phytoplasma-free colony of this insect but was unable to do so (Tsai and Thomas 1980, Tsai and Maramorosch 1988). Tsai found that nymphal stages of this leafhopper live on palm roots in the soil. Repeated attempts to obtain a permanent colony of the leafhoppers have been unsuccessful for reasons as yet unknown. The related species *M. adipodoumeensis*, reported as a presumed vector of the lethal yellowing phytoplasma of coconut palms in East Africa (Eden-Green 1994, Maramorosch 1999) has defied all attempts at cultivation. My own attempts, as well as those of Albert Hartzell and L.O. Kunkel in the 1920s and 1930s (personal communication) to maintain a colony of *Macropsis trimaculata*, the vector of peach yellows phytoplasma, have failed, primarily because the insects have but one generation a year in nature and no proper evergreen host was found to support nymphal instar stages and adults throughout the year.

A different pitfall occurred while I was growing large numbers of leafhoppers for a serial passage of the aster yellows phytoplasma at Rockefeller University in New York city. All insects from the stock colony maintained in a greenhouse, that were brought by me personally to the coldroom in another building, died within a few hours. The cause was discovered with considerable difficulty. I carried the plants from the greenhouse through corridors of 3 adjoining buildings and one elevator to my laboratory on the third floor. It was in the elevator, where unknown to me, an outside contractor had sprayed an insecticide to exterminate cockroaches earlier that day. Droplets in the air or vapors still present in the elevator several hours later were sufficient to kill all leafhoppers that were caged on the plants.

When I worked with leafhoppers at Rockefeller University in New York City, I wanted to construct cellulose nitrate cages for my work, but was told that the fire code of the city did not permit shipping cellulose nitrate material because they presented a fire hazard. The purchasing office suggested using cellulose acetate. I obtained sheets that appeared indistinguishable from nitrate sheets and I constructed several hundred larger and smaller cages for my experiments. Insects were confined to caged plants but hundreds of crimson clover plants soon developed lesions on leaflets and the plants died within 2 weeks. Insects that were caged on the plants survived, if removed before they starved, and I concluded that for some mysterious reason the new cages were killing the plants. A spectrometric analysis of the cellulose acetate cages revealed that they contained traces of diethyl phthalate, the plasticizer used to manufacture the 2 mm thick plastic sheets. When the toxic effect was tested on China aster (*Callistephus chinensis*) and tobacco (*Nicotiana tabacum*) seedlings, aster plants were found less susceptible, while tobacco proved more sensitive (Maramorosch 1952b). When cellulose acetate cages were soaked for a whole week in running tap water, their toxicity was not diminished and it became apparent that, as long as the cage material retained its sheet characteristics, the toxicity persisted. When pieces of the plastic material were soaked in ethyl alcohol, they lost their toxicity but also deformed and were no longer suitable for making cages. Diethyl phthalate is also toxic to fish and when goldfish in as fishbowl were exposed to small pieces of the material, they died in less than one hour. Later I found out that aquarium supply stores knew about this toxicity and that no acetate plastic was used in aquaria with tropical fish. After 4 months of extensive tests hundreds of plants were lost. The description of the toxicity of cellulose acetate sheets resulted in a short paper in Science (Maramorosch 1952b) and more reprint requests were received for this report than for any of my other publications.

The story had a good ending, when the purchasing department at Rockefeller University revealed that they already had a permit for bringing cellulose nitrate to the buildings. It turned out that the Nobel laureate Herbert Gasser, who headed the university at that time, obtained a permit to buy huge amounts of cellulose nitrate movie film to record his neurophysiology experiments.

Conclusion

Selections of proper host plants, cages, temperature, humidity and light conditions assured the successful rearing of several leafhopper and planthopper vectors. Insects that have several generations per year are usually easily reared in an insectary. Certain species have been maintained for many years but others have not yet yielded to attempts to rear them in captivity. Insectary work is labor intensive. Artificial diets have been developed for a few species of leafhoppers and planthoppers but until now most vector species require a constant supply of host plants. Insect vector work is being carried out in entomology and plant pathology departments at universities, experiment stations, and at international institutes for agricultural research.

References

Black, L.M. 1950. A plant virus that multiplies in its insect vector. Nature 166: 852–853.

Black, L.M. and M.K. Brakke. 1952. Serial passage of the wound tumor virus through its leafhopper vector. Phytopathology 42: 269–273.

BOCK, K.R. 1974. Maize streak virus. *In*: Descriptions of Plant Viruses. Commonw. Mycol. Inst., Assoc. Appl. Biol., 133. Kew, Surrey, England. 4 pp.

Brooks, Marion A. 1985. Nutrition, cell culture, and symbiosis of leafhoppers and plant hoppers. *In*: L.R. Nault and J.G. Rodriguez (eds.). The Leafhoppers and Planthopppers. John Wiley and Sons. 195–216.

Carter, W. 1927. A technic for use with homopterous vectors of plant disease with special reference to the sugar-beet leafhopper, Euscelis tenellus. Baker. J. Agric. Res. 34: 449–451.

Dobrowski, Z.T. 1987. Some parameters affecting suitability of Cicadulina species for resistance screening to maize streak virus (MSV). Insect Sci. Appl. 8: 757–764.

Decker, G.C. and H.G. Cunningham. 1967. The mortality of the potato leafhopper and some related species when subjected to prolonged exposure at various temperatures. J. Econ. Ent. 60: 373–379.

Eden-Green, S.J. 1997. History, world distribution and present status of lethal yellowing-like diseases of palms. *In*: S.J. Eden-Green and F. Ofori (eds.). Proceedings of an International Workshop on Lethal Yellowing-like Diseases of Coconut. Elmina, Ghana, 1995. Chatham, U.K. Natural Resources Institute. Hobbs-The Printers, Ltd. Totton, Hampshire, U.K. 9–25.

Fukushi, T. 1933. Transmission of the virus through the eggs of an insect vector. Roc. Imp. Acad. (Tokyo) 9: 457–460.

Fukushi, T. 1940. Further studies on the dwarf disease of the rice plant. J. Univ. 45: 8–154.

Koyama, K. 1973. Rearing of *Inazuma dorsalis* and *Nephothettix cincticeps* on a synthetic diet. Jpn. J. Appl. Entomol. Zool. 17: 49–53.

Maramorosch, K. 1951. Handy insect- vector cage. J. New York Entomol. Soc. 59: 49–50.

Maramorosch, K. 1952a. Direct evidence for the multiplications of aster-yellows virus in its insect vector. Phytopathology 42: 59–64.

Maramorosch, K. 1952b. Toxicity of cellulose acetate sheets in plants and fish. Science 115: 236.

Maramorosch, K. 1955. Transmission of blueberry stunt virus by *Scaphytopius magdalensis*. J. Econ. Entomol. 48: 106.

Maramorosch, K. 1956. Semiautomatic equipment for injecting insects with measured amounts of liquids Containing viruses or toxic substances. Phytopathology 46: 188–190.

Maramorosch, K. 1999. Suggestions for lethal yellowing and cadang-cadang disease prevention. Acta Horticulture 486: 141–148.

Maramorosch, K. and N. Jernberg. 1970. An adjustable multiple-insect holder for microinjection. J. Economic Entomol. 63: 1216–1218.

Mitsuhashi, J. 1974. Methods for rearing leafhoppers and planthoppers on artificial diets. Rev. Plant Prot. Res. 7: 57–67.

Mitsuhashi, J. 1979. Artificial rearing and aseptic rearing of leafhopper vectors. *In*: K. Maramorosch and K.F. Harris (eds.). Plant Disease Agents. Academic Press. 369–412.

Mitsuhashi, J. and K. Koyama. 1971. Rearing of planthoppers on a holistic diet (Delphacidae, Deltocephalidae). Entomol. Exp. Appl. 14: 93–98.

Mitsuhashi, J. and K. Koyama. 1975. Oviposition of smaller brown leafhopper, *Laodelphax striatellus,* into various carbohydrate solutions through a Parafilm membrane (Hemiptera, Delphacidae). Appl. Entomol. Zool. 10: 123–129.

Mitsuhashi, J. and K. Maramorosch. 1963. Aseptic culture of four virus transmitting species of Leafhopper (Cicadellidae). Contrib. Boyce Thompson Inst. 22: 165–173.

Nielson, M.W. 1985. Leafhopper systematics. *In*: L.R. Nault and J.G. Rodriguez (eds.). The Leafhoppers and Planthoppers. John Wiley and Sons. 11–39.

Tsai, J.H. and K. Maramorosch. 1988. Lethal yellowing of palms. *In*: S.P. Raychaudhuri and N. Rishi (eds.). Mycoplasma Diseases of Woody Plants. Malhotra Publ. House, Delhi. 29–39.

Tsai, J.H. and D.L. Thomas. 1980. Transmission of lethal yellowing mycoplasma by Myndus crudus. *In*: K. Maramorosch and S.P. Raychaudhuri (eds.). Mycoplasma Diseases of Trees and Shrubs. Academic Press. 211–229.

Weigl, R. 1947. Immunization against typhus fever in Poland during World War II. Texas Rep. Biol. Med. 5: 177–179.

10

Maintenance of *Piesma quadratum* (Fieb.)

Gerhard Proeseler

ABSTRACT

Piesma quadratum is the vector of beet leaf curl virus. The natural habitat of the bugs in infested beet fields and in special biotypes is briefly described. Similarly, references to typical morphological characteristics for identification of this species and various aspects of its biology are presented. For continuous laboratory maintenance of the bugs, long day illumination, temperatures between 20–25°C, and prevention of water condensation in the cages are important. In this chapter, the feeding behavior and the characteristics of postembryonic development are discussed.

Key Words: *Piesma quadratum*, Piesmatidae, beet leaf curl virus

Introduction

The beet leaf bug *Piesma quadratum* (Fieb.) (Heteroptera: Piesmatidae) is the vector of the beet leaf curl virus, which belongs to the plant rhabdovirus group. The virus is not transmitted via inoculation of the sap but is transmissible in a persistent (propagative) manner. Typical symptoms of diseased plants are vein clearing of youngest leaves, vein deformation, leaf curl, and growth inhibition (Fig. 1).

Korter Strasse 19, Aschersleben, D-06449, Germany.

Fig. 1. Beet leaf curl disease in a beet plant (Original).

In the past, the virus specially affected sugar beet plant that grew in sandy soils of Central Europe, but at present it is only important in some regions of Poland (Proeseler 1980, 1983).

Maintenance Methods

In areas where leaf curl virus occurs, *P. quadratum* adults (Fig. 2A) can be collected during most of the year on leaves of sugar or fodder beets plants, except in May or at the beginning of June. Presence of chlorotic spots on the leaves, caused by suction feeding of the bugs and later leaf curl diseased plants are a signal for the occurrence of the insects in a sugar beet field. During winter, the adults can be collected from the substratum formed by fallen dry leaf litter and the grasses growing within infested sugar beet fields. Other natural biotopes of *P. quadratum* include halophytes growing in extremely dry locations, such as railway embankments. Here, *P. quadratum* shares its habitat with mainly plants belonging to family Chenopodiaceae, and with some other non-vector species of Piesmatidae, such as *P. capitatum* Wolff, *P. masculatum* Lap. and *P. salsolae* Beck. These species can be differentiated from each other by using morphological characteristics (Heiss and Pericart 1983). The adults of *P. quadratum* are 3.0–3.5 mm in length, and generally males are smaller than females. Adults exhibit a range of colors that is directly related to their age after emergence. Immediately following the last molt, the adult appears green or grayish green and approximately ten days later they attain their typical darker grey

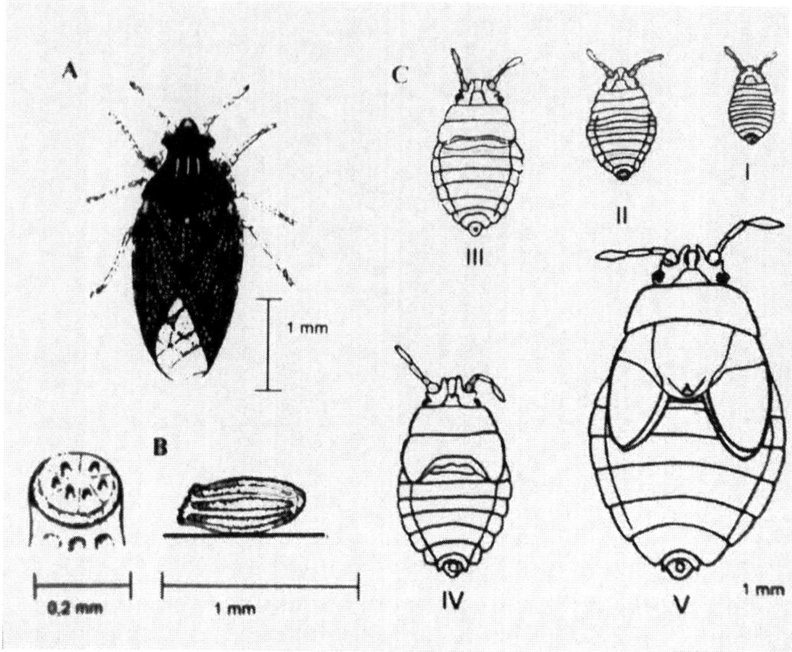

Fig. 2. *Piesma quadratum*. **A.** adult (after Proeseler 1966). **B.** egg (after Ext 1923). **C.** larval stages I–V (after Wille 1929).

or grayish brown color. The eyes are red. Other morphological characteristics of this species are specially three wedges on the pronotum and its edges are straight and lack tails.

The continuous culture of *P. quadratum* can be maintained under controlled conditions in an insectarium or a climate chamber. The recommended conditions are temperature between 20–25 cm at least 1500 lux illumination for 16–18 hr and relative humidity of 50–70%. These conditions are necessary to prevent onset of adult hibernation and their refusal to feed late in summer or at the beginning of autumn. At higher temperatures, the adults rarely fly and mostly move by crawling. Various types of cages can be used to rear or breed the bugs and the choice or selection of the cages depends on the density of insect per cage and the final aim of the experiments (Figs. 3 and 4).

Generally it is important to provide adequate ventilation or circulation of the air to avoid excessive humidity and water condensation. Therefore one or more sites of the cages have to be covered with fine nylon screens that allows the movement of air (Figs. 3 and 4). Plants of *Beta vulgaris* are suitable as feeding source. The sucking of the bugs causes typical chlorotic and later necrotic spots on the leaves (Fig. 5).

Fig. 3. Insect cages useful for experiments with the beet leaf bug (Original).

Fig. 4. Insect cage useful for experiments with the beet leaf bug showing nylon screen on two sides (Original).

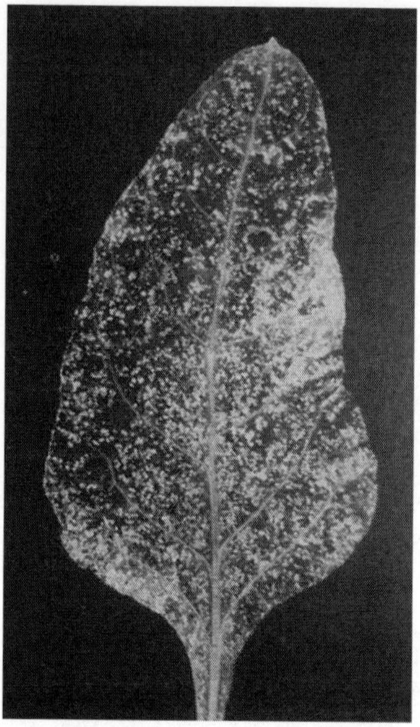

Fig. 5. Spots caused by suction feeding of *Piesma quadratum* on a beet leaf (Original).

The bugs penetrate into the plant tissue mainly intercellularly with their stylets. The salivary sheaths always reach the region of phloem. On the way to the phloem the content of parenchyma cells can be taken up, which causes the described chlorotic and necrotic spots of the leaves (Ehrhardt and Schmutterer 1964). The feeding in the phloem is responsible for the virus transmission. Large virus particle concentrations are found in phloem parenchyma of leaves (Eisbein 1976).

Each female deposits on average 100–160 eggs. The eggs are irregularly oviposited mainly along the veins of either the upper or the lower surface of he leaves and also on the pedicel of the leaves. The eggs (Fig. 2B) are yellow and elongated. Their average size is 0.65 mm by 0.27 mm, and they are rounded at one end. The opposite end is obliquely flattened and covered on its circular surface with small protrusions. The eggshell or chorion is longitudinally deeply ribbed, before the poles (Wille 1929).

At 25°C eggs hatch approximately one week after the oviposition. The first larval instar is yellow in color like the eggs and the later instars are green. All immature stages have red eyes like the adults (Fig. 2C). The postembryonic development encompassing 4–6 larval instars is completed within 30–40 days. There are no significant sex-linked differences in the duration of the larval development. Under environmental conditions that are similar to those in nature, only five larval instars can be observed

(Korcz et al. 1998, Wille 1929). Due to the absence of transovarian transmissions of beet leaf curl virus to insect progeny, it is comparatively easy to establish a virus-free colony. The eggs have to be removed carefully with a fine hairbrush from infected host plants to virus-free plants. The hatched nymphs are then used for the establishment of a virus-free stock colony.

References

Ehrhardt, P. and H. Schmutterer. 1965. Untersuchungen zum Anstich- und Saugverhaltensowie zur Nahrungsaufnahme der Rübenblattwanze *Piesma quadrata* Fieb. (Heteropt. Piesmidae). Z. Angw. Entomologie 56: 41–55.

Eisbein, K. 1976. Untersuchungen zum electronenmikroskopischen Nachweis des RübenbKraüsel-Virus (beta virus 3) in *Beta vulgaris* and *Piesma quadratum* Fieb. Arch. Phytopathol. und. Pflanzenschutz 12: 299–313.

Ext, W. 1923. Zur Biologie und Bekämpfung der Rübenblattwanze *Zosmenus capitatus*. Wolff. Arb. Biol. Reichsanst. Land-u. Forstwirtsch. 12: 1–30.

Heiss, E. and J. Pericart. 1983. Revision of palaearctic Piesmatidae (Heteroptera). Mitt. Münch. Ent. Ges. 73: 61–171.

Korcz, A., P. Olejarski and R. Paradowska. 1998. Research investigations on biology, ecology, occurrence and harmfulness of *Piesma quadratum* Fieb. (Heteroptera: Piesmidae) in Poland. Arch. Phytopath. Pflanz. 31: 323–334.

Proeseler, G. 1966. Beziehungen zwischen der Rübenblättwanze *Piesma quadratum* Fieb. und dem Rübenkräuselvirus. 1. Virusübertragungsversuche und Zucht des Vektors. Phytopathol. Z. 56: 91–211.

Proeseler, G. 1983. Piesmids. *In*: K.F. Harris and K. Maramorosch (eds.). Vectors of Plant Pathogens. Academic Press. 97–113.

Wille, J. 1929. Die Rübenblättwanze *Piesma quadratum* Fieb. Monographien zum Pflanzenschutz 2. Springer Verlad, Berlin. 118 pp.

11

Rearing Thrips Vectors

D.V.R. Reddy,[1,*] *G. Kurub*[2] and *V. Muniyappa*[3]

ABSTRACT

Several thrips species have so far been reported as vectors of tospoviruses. We describe in this chapter methods developed to rear *Thrips palmi* and *Scirotothrips dorsalis* utilizing detached leaves. The method facilitated maintenance of thrips colonies as well as selection of first instar larvae of approximately the same age, essential for obtaining maximum transmission.

Key Words: Tospoviruses, *Frankliniella fusca, F. occidentalis, F. tencicornis, Scirtothrips dorsalis, Thrips palmi, T. setosus, T. tabaci,* Virus transmission, thrips, detached leaves, aspirator

Introduction

Tospoviruses have a wide host range, which includes food, vegetable, ornamental and fruit crops. They are vectored by thrips (Thysanoptera). Seven thrips species, *Frankliniella fusca, F. occidentalis, F. tencicornis, Scirtothrips dorsalis, Thrips palmi, T. setosus,* and *T. tabaci* have been reported as vectors. Thrips are small (less than one millimeter in length) and difficult to handle in experiments on virus transmission. Many reports have been published on rearing of thrips utilizing whole plants or detached plant

[1] International Crops Research Institute for the Semi-Arid Tropics, Patancheru 502 324, Andhra Pradesh, India.
[2] Acharya N. G. Ranga Agricultural University, Regional Agricultural Research Station, Tirupati 517 502, Andhra Pradesh, India.
[3] Department of Plant Pathology, University of Agricultural Sciences, GKVK Campus, Bangalore, India 560 065.
* Corresponding author

parts, enclosed in different containers which include lamp globes, jars or celluloid cages (Samuel et al. 1930, Linford 1932), microcages (Smith 1932), microsandwich cages (Wright and Blodgett 1948), plastic cylindrical cages (Sakimura 1961) , lantern globes and glass vials (Amin et al. 1981). Various host plants utilized for rearing thrips include *Emilia sonchifolia, Callistephus chinensis* and *Stellaria media* (Sakimura 1961); potato, eggplant, cucumber, and watermelon (Calilung 1990); lettuce (Mau et al. 1990) and green pods (Ullman et al. 1992). In this chapter we describe a simple method for rearing of *S. dorsalis* and *T. palmi* utilizing detached leaflets of peanut or groundnut (*Arachis hypogaea*) under laboratory conditions.

Rearing and Maintenance

Materials Required

Incubator, stereo-binocular microscope, refrigerator, glass vials (3 x 1 cm), corks, camel hair brush, an aspirator, petri dishes, distilled water, hot air oven (for sterilization of vials and petri dishes) and a table lamp with fluorescent tube light are required.

Collection of Thrips

Initially thrips are field collected from a host known to support their multiplication. Mungbean is suitable for *T. palmi* and peanut for *S. dorsalis*. Leaves with a portion of stem containing thrips are enclosed in a glass vial (3 x 1 cm) containing filter paper or facial tissues and brought to the laboratory. They are tapped on a piece of black paper. They are collected with an aspirator (Fig. 1) and immobilized in the aspirator by leaving it in a refrigerator for approximately 15 min. They are subsequently transferred to a leaflet floated on ice cold water and examined under a binocular microscope. Different species can be separated using their distinguishing characteristics (Palmer 1990, Reddy et al. 1991). Characteristics used to separate *S. dorsalis* and *T. palmi* are listed in Table 1. They are collected using the aspirator and immobilized by leaving them in a refrigerator for about 15 min.

Rearing on Detached Peanut Leaflets

Glass vials (3 x 1 cm) are thoroughly washed and sterilized. Thrips immobilized in the aspirator by refrigeration, following their identification, are transferred to glass vials and closed with parafilm. The vials are placed in an inverted position to facilitate thrips movement to bottom of the vials. A fully expanded peanut leaflet is introduced into each vial after removing

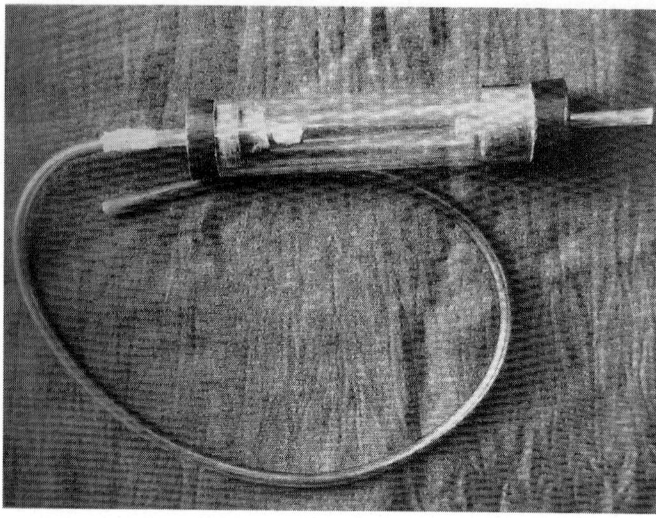

Fig. 1. Aspirator for collection of thrips.

Table 1. Distinguishing features of *Thrips palmi* and *Scirtothrips dorsalis*.

Characteristics	*T. palmi*	*S. dorsalis*
Larvae	Whitish in color	Whitish in color, both larvae and adults are active, moving in darted fashion
Adult female's color and length	Straw yellow to pale to pale brown, 0.9 mm long	Yellow in color, 0.7 mm long
Antennae	Seven segmented	Eight segmented
Pronotum	Two pairs of setae on the posteriolateral margin, no setae on the anteriolateral margin	No setae. Dark patches on the dorsal side of abdominal tergites
Wing characteristics	Forewings with broken rows of vein setae	Forewings with few small setae on the veins, hind wings with 2 setae

the parafilm, and then closed with a cork (Fig. 2). Vials were held in an incubator adjusted to 12 hr of light at 25°C and 12 hr of darkness at 22°C. After allowing 1 d for oviposition, the thrips are dislodged from leaflets by gently tapping and collected on a black paper after inverting the vials. Thrips are transferred to a new set of glass vials containing peanut leaflets and allowed again 1 d for oviposition.

This procedure is repeated for ten consecutive d, during which time 90% of the total egg laying is completed. Exposed leaflets are transferred to new glass vials once in two d and are observed for larval emergence from the 7th d after oviposition. These newly emerged larvae are ideal for

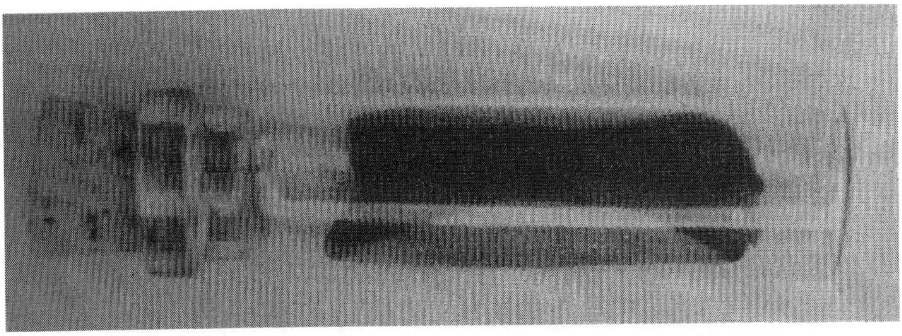

Fig. 2. Rearing of thrips in a glass vial on detached peanut leaflet.

utilization in transmission tests. For colony maintenance they are transferred to glass vials containing peanut leaflets and held in an incubator (as described above) until they became adults. The method described facilitated collection of especially 1 d old larvae which are found to be ideal for use in transmission tests.

Ullman et al. (1992) reported a very simple procedure for rearing *F. occidentalis*. The thrips are reared on green pods of *Phaseolus vulgaris* (cv. Green crop) placed in 227-g plastic deli cups with snap-on lids. A hole of 7.50 cm diameter in the lid, covered with nylon mesh, provided ventilation.

Precautions

- Use of sterilized glass vials and petri dishes minimized risk due to entomopathogenic fungi.
- Only fully emerged peanut leaflets are found to be suitable.
- Separate camel hair brush should be used to remove dead larvae from field collected samples to avoid transfer of microorganisms pathogenic to thrips.
- Moisture should not be allowed to accumulate in the glass vials.
- Aeration of glass vials for brief periods, once in two days, is found to be helpful.

A major disadvantage for using the glass vials to maintain colonies appeared to be mortality, which often exceeded 20%. Therefore for continuous maintenance of a large number of thrips, at least 75 adults were required each time they were used for oviposition.

Conclusion

The method described facilitated maintenance of two different thrips species and especially availability of 1 d larvae, which are most suitable for virus acquisition.

Acknowledgement

This article was submitted as Journal Article No. 2313 to ICRISAT.

References

Amin, P.W., D.V.R. Reddy, A.M. Ghanekar and M.S. Reddy. 1981. Transmission of tomato spotted wilt virus, the causal agent of bud necrosis disease of peanut, by *Scirtothrips dorsalis* and *Frankliniella schultzei*. Plant Disease 65: 663–665.

Calilung, V.J. 1990. Management of thrips and mites attacking potato in the low land. Study 1. Identity, biology, host range and natural enemies of potato thrips. Terminal Research Project Report. 25 pp.

Linford, M.B. 1932. Transmission of pineapple yellow spot virus by *Thrips tabaci*. Phytopathology 22: 310–324.

Mau, R.F., R. Bautista, D.E. Ullman, L. Gusukuma-Minuto, J.J. Cho and D. Custer. 1990. Factors affecting the epidemiology of TSWV in field crops: Comparative Virus acquisition efficiency of vectors and suitability of alternate hosts to *Frankliniella occidentalis*. Proceedings of ARS-US Department of Agriculture Workshop 87: 21–27.

Palmer, J.M. 1990. Identification of the common thrips of tropical Asia (Thysanoptera: Insecta). Trop. Pest Management 36: 27–49.

Reddy, D.V.R., J.A. Wightman, R.J. Bashear, B. Highland, M. Black, P. Sreenivasulu, S.I. Dwivedi, J.W. Demski, D. McDonald, J.W. Smith, Jr. and D.H. Smith. 1991. Bud necrosis: a disease of groundnut caused by tomato spotted wilt virus. Information Bulletin No. 31. ICRISAT, Patancheru 502 324, A.P., India. 20 pp.

Sakimura, K. 1961. Techniques for handling thrips in transmission experiments with the tomato spotted wilt virus. Plant Dis. Reptr. 45: 766–771.

Samuel, G.J., J.S. Bald and H.A. Pittman. 1930. Investigations on "spotted wilt" of tomatoes. Aust. Council Sci. Indi. Res. Bull. 44: 64.

Smith, K.M. 1932. Studies on plant virus diseases: IX. Further experiments with a ring Spot virus: Its identification with spotted wilt of the tomato. An. Appl. Biol. 19: 305–330.

Ullman, D.E., J.J. Cho, R.F.L. Mau, D.M. Westcot and D.M. Custer. 1992. A midgut barrier to tomato spotted wilt virus acquisition by adult western flower thrips. Phytopathology 82: 1333–1342.

Wright, C.M. and F.M. Blodgett. 1948. A detached leaf method for vector studies with the tomato-spotted-wilt virus (Abst.). Phytopathology 38: 28.

12

Whiteflies as Subjects of Laboratory Research

David N. Byrne

ABSTRACT

Whiteflies (Aleyrodidae) are truly unique insects, both agronomically and biologically. It spite of this, only a small number of the 1,200 described species have been investigated to any extent. A small number (less than 20) species have our attention. These are primarily those that are polyphagous and associated with herbaceous hosts. Other studied groups are the few attack trees, usually citrus or ornamentals. Many of these pestiferous species are maintained in laboratory colonies across the US. Those discussed in this chapter all seem to have a special set of requirements that are necessary to ensure successful experimentation and rearing.

Introduction

There are a number of significant reasons why many more of the 1,200 described species of whiteflies (Homoptera: Aleyrodidae) (Mound and Halsey 1978) deserve attention. Nevertheless, few have been the subjects of careful examination. This is likely because many of the unstudied species are found in remote habitats or because they lack economic impact. Most that have been examined do have economic importance. Consequently, one

Department of Entomology, The University of Arizona, Tucson, AZ 85721, USA.

of the best studied species is the cotton, sweet potato or tobacco whitefly, *Bemisia tabaci* (Gennadius).[1] This is because *B. tabaci* is one of the principal insect pests found on plants grown for food and fiber (Byrne et al. 1990). Most of these hosts are herbaceous. *B. tabaci* causes problems as a result of direct feeding damage (Riley and Palumbo 1995) and by acting as a vector of several viral pathogens, e.g., squash leaf curl virus, tomato yellow leaf curl virus and lettuce infectious yellows virus (Brown and Nelson 1986, Duffus 1996). Additionally problems resulting from the deposition of *B. tabaci* honeydew (excreta) and the associated sooty mold fungi (i.e., stickiness and staining) have become critical for growers (Hector and Hodkinson 1989). Finally, its feeding results in a plant disorder known as squash silverleaf with certain members of Cucurbitaceae (Costa et al. 1993). Other species, e.g., the greenhouse whitefly, *Trialeurodes vaporariorum*, is also considered significant crop pests of herbaceous crops, in some situations (Johnson et al. 1992).

Several other species are also significant pests of woody, perennial hosts. The woolly whitefly, *Aleurothrixus floccosus*, the citrus blackfly, *Aleurocanthus woglumi*, and the bayberry whitefly, *Parabemisia myricae*, have been major pests of citrus (Flint 1984). More recently, the giant whitefly, *Aleurodicus dugesi* (Zolnerowich and Rose 1996), and the ash whitefly, Siphoninus phillyreae (Gould et al. 1992), have become important pests of ornamentals.

Because of their pest status, *B. tabaci*, and the other species mentioned, are often the ones maintained in laboratory cultures. These species also are those most often found in the literature. As a consequence, much of the discussion in this chapter concerns these few whiteflies.

In additional to their economic importance, whiteflies are exceptionally interesting biologically. They have many features, which together make whiteflies unique among insects. Among these is the possession of a dorsal anus (Gill 1990) (Fig. 1). There are other insect orders that have this morphological characteristic, such as members of Psylliidae. Those structures in psyllids, however, are thought to be the result of displacement of the anus by enlarged reproductive organs. The suggested purpose of the dorsal anus in whiteflies is that it is positioned to facilitate the expulsion of excess honeydew. Whiteflies feed by inserting their mouthparts into the phloem tissue of plants. From there they draw large amounts of sap. Necessary nutrients are extracted while the excess is extruded as honeydew droplets. Honeydew is made up of a complex of compounds, but much of its composition consists of carbohydrates such as sucrose, glucose, fructose and

[1] The nomenclature for this taxon is unclear. Some assign the binomen Bemisia argentifolii to the species found in the Southwest (Bellows et al. 1994). We feel, however, that it is better to consider it as a complex of strains and use Bemisia tabaci as the scientific name.

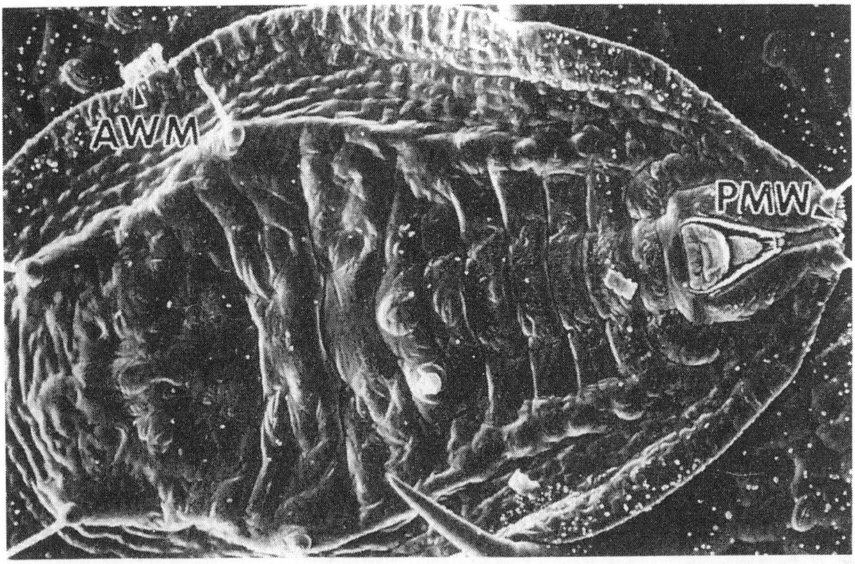

Fig. 1. Early fourth instar nymph of *Bemisia tabaci*. Note the dorsal anus near the cauda. Taken from Rossel et al. 1997.

trehalulose (Byrne and Miller 1990). As a result, honeydew is usually quite viscose (Hector and Hodkinson 1989). If whiteflies were more mobile insects such as aphids, they could release a droplet and move away. Since most whitefly immatures are sessile, they partially avoid problems associated with sticky honeydew by sending away these droplets. Honeydew exits the whitefly digestive track and fills its vasiform orifice (the well-like structure seen in Fig. 1) (Gill 1990). A spatula-shaped structure, the lingula, dips into the liquid and flips it away from the insect (Azab 1971).

Another feature associated with whitefly nymphs and adults are various forms of extracuticular wax secretions (Byrne and Hadley 1988). In nymphs these can take a number of forms, but usually appear as plumes, columns or setae-like projections. These wax secretions take a different shape in the adult stage. They are usually found in the form of short, 1 μm, curled threads. These are extruded from wax plates having been produced by wax glands (Byrne and Hadley 1988). These structures are found on the adult abdomen (Fig. 2). These threads are broken off by the hind tibiae and distributed over the entire body, with the exception of the eye surface. If adults are anesthetized, they will continue to produce wax filaments that are not removed from the wax plates by hind legs (Fig. 3). Only speculation exists concerning the function of these particles.

The final prominent whitefly features discussed here are the intermediate forms of the fourth nymphal instar. This is commonly referred to as the pupa. Whiteflies entering their fourth instar function in much the

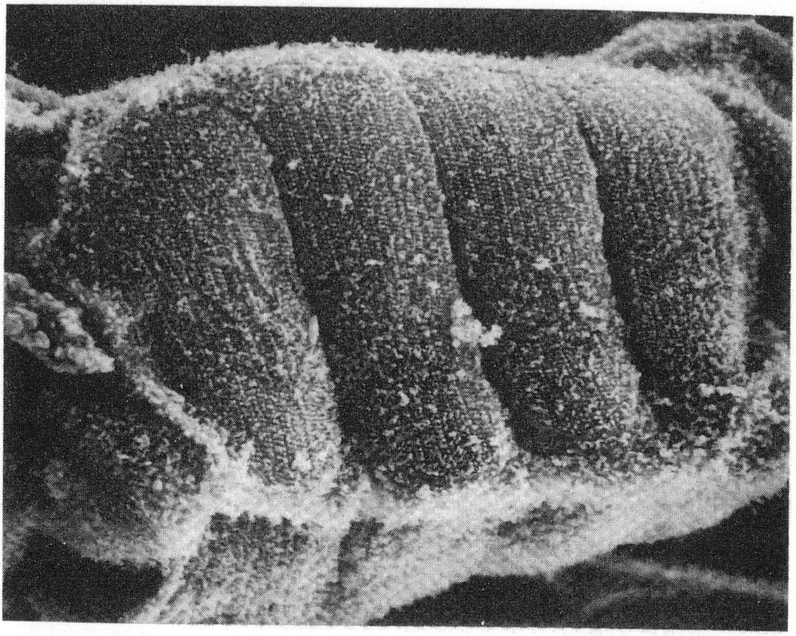

Fig. 2. Scanning electron micrograph of the abdominal wax plates of a male *Bemisia tabaci*. From Byrne and Hadley 1988.

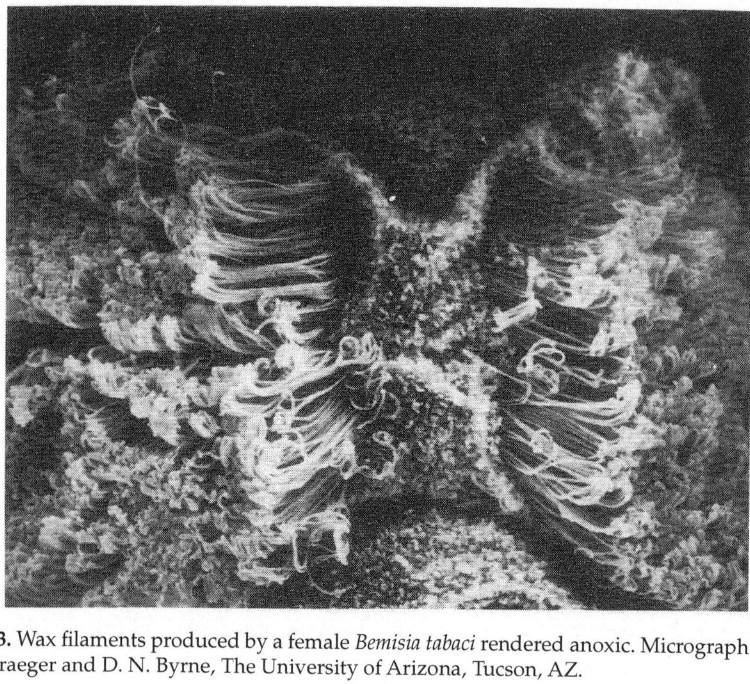

Fig. 3. Wax filaments produced by a female *Bemisia tabaci* rendered anoxic. Micrograph by E. A. Draeger and D. N. Byrne, The University of Arizona, Tucson, AZ.

same way as previous nymphal instars, i.e., the mouthparts are inserted into phloem tissue and they feed on their host plant (Gill 1990). They appear as flattened and translucent insects. They then assume an expanded and opaque white appearance. Apolysis takes place and they no longer appear to be feeding. Later the pharate adult form is clearly present beneath the old cuticular test.

Eventually adult structures such as red eyes and cuticular pigmentation can be seen. The term pupa is inappropriate, however, as whiteflies do not have complete metamorphosis. Borror et al. (1992) refer to this form of metamorphosis as intermediate.

Not a great deal is known about the taxonomy of Aleyrodidae. This likely results from the fact that they are considered tropical in origin and so less well studied than more temperate families. The majority of the over 1,200 described species are oligophagous or monophagous and associated with woody, perennial hosts (Mound and Halsey 1978).

Maintenance Methods

Rearing whiteflies in the Southwest of the United States (US) seems to present fewer problems than in other parts of the country because weather conditions appear to be favorable and whitefly natural enemy problems relatively easy to overcome. It is interesting to learn how unique the requirements are for maintaining the various whitefly species in different parts of the US.

Eggs

An expanded understanding of the whitefly life stages is critical when determining methods for maintaining laboratory colonies or designing experiments. Whitefly eggs are generally pyriform or ovoid and possess a pedicle (Fig. 4) (Poinar 1965).

The pedicel is the part of the egg that is inserted into plant tissue. It is a peg-like extension of the egg chorion. At least 13 species insert their pedicels into stomata (Poinar 1965, Paulson and Beardsley 1986). Most of these species attack grasses and sedges that may contain large amounts of silica and lignin. These leaf surfaces are likely hard to penetrate so that insertion into stomata may be less difficult.

Most whiteflies, including *B. tabaci* and *T. vaporariorum*, insert their pedicels into a slit made with the ovipositor in the leaf tissue. Several authors have speculated that whiteflies use the pedicel as a means of absorbing water from plant tissue (Hinton 1981). More definite proof was offered by Byrne et al. (1990) who demonstrated that approximately 50% of the mass

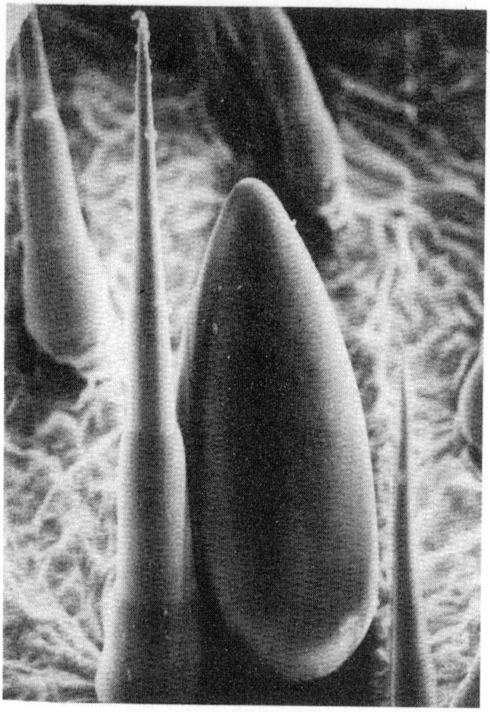

Fig. 4. Recently deposited eggs of *Bemisia tabaci* (Gennadius) on melon, *Cucumis melo* L. Picture taken by E.A. Draeger and D.N. Byrne, The University of Arizona.

of a mature egg belonging to *T. vaporariorum* is obtained from the host plant. From this it should be concluded that maintaining vigorously growing host plants is essential when examining the properties of whitefly eggs.

Rates of oviposition vary greatly between and within whitefly species. Azab et al. (1971) examined oviposition rates of *B. tabaci* reared on cotton in Egypt in an open-air insectary. The maximum number of eggs per female varied from 48 to 394. During July and August, when the daily maximum temperature was 28.5°C, the mean rate of oviposition was 252 eggs per female. During October and November, the mean was 204 (daily maximum temperature 22.7°C). In December and January (daily maximum temperature 14.3°C), the mean rate was 61 eggs. *B. tabaci* on cotton in Arizona laid no eggs at 14.9°C, but produced a significant number of eggs at 26.7°C and 32.2°C (Butler et al. 1983). These results occurred under conditions of constant temperature and light. I interpret Gameel's (1974) results on *B. tabaci* in Sudanese cotton fields to indicate a lifetime oviposition rate of 160.4 eggs per female during early fall. Von Arx et al. (1983) calculated the oviposition rate of a Sudanese strain to be 127.5 eggs. However, Dittrich et al. (1985) found that Sudanese whiteflies have an oviposition rate on cotton of 344 eggs. Obviously, when measuring oviposition rate as a trait, this amount of variation must be taken into account.

Oviposition rates also vary greatly in other species and are affected by environmental conditions and host plant. Burnett (1949) reports that the oviposition rate of *T. vaporariorum* was 0 at 9°C, 319.5 at 18°C and 5.5 at 33°C. Burnett (1949) also reported that oviposition rate for *T. vaporariorum* increased fivefold when insects were placed on a new leaf. The oviposition rate of *A. woglumi* also varied, ranging from 8.3 to 39.6 eggs per female (Dowell and Steinberg 1990). This was thought to be largely a function of the nymphal host plant.

Ovipositional habits also differ between species (Byrne and Bellows 1991). For example, *P. myricae* rarely oviposits anywhere but on the youngest leaves. *Tetraleurodes stanfordi*, the Stanford whitefly, lays its eggs on both sides of the leaf. This makes placement of clip cages particularly important. Alternatively one may consider using a sleeve cage with a drawstring.

Other whitefly biological traits display variation. The ratio of male to female *B. tabaci* adults constantly changes throughout the course of the year. Some have reported ratios of six females to every one male in the laboratory. Adult field populations have been reported as having sex ratios of 2 females:1 male (Morrill and Back 1911). It was cautioned that this may actually reflect a 1:1 primary sex ratio because female live longer in the population as adults.

First Instar Nymphs

First instar nymphs exit eggs from slits or cracks that form in the apical end. In *B. tabaci* the nymph seems to bend in the middle while trying to touch the leaf surface with its legs (D.N. Byrne, personal observation). Once they are clear of the old chorion, these nymphs, called crawlers, move freely over the leaf surface. They have fully functional legs and antennae (Fig. 5). It was once thought that movement by *B. tabaci* crawlers was somewhat limited. Summers et al. (1996), however, observed that crawlers move from senescing leaves on which they emerge to new leaves higher up on the plant. They have been observed moving distances of greater than 200 mm. Distance moved seems to be a function of host. Movement between plants was also reported. Movement between plants has also been observed in other species such as *A. woglumi* (Dowell et al. 1978). These observations emphasize the importance of using restraining devices when conducting life history studies using the first instar nymphs.

Fig. 5. Electronmicrograph of a newly emerged first instar nymph, crawler, of *Bemisia tabaci*. Note functional legs, antenna and dorsal vasiforn orifice (near rear). Micrograph courtesy of Allen C. Cohen, USDA/ARS, Mississippi State, MS.

Second and Third Instars

Most second and third instar nymphs have a body shape that is oval or elongate-oval (Byrne and Bellows 1991). Some species, however, have circular or nearly heart-shaped second and third instar nymphs.

Differentiating between second and third instar nymphs of *B. tabaci* often causes problems. Our laboratory members[2] often rely on size as an indicator of which instar is under consideration. Because there is considerable overlap in size, absolute size can not be used as a criterion when determining the nymphal instar. It can, nonetheless, be helpful. In the population measures, the median size of the crawler stage is near 0.24 mm, second instar nymphs near 0.34 mm, third instar nymphs near 0.43 mm, and early fourth nymphs near 0.67 mm. When generating this type of data, however, it is important to be mindful of the host plants used and their condition.

Fourth Instar Nymphs or Pupae

What takes place during the whitefly's fourth nymphal instar is distinct from what occurs in holometabolous families. Nichols and Tauber (1977),

[2] See acknowledgements.

discussing *T. vaporariorum*, stated that the fourth nymphal instar has three morphologically distinct forms. The early fourth instar is flattened and translucent. This stage, according to Gill (1990), feeds and so clearly is not a pupa. The next form, the transitional substage, is expanded and opaque-white with dorsal and lateral waxy, spinelike processes. At this point, apolysis is likely taking place. As stated, the last stage (particularly in *B. tabaci*), the red eyes and the yellow body pigment of the adult are apparent (Fig. 6). Apolysis seems to be complete and the adult cuticle is laid down. J. R. Nechols (personal communication) states that even after careful histological examination, he was unable to observe a distant stadium that intervened between the last nymphal stage and the adult. For purposes of convenience, many reserve the term pupa for the nonfeeding portion of the last nymphal stadium.

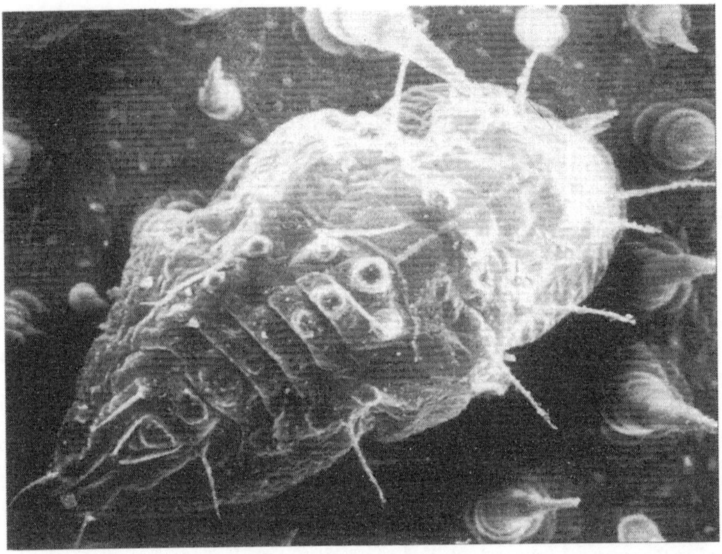

Fig. 6. Later stages of the fourth instar nymph (pupa) of *Bemisia tabaci*. Note the appearance of adult characteristics, particularly the eyes. Micrograph taken by E.A. Draeger and D.N. Byrne, The University of Arizona.

Adults

While I have already discussed some of the difficulties associated with manipulating adults, other words of caution are important. When measuring whitefly life history traits it is particularly essential to ensure that all treatments are exposed to the same conditions. As an extreme example, Blackmer et al. (1998) found that female adults reared on vigorously growing melon, *Cucumis melo* L., plants weighed significantly more than individuals reared on senescing plants (Table 1).

Table 1. Weight variation, in µg, of *Bemisia tabaci* (Gennadius) resulting from being reared on hosts, *Cucumis melo* L., of different quality (Blackmer and Byrne 1999).

	Male	**Female**
Senescing host	20.5 ± 0.5	33.0 ± 1.5
Vegetative host	23.7 ± 0.4	44.5 ± 0.6

While host quality has an influence on adult mass, so does host species. In one study on vegetative cotton hosts males weighed 17.08 ± 0.5 µg, which females weighed 35.05 ± 0.9 µg (Blackmer and Byrne 1993). Compare to Table 1 data.

Size

The small size of whiteflies can present interesting problems when they are used as research subject and warrants being considered separately. Gill (1990) reports that the largest genus is *Aleurodicus* spp. It has an adult body length in excess of 2 mm and a wing expanse of more than 3.5–4.0 mm. The dimensions of most whitefly adults can arbitrarily be placed in three categories (Table 2). *B. tabaci* is among the smaller whitefly species. *T. vaporariorum* illustrates the midrange and *D. citri* the upper end.

Sexual dimorphism in adult body dimension is usual for members of Aleyrodidae. Males are generally smaller (Table 2). Sexual dimorphism has also been found in *D. citri* (T.S. Bellows, unpublished) as well as *D. chittendeni* Laing (Wilson 1929) and *Tetraleurodes acaciae* (Quaintance) (Dowell 1983).

Table 2. Body measurements (mm) of three species of Aleyrodidae (n = 5).

Species	**Body length**	**Wing expanse**
Bemisia tabaci		
Male	0.85 ± 0.05	1.81 ± 0.06
Female	0.91 ± 0.04	2.13 ± 0.06
Trialeurodes vaporariorum		
Male	0.99 ± 0.03	2.41 ± 0.06
Female	1.06 ± 0.04	2.65 ± 0.12
Dialeurodes citri		
Male	1.20 ± 0.03	2.49 ± 0.07
Female	1.20 ± 0.05	3.01 ± 0.05

Handling Laboratory Colonies

Host Plants

The choice of host plants on which to rear whitefly colonies is critical. There is often a trade off between the ideal host plant for whiteflies and the plant's susceptibility to other pests. For example, poinsettia, *Euphorbia pulcherrima* (Willd.), will support whitefly colonies, but populations tend to be less robust. On the positive side, *E. pulcherrima* is attacked by almost no other insect found in greenhouse in the Southwest.

Other plant species, while excellent hosts for certain whiteflies, require a great deal of special maintenance. We know that *P. myricae* lay their eggs only on very young citrus leaves. To maintain colonies of these insects, G.P. Walker (personal communication) recommends using lemon or rough lemon, *Citrus* spp., as a host. To ensure very young leaves, Walker suggests that potential hosts be stripped of their foliage, which will prompt a flush of new growth. Stripping leaves, however, stresses plants. Rough lemon seems to respond better to the stripping of its leaves than does lemon.

Walker recommends a similar strategy of leaf stripping when maintaining colonies of *A. floccosus*. The requirements for very young leaves is not so stringent for this species, however, and plants can be allowed to recover for 2 to 3 wk after having their leaves stripped before being placed into laboratory colonies of *A. floccosus*.

T.S. Bellows and C.T. Meisenbacher (personal communications) have been largely responsible for developing the techniques necessary to maintain colonies of the nesting whitefly, *Paraleyrodes minei* Iaccarino (Bellows et al. 1998). They found these insects would not readily colonize citrus plants until they have already been infested with other whitefly species, usually *D. citri*. If plants are first colonized with *D. citri* (until these whiteflies pupate, usually 3 to 4 wk) they are a much better host for *P. minei*.

J.E. Polston (personal communication) has specific requirement for her colonies of *B. tabaci*, which she rears on herbaceous hosts. Her plant species of choice is lima bean, *Phaseolus lunatus* L. She prefers these to green bean, *P. vulgaris* L. In Florida *P. lunatus* is less subject to root pathogens. Additionally, *P. vulgaris* is more prone to become rangy under the low light conditions of growth chambers.

Our laboratory uses *C. melo* to maintain colonies of *B. tabaci*. These plants are excellent hosts, in part because of the leaf architecture; they have numerous minor veins that are close to the surface (Cohen et al. 1996). The watering scheme for these plants needs constant attention. They do, however, readily show signs of water stress so watering regimen can easily be adjusted.

Overwatering plants can also cause problems. This is particularly true when germinating seedlings. Young plants need aerated soil to establish strong root systems. Watering schemes must be adjusted for each condition. The key, as with most greenhouse operations, is to be extremely attentive.

In terms of a planting scheme for *C. melo* seeds, we plant four seeds per pot and then thin until the two healthiest plants remain. It is important to follow planting instructions for each type of host plant to ensure proper depth of planting.

Another precaution recommended by L. Osborne (personal communication) is that when using hibiscus, *Hibiscus* sp., although it is an excellent host for *B. tabaci*, you should avoid plants that have been treated with the plant growth regulator chlormequat chloride, Cycocel®. This material has an adverse effect on whitefly populations.

In addition to the above specific considerations, researchers must be mindful of more general greenhouse conditions for plants. Those recommended by R.E. Santangelo (personal communication) for eggplant, *Solanum melongena* L., grown in the Southwest seem to be optimal for *C. melo*, cotton, *Gossypium hirsutum* L., *E. pulcherrima*, and several other greenhouse hosts. These are a temperature near 27°C during daylight hours and no lower than 15°C during the night. Relative humidity should remain between 27 and 30% during the day, while they can be from 50 to 60% during the night. Conditions in other parts of the US are likely to vary, but these recommendations may serve as starting points. R. E. Santangelo (personal) also fertilizes his plants daily with a solution of Peters® EXCEL® Water Soluble Fertilizer. Its 20-20-20 mixture has adequate levels of nitrogen, phosphorus and potassium.

There are others matters to consider when rearing whitefly parasitoids. R.E. Santangelo (personal communication) is responsible for rearing a large number of whitefly parasitoids, primarily *Eretmocerus* spp. To do so, he utilizes tobacco, *Nicotiana* spp., as a greenhouse host to maintain parasitoid-free whitefly colonies. For reasons that are unclear, *B. tabaci* reared on this host plant are not subject to parasitism. For his purposes, the tobacco plant also has an appropriate architecture. To collect large numbers of parasitoid pupae, he needs a broad, flat leaf. Individual plants are used for up to 3 months. As a result, plants must be managed so that they maintain vigorous growth throughout.

Pest Control

Aphids

Several whitefly researchers reported carefully trimming off flower buds on their host plants as a means of controlling aphid numbers. Others routinely wash plants with soap and water. Various proportions of soap to water have been suggested, but phytotoxicity is a concern when using high concentration of soap under conditions of high temperatures. It is suggested that a few plants be washed first as a means of checking for potential damage. C.T. Meisenbacher (personal communication) warns that if you attempt to rid plants of aphids in *A. dugesii* colonies using this method, you will wash off an unacceptable number of whiteflies.

Mites

Mites are a problem in many parts of the US. In our colonies these arthropods do not respond to miticides. We have used a number of different compounds without success. Mites are most often a problem in the Southwest during the summer. To control their populations we turn off cooling systems for a three-week period and subject them to extreme ambient temperatures. These have proven to be very effective strategy. During winter months, we will abandon greenhouses entirely, periodically placing a single plant inside the greenhouse to check for still-existing mite problems. G.P. Walker (personal communication) will dip his leaf stripped citrus plants in an oil solution to rid them of mites.

Ants

We also experience problems with ants, primarily *Solenopsis* spp., in our whitefly colonies. These are troublesome because their presence is associated with aphid colonies, which are tended by ants. We have also had to control ant populations because people working in our whitefly colonies were found to be sensitive to stings. To control ants, we use the bait form of hydramethylnon (Amdro®). Since the ant species we are trying to control is relatively small, we grind the bait in a mortar because these ants are incapable of carrying away the formulated product. We usually try to 'spot treat' the ant nest site.

Lepidopterans

Lepidopteran larvae such as black cutworm, *Agrotis ipsilon* (Hufnagel), and alfalfa loopers, *Autographa califonica* (Speyer), are occasionally a problem. When they are present, their damage is quite apparent so they can be easily controlled by hand picking. We have never resorted to pesticides for their control, although others have.

Parasitoids

Because our laboratory regularly conducts behavior studies, we want to avoid experimenting with whitefly populations that are maintained in the laboratory for too long. We are concerned about problems associated with adaptation to greenhouse conditions and in-breeding. To ensure that our colonies do not suffer from these problems, on an annual basis we collect whiteflies from the field using a D-Vac® Vacuum Sampler. Large numbers, i.e., > 300,000 individuals are released into the colonies. We are certain that along with these new whiteflies, their predators and parasitoids are being introduced. We have never had problems, however. This is presumably because conditions are so favorable for the propagation of *B. tabaci* that whitefly populations do well in spite of the level of parasitism they experience.

Rates of parasitism within a colony can be estimated by inspecting spent pupal cases. Those that have been parasitized will have a round exit hole produced by the exiting wasp. Pupal cases from which adult whiteflies have emerged will have the characteristic t-shaped slit shown in Fig. 7.

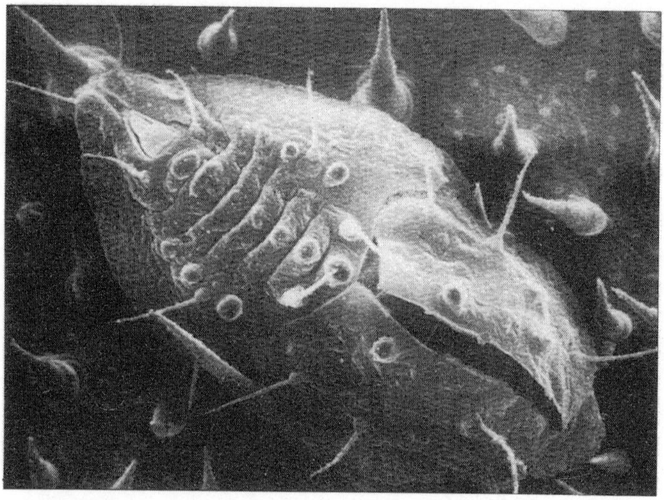

Fig. 7. Pupal case from which an adult *Bemisia tabaci* has emerged.

While we have been able to adopt a rather cavalier attitude in our whitefly colonies relative to parasitoids, R.E. Santangelo can not. Because he is concerned about releasing large numbers of specific parasitoids, he must maintain colonies free of other parasitoids.

Problems do exist elsewhere, as in Florida. L. Osborne (personal communication) reported they must strip the hibiscus they are using as host plants to avoid parasitism by *Eretmocerus* spp.

Unwanted Whiteflies

For obvious reasons, most researchers want to maintain colonies that are free of unwanted whitefly species. R. Santangelo (personal communication) is concerned when he finds *T. vaporariorum* in his *B. tabaci* colonies. Fortunately, *T. vaporariorum* (personal communication) is more susceptible to high temperatures than *B. tabaci.* To rid his colonies of the unwanted whiteflies, he raises the greenhouse temperature to between 29 and 33°C for a period of approximately 2 wk. This does not effect host plants if temperatures are quickly reduced after that time.

R.E. Santangelo (personal communication) also uses a mesh screen over the air intake system of greenhouses to prevent entry by unwanted parasitoids and whiteflies. This seems to be effective, but no empirical information is available from his greenhouses. J.E. Polston (personal communication) uses a similar method in Florida. She takes the further precaution of keeping her colonies in cages inside her screened air-conditions rearing facility. Her concern is heightened because she must entirely exclude viruliferous *B. tabaci* from colonies of the same species that are going to be used in transmission studies.

K.C. Bennett (personal communication) works with *T. vaporariorum* in New York, because these insects are among the principal ornamental pests. She minimizes plant and human traffic into her greenhouses primarily to exclude *B. tabaci.*

Fungal Pathogens

In more humid situations, such as those experienced in the Southeast, fungal pathogens such as *Paecilomyces fumosoroseus* (Wize) Brown & Smith, cause considerable problems in whitefly colonies (Vidal et al. 1998). Because of fungal pathogens whitefly colonies are difficult to rear there in traditional greenhouses. Colonies must be maintained in air-conditioned rooms. If greenhouses are used, an adequate system to dehumidify the air must be maintained to avoid fungi. During winter months dehumidifier can be inadequate and new host plants must be constantly rotated into the colony.

Sooty mold, resulting from a variety of fungi such as *Capnodium* spp., is a problem in most all whitefly colonies. Whitefly honeydew provides a rich medium for the growth of these fungi (Byrne and Miller 1990). Problems are most prevalent in the Southwest during summer rainy seasons when both humidity and temperatures are relatively high. We prevent problems in our *C. melo* colonies by replacing approximately 20% of the plant population every week. As a consequence, no *C. melo* plant remains in our whitefly greenhouse for an extended period of time. This is, in a large part, a reflection of the fact that melon plants grow quickly and can be readily available to insert into greenhouse plantings.

In situations where plants must remain in colonies for more than 2 months, e.g., *Nicotianna* spp., this strategy is not possible. Workers must be content to wipe off tops of leaves with soap and water.

Powdery mildew, caused by *Sphaerotheca* spp. and other species, may attack host plants as well. Maintaining colonies in air-conditioned rooms in more the humid environments of the Southeast can do much to control these pathogens.

In the drier conditions found in the Southwest, soil-borne fungal pathogens are more likely to present problems. Is this part of the US we steam autoclave soil before we use it to present problems such as those associated with damping off, *Pythium* spp. Unfortunately, for reasons we can only guess, soil that has been autoclaved more than once does not produce vigorous host plants. We have had considerable luck in using new lots of soil two times before having to discarding them. R.E. Santangelo (personal communication) informs me that he is only able to use soil once. He feels the *S. melongena* he uses are particularly susceptible to a number of fungal pathogens such as *Verticillium* spp., *Pythium* spp. and *Fusarium* spp. Even after sifting the soil, a great many old root hairs remain in the recycled soil, so that problems persist. He has tried several types of fungicide drenches, but the overall vigor of the plants remained poor. Using new, freshly mixed soil eliminates a number of these fungal problems. This is costly, but stronger plants can support greater numbers of whiteflies and produce more wasps.

Handling Individual Whiteflies

Manipulating individuals of any whitefly species should be done with as much care as possible. Species that primarily infest herbaceous host such as *B. tabaci* and *T. vaporariorum* present a unique set of problems.

Whiteflies that have been transferred to new hosts using a pooter may not respond well when measuring life history traits. Adopting a method used by Walker and Aitken (1993), we allow whiteflies to walk into and out of a glass tube with a diameter of approximately 3.0 mm when being transferred to new hosts. This is thought to be less disruptive than forcing them through the narrow end of a pooter into its larger chamber.

If concerned about such traits as oviposition rates, it is best to confine adults to sleeve cages. *B. tabaci* tend to deposit a few eggs on the leaves upon which they emerge as adults and then move to younger leaves if they are available. Also, the number of eggs laid is often dependent upon the age and mating status of the adult. When comparing oviposition rates between host plants, it best to used females that are of the same age (preferably 3 d) and experience, relative to exposure to males.

We have found that the rate of oviposition seems to be closely tied to host experience. If you are trying to determine whitefly traits for a given plant host then the insects must be raised in that host for at least three generations. Costa et al. (1996) suggest that this reflects a period of entrainment. When whiteflies are first moved on to a new host, the first generation lacks some vigor and subsequent generations are likely to provide data that more closely approximates true oviposition rates.

When conducting transmission studies, J.E. Polston also prefers to acclimatize whiteflies to the host of the virus being examined. For example, if the experiments involve tomato mottle virus, whiteflies must be reared on tomatoes for at least three generations. The architecture of the tomato plant, with its narrow leaves, can make this difficult host plant. This is particularly true if you must utilize clip cages.

Certain species that are associated with woody perennials are difficult for other reasons. T.S. Bellows and C.T. Meisenbacher (personal communications) have a great deal of experience in maintaining colonies of *A. dugesii*. These they rear on *H. esculentum*. When moving these insects between plants, it is best not to use an aspirator. This is thought to be because they have relatively long mouthparts. Trying to move them from leaf to leaf without allowing them to fully extract these mouthparts results in high levels of mortality. Even shaking leaves until the whiteflies appear more mobile before moving them can result in less than 15% survival. It is best to transfer highly infested leaves in open greenhouse, not individual cages, and allow them to move at will. T.S. Bellows and C.T. Meisenbacher further inform us that *A. dugesii* will not move between plants in the presence of any air movement. It is best to turn off even air handling systems in greenhouses when trying to coax these insects to move from one plant to another.

R.E. Santangelo (personal communication) transfers whiteflies between hosts by placing partially matured *S. melongena* into his colonies of whitefly-infested *Nicotiana* sp. He moves whitefly adults to the new host using the blower from a Dayton® Wet/Dry Vacuum. *S. melongena* are removed from the *Nicotiana* colonies and kept for 1 d in an isolation greenhouse. After a day, the blower is used to remove the adults. This results in a colony consisting of eggs of approximately the same age. When the colony is made up of nymphs of the proper instar, here second, parasitoids are introduced.

Conclusion

Conversations with a number of individuals, working with several species of whiteflies, have led to the conclusion that whiteflies have a special set of requirements depending on the questions being address or the insect under evaluation. While able to examine only a few of the whitefly species and their associated host plants, I hope I have provided an insight into some of the problems associated with maintaining these insects. We must each adopt our rearing and handling techniques to suit our own purposes.

Acknowledgements

Throughout this chapter I have used the first person, plural pronoun when discussing techniques used in our laboratory. This is to reflect the contributions of the individuals here at the University of Arizona who have the actual responsibility for maintaining our whitefly colonies. In particular I would like to give thanks to Erich Draeger, Felisa Blackmer and Mary Holloran.

References

Avidov, Z. 1956. Bionomics of the tobacco whitefly (*Bemisia tabaci* Gennad.) in Israel. Ktavin 7: 25–41.

Azab, A.K., M.M. Megahed and D.H. El-Mirasawi. 1971. On the biology of *Bemisia tabaci* (Genn.). Bulletin. Societe Entomogique d'Egypte 55: 305–315.

Bellows, T.S., C. Meisenbacher and D.H. Headrick. 1998. Field biology of *Paraleyrodes minei* (Homoptera, Aleyrodidae) in southern California. Environmental Entomology 27: 277–281.

Bellows, T.S., T.M. Perring, R.J. Gill and D.H. Headrick. 1994. Description of a species of *Bemisia* (Homoptera, Aleyrodidae). Annals of the Entomological Society of America 87: 195–206.

Bemis, F.E. 1904. The aleyrodids, or mealy-winged flies, of California, with references to other America species. Proceedings of the US National Museum 27: 471–573.

Blackmer, J.L. and D.N. Byrne. 1993. Flight behavior of *Bemisia tabaci* in a vertical flight chamber, effect of time of day, sex, ages and host quality. Physiological Entomology 18: 223–232.

Blackmer, J.L. and D.N. Byrne. 1999. Changes in amino acids in *Cucumis melo* in relation to life-history traits and flight propensity of *Bemisia tabaci*. Entomol. Exp. Appl. 93: 29–40.

Borror, D. J. C. A. Triplehorn and N.F. Johnson. 1992. An Introduction to the Study of Insects. Harcourt Brace College Publishers, New York. 875 pp.

Brown, J.K. and M.R. Nelson. 1986. Whitefly-borne viruses of melons and lectuce in Arizona. Phytopathology 76: 236–239.

Burnett, T. 1949. The effect of temperature on an insect host-parasite population. Ecology 30: 113–134.

Butler, G.D., T.J. Henneberry and T.E. Clayton. 1983. *Bemisia tabaci* (Homoptera, Aleyrodidae), development, oviposition and longevity in relation to temperature. Annals of the Entomological Society of America 76: 310–313.

Byrne, D.N. and N.F. Hadley. 1988. Paeticulate surface waxes of whiteflies, morphology, composition and waxing behavior. Physiological Entomology 13: 267–276.

Byrne, D.N. and T.S. Bellows. 1991. Whitefly biology. Annual Review of Entomology 36: 431–457.

Byrne, D.N. and W.B. Miller. 1990. Carbohydrate and amino acid composition of phloem sap and honeydew produced by *Bemisia tabaci*. Journal of Insect Physiology 36: 433–439.

Byrne, D.N., T.S. Bellows and M.P. Parrella. 1990. Whiteflies in agricultural systems. pp. 227–261. *In*: D. Gerling (ed.). Whiteflies, Their Bionomics, Pest Status and Management. Intercept Limited, Andover, UK.

Cohen, A.C., T.J. Henneberry and C.C. Chu. 1996. Geometric relationships between whitefly feeding behavior and vascular bundle arrangements. Entomologia Experimentalis et Applicata 78: 135–142.

Costa, H.S., D.E. Ullman, M.W. Johnson and B.E. Tabashnik. 1993. Antibiotic oxytetracycline interferes with *Bemisia tabaci* (Homoptera, Aleyrodidae) oviposition, development, and ability to induce squash silverleaf. Annals of the Entomological Society of America 86: 740–748.

Dittrich, V.S. and O.I. Gameel. 1974. Some aspects of the mating and oviposition behavior of the cotton whitefly *Bemisia tabaci* (Genn.). Revue de Zoologie Africaine 88: 784–788.

Dittrich, V., S.O. Hassan and G.H. Ernst. 1985. Development of a new primary pest of cotton on the Sudan, *Bemisia tabaci* (Gennadius) in cotton fields in Israel. Ecological Applications 5: 221–233.

Dowell, R,V. 1983. Biology of *Tetraleurodes acaciae* (Quaintance) (Homoptera, Aleyrodidae). Pan-Pacific Entomologist 58: 312–318.

Dowell, R.V. and B. Steinberg. 1990. Influence of host plant on fecundity of *Aleurocanthus-woglumi* Ashby (Homoptera, Aleyrodidae). Pan-Pacific Entomologist 66: 62–65.

Dowell, R.V., G.E. Fitzpatrick and F.W. Howard. 1978. Activity and dispersal of the first instar larvae of the citrus blackfly. Journal of the New York Entomological Society 86: 121–122.

Duffus, J.E. 1996. Whitefly-borne viruses. pp. 255–263. *In*: D. Gerling and R.T. Mayer (eds.). Bemisia, 1995 Taxonomy, Biology, Damage, Control and Management. Intercept Limited, Andover, UK.

Flint, M.L. 1984. Integrated Pest Management for Citrus. University of California IPM Publication No. 3303. 144 pp.

Gameel, O.I. 1974. Some aspects of the mating and oviposition behavior of the cotton whitefly *Bemisia tabaci* (Genn.). Review Zoologique Afrika 88: 784–788.

Gerling, D., A.R. Horowitz and J. Baumgartner. 1986. Autocology of *Bemisia tabaci*. Agriculture, Ecosystems and Environment 17: 5–19.

Gill, R.J. 1990. The morphology of whiteflies. pp. 13–46. *In*: D. Gerling (ed.). Whiteflies, Their Bionomics, Pest Status and Management. Wimborne, UK, Intercept.

Gould, J.R., T.S. Bellows and T.D. Paine. 1992. Population dynamics of *Siphoninus phillyreae* in California in the presence and absence of a parasitoid, *Encarsia partenopea*. Ecological Entomology 17: 127–134.

Hector, D.J. and I.D. Hodkinson. 1989. Stickiness in Cotton. C.A.B. International, Oxon, UK.

Hinton, H.E. 1981. Biology of Insect Eggs. Volumes 1 and 2. Pergammon Press, New York.

Husain, M.A. and K.N. Trehan. 1933. Observations on the life history, bionomics and control of the white-fly of cotton (*Bemisia gossypiperda* M & L). Indian Journal of Agricultural Science 3: 701–753.

Johnson, M.W., L.C. Caprio, J.A. Coughlin, B.E. Tabashnik be, J.A. Rosenheim and S.C. Welter. 1992. Effect of *Trialeurodes vaporariorum* (Homoptera, Aleyrodidae) on yield of fresh-market tomatoes. Journal of Economic Entomology 85: 2370–2376.

Laird, E.F. and R.C. Dickson. 1959. Insect transmission of the leaf-crumple virus of cotton. Phytopathology 49: 324–327.

Morrill, A.W. and E.A. Back. 1911. Whiteflies injurious to citrus in Florida. USDA Bureau of Entomology Bulletin #92. 109 pp.

Mound, L.A. 1983. Biology and identification of whitefly vectors of plant pathogens. pp. 305–313. *In*: R.T. Plumb and J.M. Thresh (eds.). Plant Virus Epidemiology. Oxford, Blackwell Scientific.

Mound, L.A. and S.H. Halsey. 1978. Whitefly of the World, a systematic catalogue of the Aleyrodidae (Homoptera) with host plant and natural enemy data. John Wiley and Sons, New York. 340 pp.

Nechols, J.R. and M.J. Tauber. 1977. Age-specific interaction between the greenhouse whitefly and *Encarsia formosa*, influence of the parasite on host development. Environmental Entomology 6: 207–210.

Paulson, G.S. and J.W. Beardsley. 1985. Whitefly (Hemiptera, Aleyrodidae) egg pedicel insertion into host plant stomata. Annals of the Entomological Society of America 78: 506–508.

Poinar, G.O. 1965. Observations on the biology and ovipositional habits of *Aleurodicus occiduus* (Homoptera, Aleyrodidae) attacking grasses and sedges. Annals of the Entomological Society of America 58: 618–620.

Riley, D.G. and J.C. Palumbo. 1995. Interaction of silverleaf whitefly (Homoptera, Aleyrodidae) with melon crop quality. Journal of Economic Entomology 88: 1726–1732.

Rosell, R.C., I.D. Bedford, D.R. Frohlich, R.J. Gill, J.K. Brown and P.C. Markham. 1997. Analysis of morphological variation in distinct populations of *Bemisia tabaci* (Homoptera, Aleyrodidae). Annals of the Entomological Society of America 90: 575–589.

Summers, C.G., A.S. Newton and D. Estrada. 1996. Intraplant and interplant movement of *Bemisia argentifolii* (Homoptera, Aleyrodidae) crawlers. Environmental Entomology 25: 1360–1364.

Van Lenteren, J.C. and L.P.J.J. Noldus. 1990. Whitefly-plant relationships, behavioural and ecological aspects. pp. 47–89. *In*: D. Gerling (ed.). Whiteflies, Their Bionomics, Pest Status and Management. Wimborne, UK, Intercept.

Vidal, C., L.S. Osborne, L.A. Lacey and Fargues. 1998. Effect of host plant on the potential of *Paecilomyces fumosoroseus* (Deuteromycotina, Hyphomycetes) for controlling the silverleaf whitefly, *Bemisia argentifolii* (Homoptera, Aleyrodidae) in greenhouses. Biological Control 12: 191–199.

Von Arx, R. Baumgartner and J. Delucci. 1983. Development biology of *Bemisia tabaci* (Genn.) (Sternorrhych, Aleyrodidae) on cotton at consistent temperature. Bulletin. Societe Entomologique Suisse 56: 389–399.

Walker, G.P. and D.C.D. Aitken. 1993. Development time of *Parabemisia myricae* (Kuwana) (Hom, Aleyrodidae) in different seasons and a degree-day model. Journal of Applied Entomology 115: 398–404.

Wilson, G.F. 1929. The rhododendron white fly. Journal of the Royal Horticultural Society 54: 214–217.

Zolnerowich, G. and M. Rose. 1996. A new species of *Entedononecremnus* (Hymenoptera, Chalcidoidea, Eulophidae) parasitic on the giant whitefly, *Aleurodicus dugesii* Cockerell (Homoptera, Aleyrodidae). Proceedings of the Entomological Society of Washington 98: 369–373.

13

Xiphinema index, Maintenance and Feeding in Monoxenic Cultures#

Urs Wyss

ABSTRACT

Grapevine fanleaf, worldwide the most important virus of grapes, is transmitted by the nematode *Xiphinema index*. Hence this species is currently one of the best studied ectoparasitic nematodes. It can be reared in monoxenic agar culture on roots of *Ficus carica*, and the results obtained from feeding studies are transferable to *Vitis vinifera*. Prerequisites to be considered for the successful establishment of such cultures are described. As *X. index* feeds on the root-tips of host plants, its reproduction success depends on the response of meristematic cells to become transformed into multinucleate nurse cells. These metabolically highly active cells are induced by esophageal gland secretions. Uninucleate and multinucleate cells are exploited by dorsal gland cell secretions that rapidly liquefy the cytoplasm of these cells in order to facilitate efficient food withdrawal. Although the structure and function of this gland cell has been assesses in great detail, future research is required to understand the mechanisms involved in multinucleate nurse cell induction and virus transmission, with special emphasis on the possible role played by subventral gland secretions.

Institu für Phytopathologie, Universität Kiel, Hermann-Rodewald-Strabe 9, D-24118, Germany.

This chapter is dedicated to my friend Bernhard Weischer, who introduced me into nematology and later on to *Xiphinema index* and associated pleasures of wine tasting.

Introduction

Grapevine fanleaf virus (GFLV) affects only grapevine and occurs in most regions of the world where grapevines are cultivated. Infection by specific virus strains causes an abnormal spreading of the leaf veins so that the leaf assumes the appearance of a fan. Other strains cause different symptoms but in all cases virus infection is associated with severe losses (Taylor and Brown 1997). Forty years ago it was confirmed for the first time that GFLV is transmitted by a root parasitic nematode. *Xiphinema index* was then shown to be the vector (Hewitt et al. 1958). This finding prompted an intensive search for other virus-vector nematodes, with the result that only a few species of the nematode orders Dorylaimida (family Longidoridae) and Triplonchida (family Trichodoridae) are able to transmit soil-borne viruses (Brown et al. 1995, Taylor and Brown 1997).

Xiphinema index is economically one of the most important representatives of the genus *Xiphinema*, currently comprising more than 200 species (Taylor and Brown 1997). This genus, together with the genera *Longidorus* and *Paralongidorus*, is placed into the family Longidoridae (longidorid nematodes) of which several members are able to transmit nepoviruses. The name of this virus group was derived from nematode-transmitted viruses with polyhedral particles. The 37 reported nepoviruses (Taylor and Brown 1997), including GFLV, have isometric particle of about 28 nm diameter and a bipartite genome, each containing single-stranded RNA. The larger molecule, RNA-1, carries genetic determinants for host range, symptom expression and seed transmissibility, whereas the smaller RNA-2 carries determinants for serological specificity and nematode transmissibility.

Longidorids are relatively large nematodes (up to 12 mm long) with long hollow needle-like styles (Fig. 1) that enable them to feed deep within plant roots. Only the anterior part of the stylet, termed odontostyle, is inserted into root cells. The posterior part, called odontophore, is formed from the cuticle of the anterior esophagus and it contains adjacent to the food canal nerve tissues thought to have a discriminating gustatory function (Robertson 1976, Robertson and Wyss 1983). Protraction of the stylet is achieved by contraction of protractor muscles attached to the rear of the odontophore. The esophagus consists of a long, narrow and flexible anterior

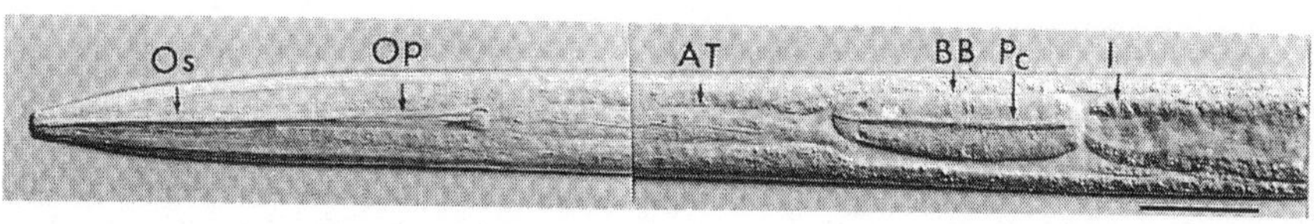

Fig. 1. Light micrograph of the feeding apparatus of a *X. index* female. **Abbreviations**: **AT**: Anterior tube of esophagus; **BB**: Basal bulb of esophagus; **I**: Intestine; **Op**: Odontophore (posterior part of stylet); **Os**: Odontostyle (anterior part of stylet); **Pc**: Pump chamber of basal bulb. Bar: 50 μm.

tube and a posterior muscular cylindrical bulb, termed basal bulb (Fig. 1), the structure and function of which will be described in detail later on.

Xiphinema index, first described by Thorne and Allen (1950) from fig, is common in most grapevine-growing regions throughout the world, where, apart from transmitting GFLV, it can also cause serious feeding damage to grapevine roots (Raski and Krusberg 1984) when present in large numbers. Affected plants suffer from a considerably distorted root system, in which root-tips have been transformed into terminal galls (Fig. 2). Sections through such galls (Fig. 3) show among and below necrotic cells many greatly enlarged multinucleate cells (Weischer and Wyss 1976, Rumpenhorst and Weischer 1978, Vovlas et al. 1978).

Xiphinema index has a restricted host range, with *Vitis* spp. and fig being the most important hosts to support a rapid population build-up. In the laboratory this species can also reproduce on other plants, including some herbaceous hosts, depending on the origin of the population tested (e.g., Coiro and Brown 1984).

In the present chapter suitable methods for studying the interactive responses between *X. index* and its host plants are described, with special emphasis on the feeding behavior and root cell modifications induced by this nematode.

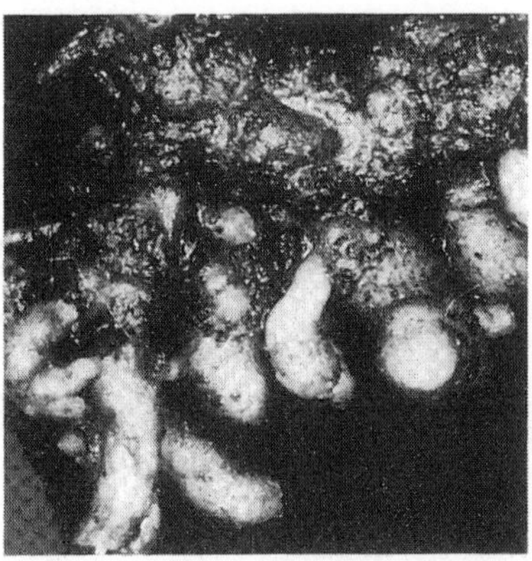

Fig. 2. Damaged root system of *Vitis vinifera*, with root-tip galls, fed upon by a large *X. index* population maintained in pot culture.

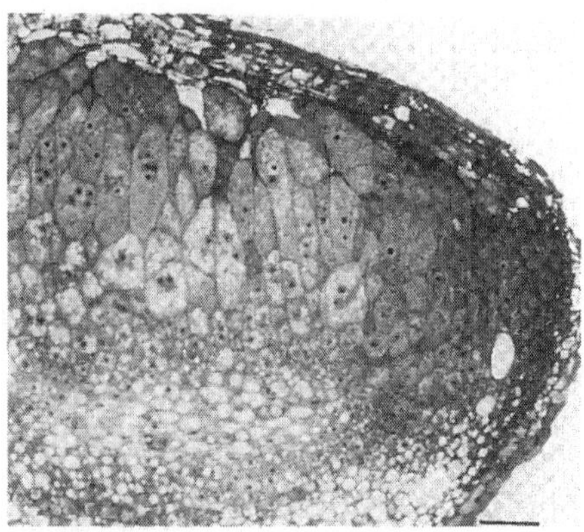

Fig. 3. Longitudinal section through one of the root-tip galls of Fig. 2, showing many greatly enlarged multinucleate cells. Bar: 50 μm.

Feeding behavior on *Vitis Vinifera*

Fisher and Raski (1967) were the first to describe in some detail the feeding behavior of *X. index* on grape. Seedlings were grown in 0.75% distilled water agar in Petri dishes. No attempt was made to work under aseptic conditions and no nutrients were added to the agar medium as germinating grape seedlings could grow for up to four weeks on the reserve food of the cotyledons. Young juvenile stages were seen to feed primarily in the piliferous region of the root without causing visible symptoms. Older juvenile stages and adult females always fed at the root-tips, and root-tip galls were evident within 24 hrs after inoculation. Weischer and Wyss (1976) observed *X. index* feeding on excised grapevine roots grown in 0.5% distilled water agar. Again no nutrients were added to the agar and the nematodes were also not surface-sterilized. On these excised roots the preferred feedings sites were in the zone of growing rootlets, just emerging rootlets and callus tissue developing from cuts. A twisting of the stylet was noticed during cell wall perforation and the nematodes fed from progressively deeper cells. The action of the esophageal bulb for food ingestion was described to some extent. A more natural condition, allowing observation only under the stereomicroscope (Weischer and Wyss 1976), can be obtained by diverting growing roots from pot cultures into water agar in Petri dishes as shown in Fig. 4.

When fine sand particles, not exceeding 1 mm Ø, are added to the agar, the nematodes feed more readily under these conditions. They are seen to aggregate in large numbers behind the tip of rapidly growing roots (Fig. 5) and they follow along this region until root growth starts to slow down. Feeding arrears are characterized by a browning and collapse of epidermal cells. When the growth of the roots is much retarded, its tip become transformed into a gall, and several lateral roots are formed, which,

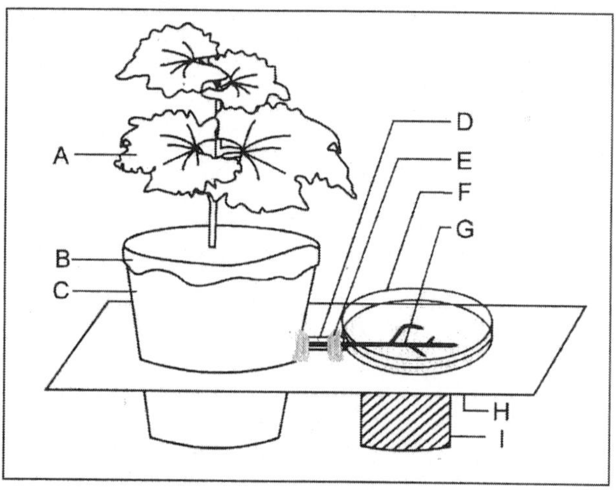

Fig. 4. Method used for feeding observations of *X. index* on extending *Vitis vinifera* roots in water agar. **A**: Young plant; **B**: Aluminium foil; **C**: Plastic beaker; **D**: Glass tube; **E**: Vaseline; **F**: Petri dish (14 cm Ø); **G**: Extending root, growing in distrilled water agar; **H**: Piece of cardboard; **I**: Wooden block to support observation chamber.

when attacked soon after emergence, develop into bud-like swellings. Sections through these swellings and root-tip galls reveal, as expected, the presence of hypertrophied multinucleate cells beneath necrotic, fed on cells (Fig. 6).

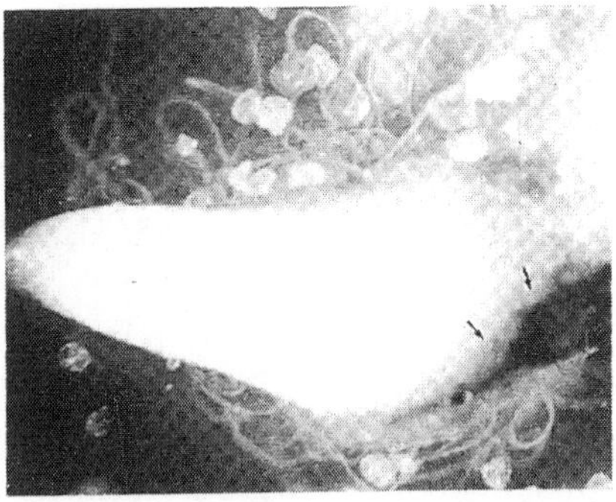

Fig. 5. *X. index* aggregating behind the tip of a rapidly growing *V. vinifera* root. Arrows indicate discoloration of surface cells at feeding sites.

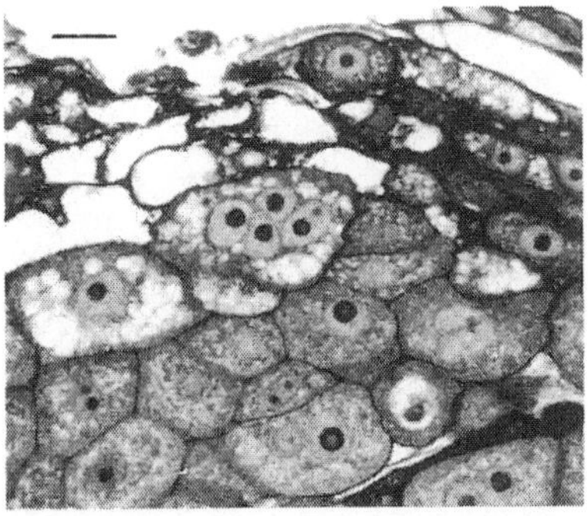

Fig. 6. Longitudinal section through part of a bud-like swelling of a lateral *V. vinifera* root fed upon by *X. index*. Bar: 20 μm.

Maintenance of Monoxenic *X. index* cultures on *Ficus carica*

As already mentioned, fig (*F. carica*) is a good host of *X. index* and due to its easier handling, compared to *Vitis vinifera*, and also due to its thin roots, this plant is an excellent model for detailed observations of the feeding behavior of *X. index* and associated root cell responses (Robertson and Wyss 1983, Wyss 1987).

The establishment of monoxenic *X. index* cultures on *Ficus carica* seedlings, first described by Wyss (1977b, 1978) for observations not exceeding several months, has been modified by Aumann (1997) and is described here in detail: Seeds, obtained from commercially sun-dried fig fruits are surface-sterilized for 1 hr with 4% (w/v) previously paper-filtered sodium hypochlorite and then transferred to Petri dishes containing 0.8% (w/v) water agar. The plates are kept at 25±2°C under long-day top-light conditions (16 hr light). The first seeds germinate after about two weeks and after another two to four weeks seedling plants with a main root length of c. 5 cm are transferred to 9 cm Ø Petri dishes with autoclaved 0.8% (w/v) agar medium prepared with Hoagland No. 2 nutrient solution (Hoagland and Arnon 1938) containing Fe-EDTA as an iron source. Hoagland nutrient medium proved to be most suitable for maintaining monoxenic cultures as any other medium containing sucrose as an energy source was seen to inhibit nematode survival. Six stock solutions are prepared with glass-distilled water as described below and are then stored at 4°C until use.

Preparation of Hoagland nutrient solution No. 2:

Stock solution	Elements	Amount (g/l)	ml used for 1 litre solution
I	KH_2PO_4	136.09	0.1
II	KNO_3	101.11	0.5
III	$Ca(NO_3)_2 \cdot 4H_2O$	236.15	0.5
IV	$MgSO_4 \cdot 7H_2O$	246.48	0.2
V	H_3BO_3 $Mn(II)Cl_2 \cdot 4H_2O$ $ZnSO_4 \cdot 7H_2O$ $Cu(II)SO_4 5H_2O$ $H_2MoO_4 \cdot H_2O$	2.86 1.81 0.22 0.08 0.09	0.1
VI	Fe-EDTA	6.605	0.1

Nematode stock cultures are started by pipetting several hundred hand picked *X. index* together with sterile water to the roots of fig plants (originally raised from aseptic cultures as described above) in 17.6 cm plastic pots filled with fine autoclaved sand. The plants are then cultivated in a growth chamber at 28±2°C under long-day (16 hr light). They are watered when required and are fertilized monthly with Hoagland No. 2 solution. Under these conditions, without competition by other nematodes, *X. index* reproduces extremely well so that after several months hundreds of nematodes can be extracted by the Baermann funnel technique from a handful of sand collected from around galled roots.

Xiphinema index reproduces by meiotic parthenogenesis (Dalmasso and Younes 1969) and hence, after extraction, it is easy to suck about 100 females and late juvenile stages from clean nematode suspensions into a fine glass pipette. These nematodes are transferred to a heat-sterilized staining block for surface sterilization. When they have settled on the bottom, the supernatant water is sucked off and about 4 ml 0.03% (w/v) sodium azide (NaN_3) are added. The nematodes are soon immobilized by this microbial inhibitor but most of them survive a 20 min (Aumann 1997) or even longer exposure (Wyss 1978) when they are thereafter thoroughly washed with sterile, glass-distilled water. The nematodes are then transferred with a sterile microneedle to a single *F. carica* plant grown in a 9 cm Ø Petri dish and placed at some distance from its growing roots. Usually some bacterial contamination develops on the surface where the nematodes have been added, but when nematodes are removed from root-tips on which they have started feeding and are transferred to new Petri dishes with fig plants, these plates stay generally clean. As the Hoagland nutrient medium contains no sucrose, the plants have to be exposed to light. This is a critical factor as *X. index* responds sensitively to higher light intensities (Wyss 1978). Good results in maintaining monoxenic cultures in 9 cm Ø Petri dishes for a period of up to one year have been obtained by using two-chamber dishes with one light-protected chamber filled with Hoagland nutrient agar medium containing growing fig roots and nematodes and a second light-exposed chamber containing the shoots (Aumann 1997).

Feeding Behavior on *Ficus carica*

The Petri dishes with feeding nematodes are placed upside down on the stage of a light microscope for observations at low magnification. Single females usually start feeding on tap roots, preferably near or in the meristematic region (Fig. 7).

Prolonged feeding at different sites along this region causes a retardation of root growth, accompanied by a conspicuous swelling of the root-tip (Fig. 8), which gradually becomes transformed into a terminal gall (Wyss 1977a, 1978, Bleve-Zacheo and Zacheo 1983). The odontostyle is usually inserted two to three cells deep into root-tips and root-tip galls before nutrients are withdrawn by a rapid pumping action of the basal esophageal bulb. *Xiphinema index* always feeds on a column of cells, feeding on each for

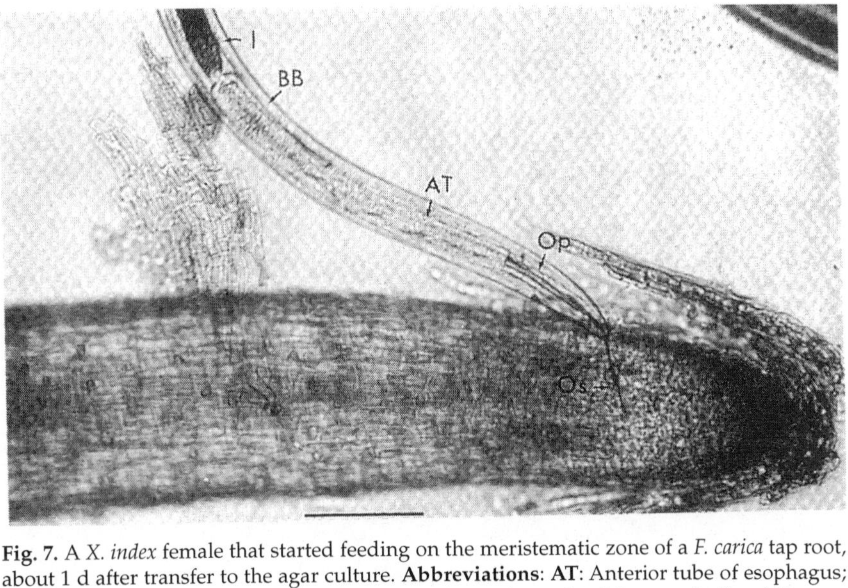

Fig. 7. A *X. index* female that started feeding on the meristematic zone of a *F. carica* tap root, about 1 d after transfer to the agar culture. **Abbreviations**: **AT**: Anterior tube of esophagus; **BB**: Basal bulb of esophagus; **I**: Intestine; **Op**: Odontophore; **Os**: Odontostyle; Bar: 100 µm.

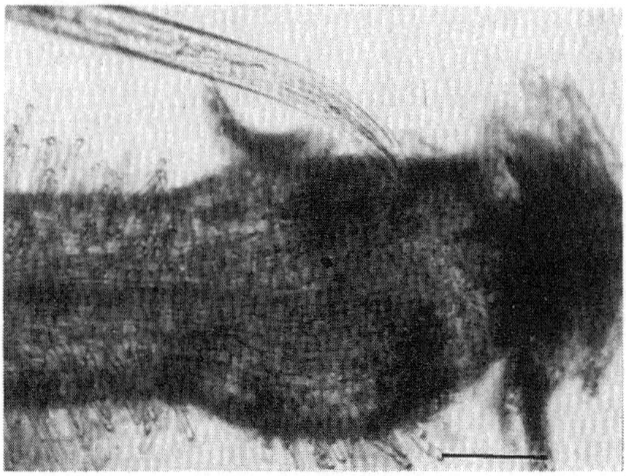

Fig. 8. Same root and female as in Fig. 7, now 22 h after the initial attack. Root-tip stopped growing and root-tip swelling is now evident. Bar: 100 μm.

several minutes and then progressing into successively deeper layers until maximal stylet protrusion.

Egg Production

Root-tip galls remain strongly attractive to individual and also several females (Fig. 9) and they are evidently the appropriate food source for egg production. Figure 10 presents two typical examples of egg production, when the feeding positions of two typical examples of egg production, when the feeding positions of two females, each kept on a single fig plant in separate monoxenic culture, were daily recorded over a period of 30 d. Females I was for instance first seen to feed three days after inoculation

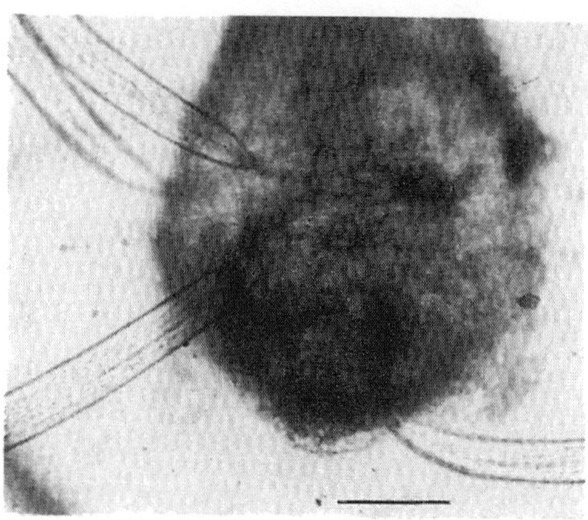

Fig. 9. Four *X. index* females feeding on a *F. carica* root-tip gall. Bar: 100 μm.

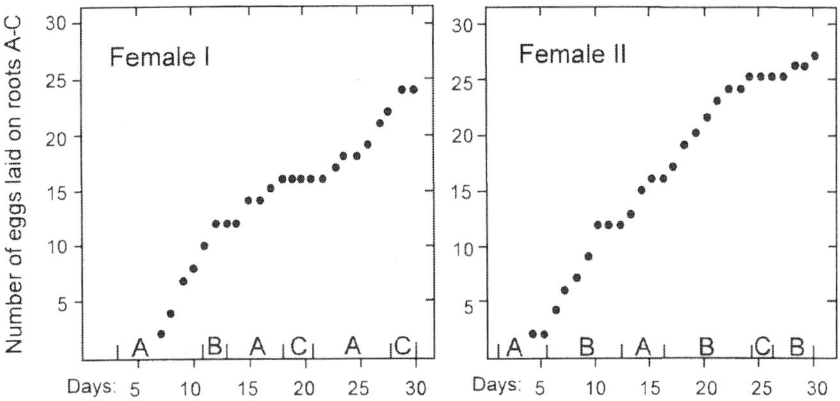

Fig. 10. Feeding positions of two *X. index* females, each kept on a single fig plant in separate monoxenic culture, showing the correlation between egg production and preferred root-tips (galls).

on root-tip A. One d later the tip was slightly swollen and proliferation continued over the following days. The first eggs were laid four d after root-tip gall initiation. Feeding on root-tip B was of short duration and did not stimulate egg production, which was resumed after the female had returned to root-tip gall A. Although feeding on a nearby rooting tip C formed a terminal gall, egg production was not initiated. This was again the case when feeding was continued on root-tip gall A, which had increased sixfold in diameter within 25 d.

A similar case is represented for females II, where root-tip gall B was the principal source for egg production. Root-tip galls are equally attractive to hatched juveniles (J1) and later developmental stages (J2–J4). At 25±1°C, the first stage juveniles hatch on average seven days after egg deposition, and the life cycle between oviposition and the last molt to the young female requires at least 77 days (Wyss 1977a).

Cellular Modifications of Feeding Sites on *Ficus Carica*

Root-tip swellings, induced by a single female 12–24 hr after its first attack, contain groups of necrotic cells surrounded by slightly enlarged binucleate meristematic cells (Fig. 11). The necrotic cells are obviously those cells from which food has been removed, whereas the binucleate state of neighboring cells is induced as a response to a trigger released by the nematode (Wyss et al. 1980).

One day later, necrotic cells are surrounded by enlarged modified cells with 4 nuclei at the plane of section (Fig. 12). The multinucleate state

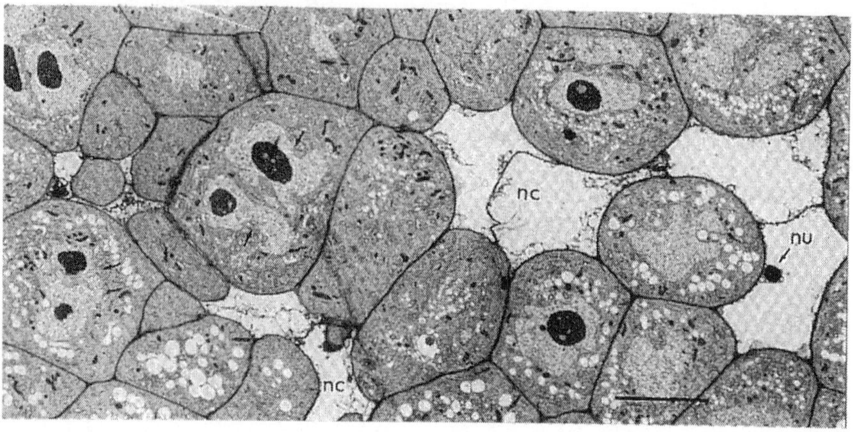

Fig. 11. EM picture of a longitudinal section through a swollen *F. carica* root-tip, 1 day after the first attack by a single *X. index* female. Necrotic empty cells (**nc**) are surrounded by binucleate cells with lobed nuclei (arrows). Degraded nucleoli (**nu**) are commonly left in necrotic cells. Bar: 10 μm.

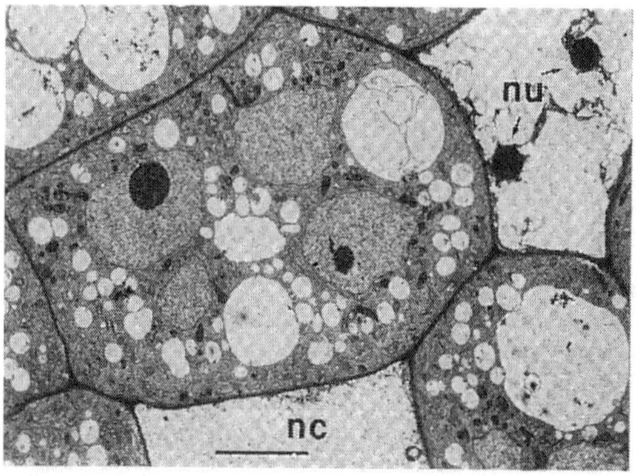

Fig. 12. EM picture of a longitudinal section through a swollen *F. carica* root-tip, 2 days after the first attack by a *X. index* female. Modified cells around necrotic cells (**nc**) with degraded nucleoli (**nu**) now contain 4 nuclei. Bar: 5 μm.

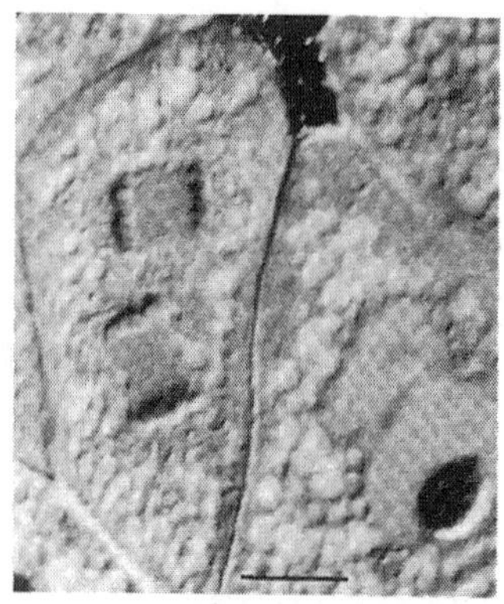

Fig. 13. Binucleate cell undergrowing synchronous mitosis, here at late anaphase. Bar: 10 μm.

of these cells arises by synchronous mitoses (Fig. 13) without cytokinesis (Rumpenhorst and Weischer 1978, Wyss 1978).

When root-tip galls supporting egg production are sectioned and examined, it becomes evident that the multinucleate cells serve as a source of food and that these cells are killed by nematode feeding. Fed on cells are empty (Fig. 14) and usually contain only remnants of degraded nucleoli (Fig. 15). These empty 'giant cells' are frequently located behind a layer of necrotic and collapsed cells (Fig. 14). As all histological studies were from roots maintained under monoxenic conditions, such damage is directly attributable to nematode feeding. Necrosis is never seen to spread from damaged cells to adjacent multinucleate cells, and there is always a clear-cut demarcation between both types of cells (Fig. 14).

Unparasitized multinucleate cells in root-tip galls are packed with nuclei (Fig. 16) and show characteristic features of a high metabolic activity (Fig. 17).

Additional cellular modifications, such as occasional cell wall stubs and wall ingrowths in multinucleate cells, induced by *X. index* in fig roots, have been described and discussed (Wyss et al. 1980). Based on a definition proposed by Bird (1979), the term coenocyte was used for a

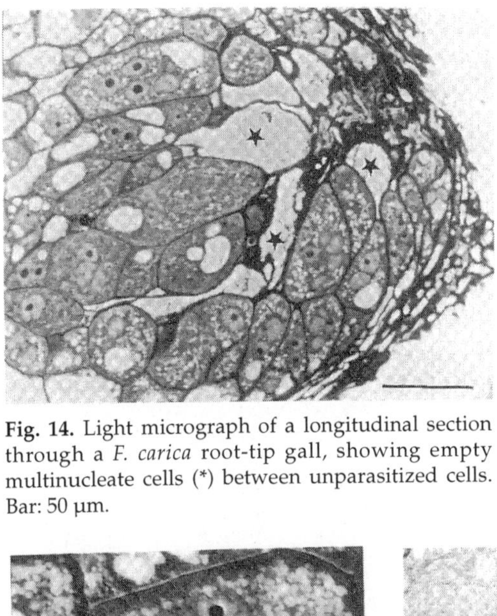

Fig. 14. Light micrograph of a longitudinal section through a *F. carica* root-tip gall, showing empty multinucleate cells (*) between unparasitized cells. Bar: 50 µm.

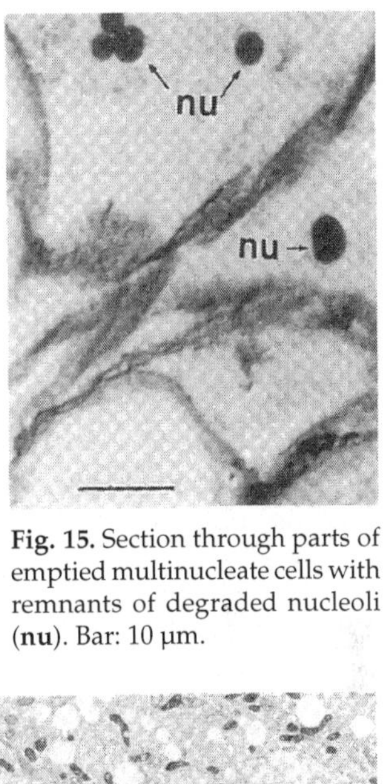

Fig. 15. Section through parts of emptied multinucleate cells with remnants of degraded nucleoli (**nu**). Bar: 10 µm.

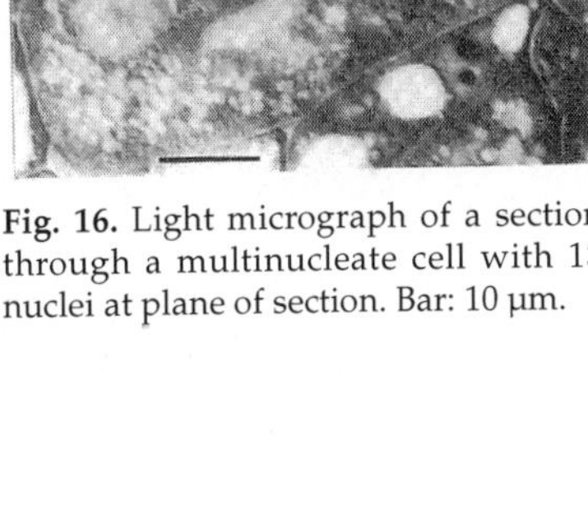

Fig. 16. Light micrograph of a section through a multinucleate cell with 13 nuclei at plane of section. Bar: 10 µm.

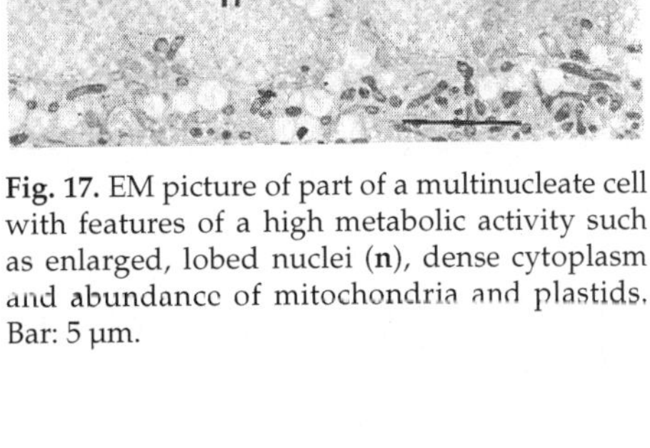

Fig. 17. EM picture of part of a multinucleate cell with features of a high metabolic activity such as enlarged, lobed nuclei (**n**), dense cytoplasm and abundance of mitochondria and plastids. Bar: 5 µm.

multinucleate mass of protoplasm formed by repeated nuclear division without cytokinesis. Thin and rather irregular cell walls were formed in older coenocytes, involving typical phragmoplast and cell plate formation. Cytokinesis in older coenocytes was interpreted as the initial stage of renewed meristematic activities during the formation of new lateral roots which commonly arise from older root-tip galls (Wyss 1978).

Bleve-Zacheo and Zacheo (1983) described similar ultrastructural cell modification induced by *X. index* in fig roots during the early stages of nematode parasitism. They summarized the histopathological changes induced by *X. index* in its host plants (Zacheo and Bleve-Zacheo 1995) and compared them with those induced by other *Xiphinema* spp. on tomato as host plant (Bleve-Zacheo et al. 1987). However, ultrastructural changes may be difficult to interprete precisely when root-tip galls by *Xiphinema* spp. on their host plants are collected at random from pot cultures or when cellular modifications are examined without preceding semi-thin sections that provide a general overview.

Responses of Non-host and Resistant Plants

When *X. index* females were added to tomato seedling roots grown in monoxenic culture, they were seen to feed on this non-host plant for a while in a similar manner as on fig. Attacked root-tips stopped growing and showed some swellings behind the elongation zone (Wyss 1978). However, root-tip galls were not formed, egg deposition was not recorded and the roots were soon abandoned. When attacked root-tips of tomato were sectioned and examined under the light microscope, multinucleate

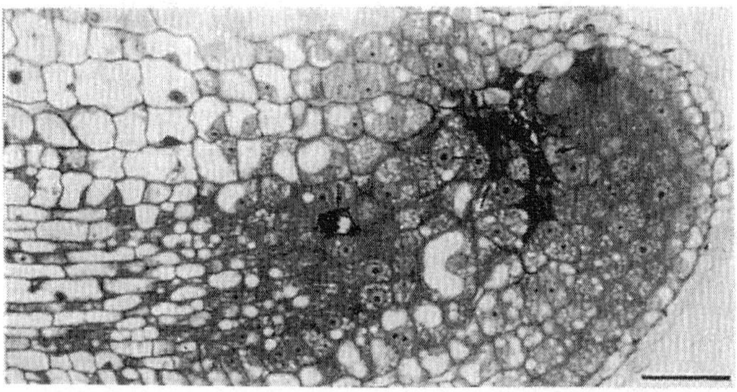

Fig. 18. Light micrograph of a longitudinal section through a tomato seedling root, a non-host, attacked by *X. index* females in monoxenic culture. Only necrotic cells are seen at feeding site (arrows). Adjacent cells are not modified. Bar: 50 μm.

cells were not detected near or distant to feedings sites, which were easily recognized as isolated groups of necrotic cells (Fig. 18).

This non-host plant does obviously not respond to nematode secretions that trigger the induction of multinucleate nurse cells. It can be concluded that this unfavorable response for nematode reproduction will generally occur in non-hosts and resistant plants. In fact it has been shown that roots of *Vitis* spp. with resistance to transmission of GFLV and lack of attraction to *X. index* respond by a necrotic, hypersensitive response to *X. index* feeding (Weischer 1982, Staudt and Weischer 1992).

Analysis of the Feeding Mechanisms of *X. index*

Preparation of Observation Chambers

So far the feeding behavior of *X. index* and associated responses of *F. carica* root-tips to feeding have been described from observations in Petri dishes under the light microscope at low magnification. In order to obtain more precise information on the feeding mechanisms, special chambers, allowing observations under 100x oil immersion objectives, must be prepared. Suitable chambers for short term observations over a few days can be prepared as shown in Fig. 19.

The simplicity of this method, suitable for general feeding observation of root parasitic nematodes (Wyss and Zunke 1985), is based on the use of very thin 80 mm Ø glass coverslips (0.13 mm thickness) on the surface of which seedling roots of host plants are grown in a thin agar layer with a nutrient medium. As already mentioned, Hoagland solution should be used for feeding studies with *X. index*. At the time when the seedling plants have established a well growing root system, nematodes are extracted from stock cultures, surface-sterilized as described before and then transferred singly to the roots with a sterile microneedle. As soon as the nematodes start feeding, the coverslip is removed from the bottom of the Petri dish with a hooked needle or a scalpel (Fig. 19H). An identical coverslip is then carefully placed on top of the seedling so that a thin observation 'sandwich' is formed, which is then placed upside down onto the stage of a light microscope. Feeding on roots growing along the surface along the surface of the coverslip can then easily be observed at high magnification under oil immersion. The chambers may be stored for a few days in Petri dishes sealed with parafilm and kept under low light until further observations.

For long term observations, special chambers should be prepared by using threaded brass rings between which the observation 'sandwiches' are fixed (Wyss 1992). All procedures for preparing the 'sandwiches' are identical to those described before, with the exception that a 80 mm Ø autoclaved rubber ring (1 mm thick and 3 mm wide) is now laid between

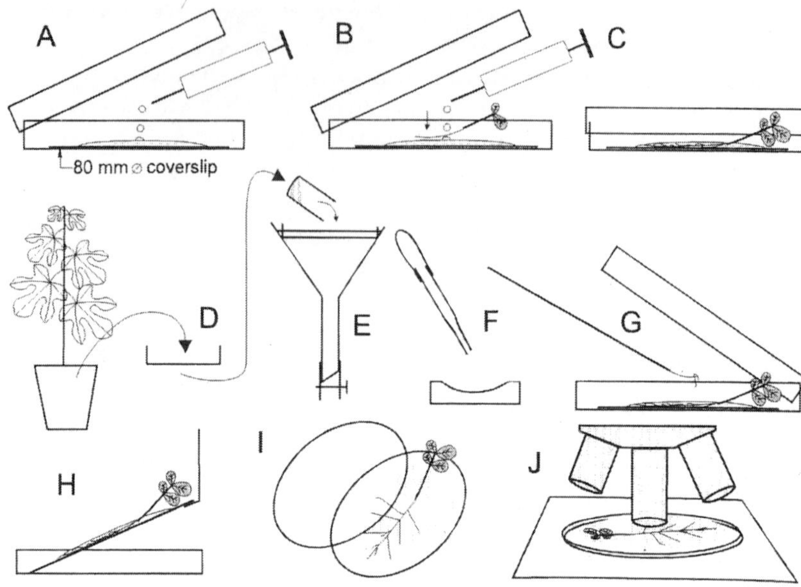

Fig. 19. Preparation of simple observation chambers to study the feedings behavior of *X. index* at high magnification under the microscope. **A**: Preparation of Petri dish for fig seedlings by placing a 80 mm Ø glass coverslip of 0.13 mm thickness on bottom of Petri dish and then adding about 5 ml 0.8% distilled water agar on cover slip; **B**: Transfer of a fig seedling onto agar layer and then adding a few drops of Hoagland solution; **C**: Storage of Petri dishes at 25°C under light; **D**: Extraction of *X. index* from fig stock cultures, followed by sieving and decantation; **E**: Final extraction in Baermann funnel; **F**: Surface sterilization of nematodes in a staining block; **G**: Transfer of hand picked nematodes to fig seedlings; **H**: Removal of coverslip with fig seedling after nematodes have started to feed on its roots; **I**: Placing an identical coverslip on top of coverslip with seedling roots and feeding nematodes, thus forming a 'sandwich' observation chamber; **J**: Placing the sandwich upside down onto the stage of a microscope for observation under a 100 x oil immersion objective. Further explanations in text.

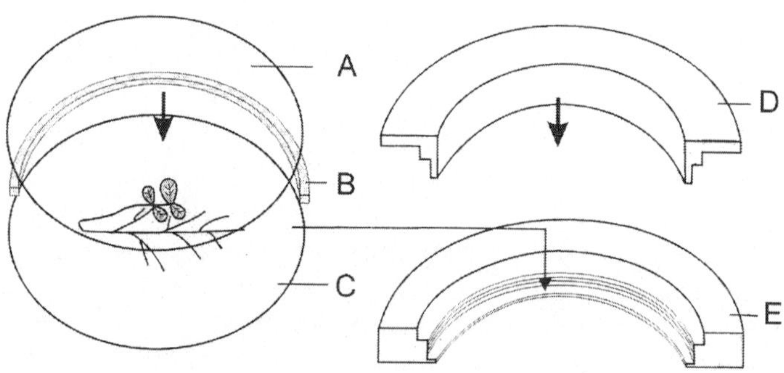

Fig. 20. Preparation and assembly of observation chambers to study the feeding behavior of *X. index* at high magnification under the microscope over longer periods. **A**: Upper glass coverslip (Ø 80 mm and 0.13 mm thick); **B**: Rubber ring; **C**: Lower glass coverslip with fig seedling in a thin agar layer with nutrient medium and nematodes feeding on roots; **D**: Upper brass ring; **E**: Lower brass ring.

the lower and upper glass coverslip. The 'sandwich' is then placed onto the brim of a threaded 90 mm Ø brass ring (Fig. 20) and fixed by twisting the upper threaded brass ring into the lower ring. These chambers are very stable and can be used over longer periods than the previous more simple chambers. Their only disadvantage is that they require specially designed brass rings.

Feeding Mechanisms

When feeding of *X. index* was examined and recorded with ciné film at high magnification (Wyss 1997b, Wyss and Inst. Wiss. Film 1977), special features in the nematode's basal bulb became evident, indicating that this pumping organ for food ingestion also harbors a complex system of ducts, to be assigned to the dorsal secretory gland cell. A few seconds after odontostyle insertion into a cell, the bulb is suddenly and ducts become distinctly dilated, especially at the anterior end of the bulb. These ducts remain dilated for a few seconds, then they narrow from back to front, forcing fluids through them. After duct depletion, the bulb persists in the stretched position for a few further seconds, then it shortness and is immediately followed by a pumping action of the basal bulb for food ingestion. Figure 21 shows the two main dilated ducts of the dorsal gland that extend along almost the entire length of the bulb on the dorsal side and are connected to a transverse collecting duct at the anterior end of the bulb.

The basal bulb pumps rapidly during food ingestion. Rates of radial muscle contractions that dilate the lumen in the pump chamber within the bulb are on average 3 per second. Dilation of this chamber is about four times slower than its rapid closure when the radial muscles relax and food is forced into the foregut of the intestine via a one-way intestinal valve (Fig.

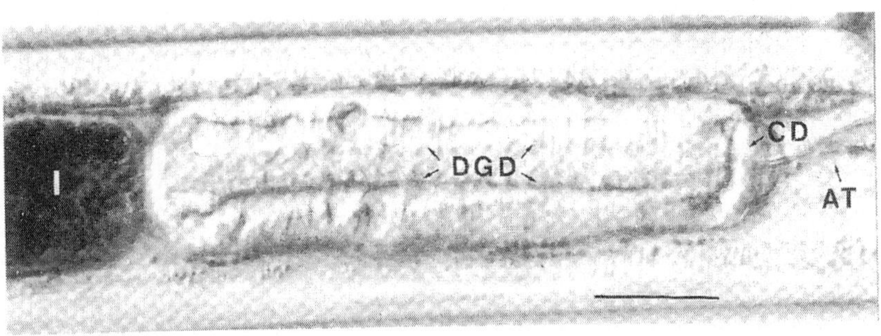

Fig. 21. Dorsal view of the basal bulb of *X. index*, showing the two main ducts of the dorsal gland cell and the transverse collecting duct at the anterior end of the bulb. **Abbreviations**: **AT**: Anterior tube of the esophagus; **CD**: Collecting duct; **DGD**: Dorsal gland ducts; **I**: Intestine. Bar: 20 μm.

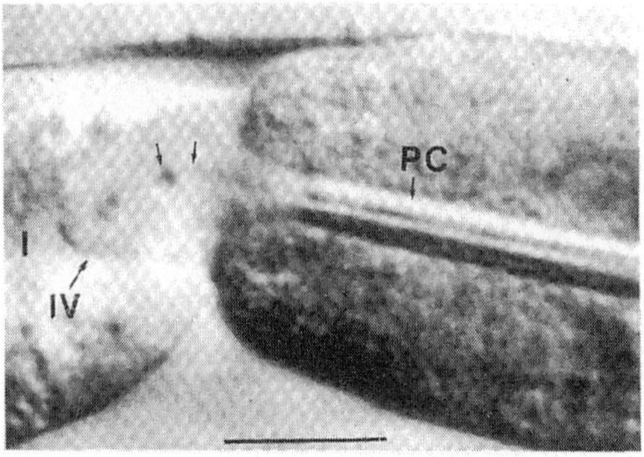

Fig. 22. Posterior end of basal bulb, showing passage of ingested food (arrows) through the one-way intestinal valve into the intestine. **Abbreviations**: **I**: Intestine; **IV**: Esophago-intestinal valve; **PC**: Pump chamber. Bar: 10 μm.

22) as the pump chamber narrows from front to back (Wyss 1977b, Wyss and Inst. Wiss. Film 1977).

Pumping is typically inter-mittent; periods of continuous bulb contractions are always interrupted by short pauses during which secretory fluids within the dorsal gland ducts are rapidly flushed toward and through the main collecting duct at the anterior end of the bulb. Whenever pumping is resumed after a very short pause, the ducts become again dilated (Wyss 1977b, Wyss and Inst. Wiss. Film 1977).

Structure and Function of the Basal Esophageal Bulb in *X. index*

The ultrastructure and possible function of the feeding apparatus in *X. index*, first described by Wright (1965) and Roggen et al. (1967), have been reviewed by Robertson and Taylor (1975), Robertson and Wyss (1983), Taylor and Brown (1997). Wright et al. (1983) described in detail the musculature of the stylet of *Xiphinema* spp. Here emphasis is placed on the basal bulb, which serves as the suction pump for food ingestion, and in addition harbors three secretory gland cells that play a decisive role in feeding/digesting processes. Figure 23 represents schematically the basic features of the bulb.

The basal bulb in adult *X. index* females is cylindrical in shape, about 100 μm long and is composed mainly of muscle, gland and nerve cells. At the most anterior end, the lumen (food canal) of the bulb is still circular in cross section (Figs. 1, 24A), but in the pump chamber that extends the bulb for most of its entire length (Fig. 23B), the lumen is triradiate with a thickened cuticle in the form of three pairs of triangular platelets to which six sets of radial muscles are attached (Figs. 23D, 24B, C). In this pump chamber

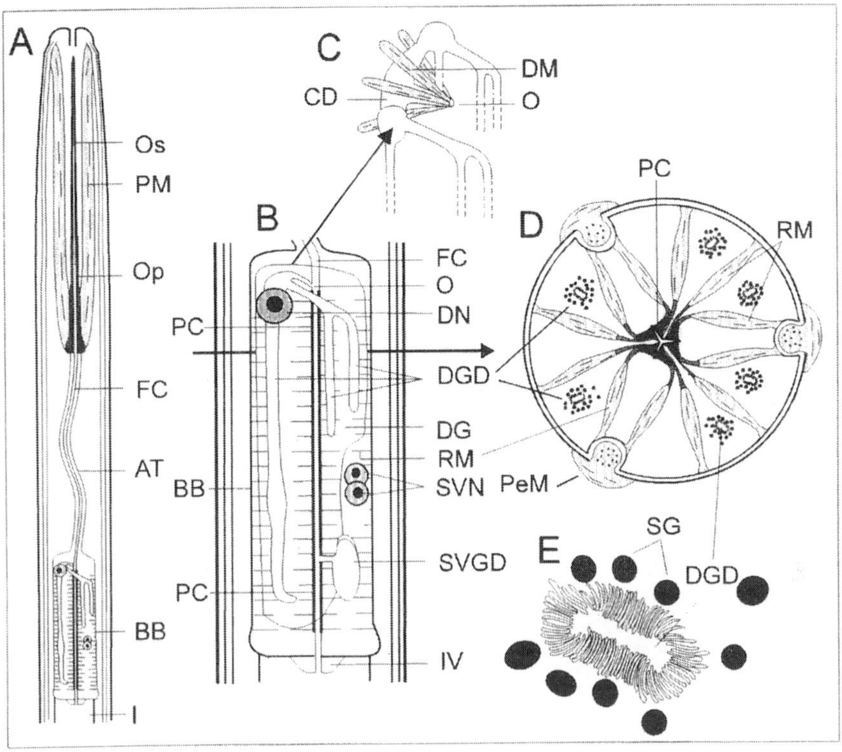

Fig. 23. Structure of the basal bulb in *X. index*, schematic representations. **A**: General overview of the feeding apparatus; **B**: Magnification of the basal bulb, showing size of the dorsal esophageal gland and its ducts (differentially shaded areas); **C**: Collecting main duct of dorsal gland ducts with orifice; **D**: Cross section through basal bulb at level indicated by arrow (modified from a drawing by Robertson and Taylor 1975); **E**: Enlargement of one of the six dorsal gland ducts. **Abbreviations**: **AT**: Anterior tube of esophagus; **BB**: Basal bulb of esophagus; **CD**: Collecting main duct of dorsal gland duct system; **DG**: Dorsal esophageal gland (shaded); **DGD**: Dorsal gland ducts (shaded); **DGD**: Dorsal gland ducts (shaded); **DM**: Dilator muscles that open main duct orifice of dorsal gland; **DN**: Dorsal gland nucleus; **FC**: Food canal; **I**: Intestine; **IV**: Esophago-intestinal valve; **RM**: Radial muscle of basal bulb; **O**: Orifice of collecting main duct of dorsal gland; **Op**: Odontophore; **Os**: Odontostyle; **PC**: Pump chamber in basal bulb; **PeM**: Peripheral muscles of basal bulb; **PM**: Protractor muscles of stylet; **SG**: Secretory granules; **SVGD**: Subventral gland ducts; **SVN**: Nuclei of the two subventral glands.

suction for food ingestion is created when the radial muscles contract and thus dilate the lumen of the food canal to a triangular cross section (Fig. 24C) as proposed by Robertson and Taylor (1975) and substantiated by Towle and Doncaster (1978) and Robertson et al. (1987) for the basal bulb of another longidorid nematode, *Longidorus caespiticola*. Three groups of peripheral muscle, attached along the bulb to the outside of its outside of its membrane (Fig. 23D), have been suggested to act as springs to conserve the volume of the bulb during pumping (Robertson et al. 1987). An analysed from ciné films (Wyss and Inst. Wiss. Film 1977, Towle and Doncaster 1978), the basal bulb of longidorid nematodes elongates upon contraction of the radial muscles and shortens again when the muscles relax. Pumping in

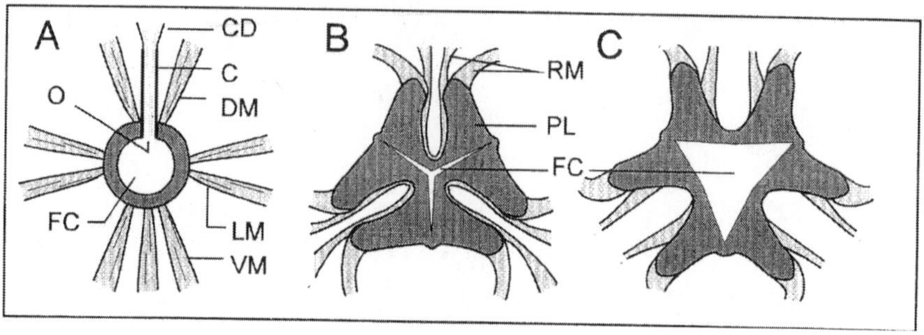

Fig. 24. Schematic representation of cross sections through the esophageal lumen (food canal) in the basal bulb of a *X. index* female. **A**: Most anterior part of the bulb in front of the triradiate pump chamber. **B-C**: Pump chamber, **B**: Chamber closed when radial muscles attached to the triangular platelets of the thickened cuticle are relaxed. **C**: Chamber dilated to a triangular (and not possible hexagonal) cross section upon contraction of the radial muscles. **Abbreviations**: C: Cuticularized part of the collecting main duct of the dorsal gland duct system; **CD**: Basal part of the collecting main duct, **DM**: Dilator muscles that open main duct orifice of dorsal gland; **FC**: Food canal; LM: Radial muscles attached to lateral side of food canal; **O**: Orifice of collecting main duct of dorsal gland into food canal; **PL**: Platelet of thickened food canal cuticle in pump chamber; **RM**: Extensions of radial muscles attached to platelets; **VM**: Radial muscles attached to ventral side of food canal.

longidorid is apparently driven entirely by the radial muscles (Robertson et al. 1987).

Structure and Function of the Dorsal Gland Cell

As previously described, observations of the basal bulb just prior and during pumping revealed the existence of a complex duct system of the dorsal gland in *X. index*. Subsequent analysis of the ultrastructure and possible function of this gland cell (Robertson and Wyss 1979) showed that it is the largest structure within the bulb, extending along the entire length on the dorsal side and half-way along the ventral side (Fig. 23B). The membrane of this gland is highly convoluted to fit between and around the radial muscles and nerves. The bulb also harbors two subventral gland cells, which, however, are smaller, occupying the posterior, ventral half of the bulb with their ducts opening into the food canal in the posterior region of the bulb (Fig. 23B).

Just anterior to its nucleus, the duct system of the dorsal gland comprises a main, transverse and fan-shaped collecting dust (Figs. 21, 23C,

25A, B) which is fed by six longitudinal ducts. Two of these extend along the whole length of the bulb in a subdorsal position (Figs. 21, 23B, D) and two pairs extend the anterior third of the bulb, one along the lateral and the other along the ventral side (Fig. 23B, D). The lining of the ducts has many longitudinal folds (Figs. 23E, 26) to allow extensive duct dilation. Numerous secretory granules surround these folds in feeding nematodes.

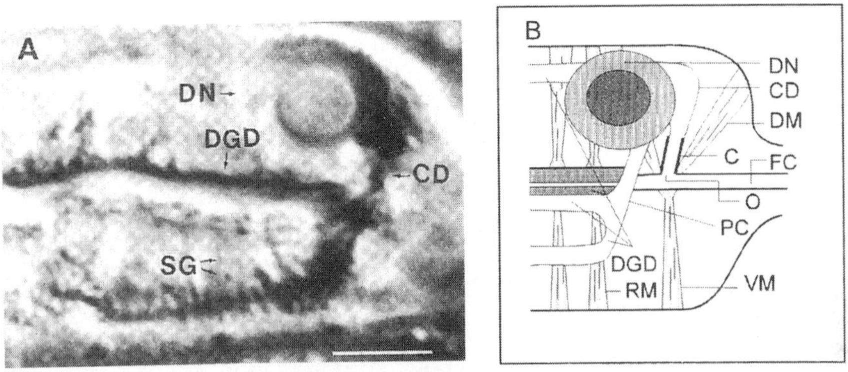

Fig. 25. Dilation of dorsal gland ducts in anterior part of bulb, a few seconds after stylet insertion into a root cell. **A**: As seen under the light microscope, Bar: 5 µm; **B**: Schematic representation. **Abbreviations**: **C**: Cuticle around base of collecting main duct; **CD**: Collecting main duct; **DGD**: Dorsal gland ducts; **DM**: One of four dilator muscles that open main duct orifice into food canal; **DN**: Dorsal gland nucleus; **FC**: Food canal; **O**: Orifice of collecting main duct; **PC**: Pump chamber; **RM**: Radial muscles attached to pump chamber; **SG**: Secretory granules; **VM**: One of three radial muscles attached to ventral side of food canal anterior to pump chamber. Lateral radial muscles (s. Fig. 24A) not shown.

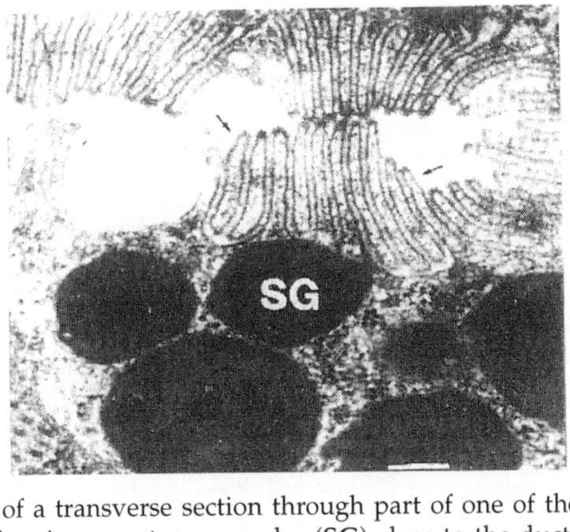

Fig. 26. EM picture of a transverse section through part of one of the ducts of the dorsal esophageal gland, showing secretory granules (**SG**) close to the duct folds (Arrows). Bar: 0,2 µm.

An intensive staining for carbohydrates was detected in the membrane lining of the ducts, whereas the membranes of secretory granules remained unstained (Robertson 1987).

Exocytosis of the secretory granules around the ducts, especially in the bulb's anterior region only occurs after odontostyle insertion into a root cell, when the bulb is suddenly stretched and the ducts become much dilated (Wyss et al. 1988, Wyss and Inst. Wiss. Film 1988). Figures 25A and B show such a moment.

The collecting main duct of the dorsal gland enters the food canal just anterior to the pump chamber (Figs. 23B, 25B), where the lumen is still circular in cross section (Fig. 24A). The orifice is a longitudinal slit, about 0.6 μm long. This slit and the basal part of the main duct is lined with a thin layer of cuticle (Figs. 24A, 25B), which may act as a one-way valve (Roberston and Wyss 1979, 1983). The slit is opened by the contraction of four dilator muscles (Fig. 23C) that extend radially to the periphery of the bulb (Fig. 25B). Radial muscles also attach to the lateral and ventral side of the food canal (Fig. 24A). These muscles appear to contract in unison to keep the food canal in position when the dilator muscles open the main duct orifice. Simultaneous contraction of all these radial muscles most likely causes the sudden stretching of the bulb at its anterior end when the orifice is opened.

As already mentioned, exocytosis of the secretory granules surrounding the ducts is only observed when the ducts dilate for the first time after odontostyle insertion into a root cell. Nerve cells found in close association with the ducts at the anterior end of the bulb are most likely involved in regulating exocytosis and they may also detect the degree of duct dilation for a controlled discharge of the secretory fluids into the food canal. This discharge is most pronounced when the bulb suddenly stops pumping to release the contents of the re-filled ducts through the main duct orifice.

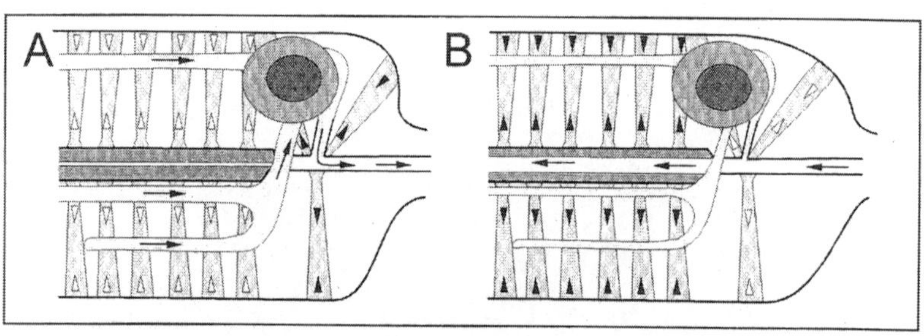

Fig. 27. Schematic representation of the function of the basal bulb muscles in *X. index* during **A:** Release of secretory fluids from the dorsal gland system, when the bulb suddenly stops pumping and **B:** At the beginning of resumed pumping, when the ducts are not yet re-filled with secretory fluids. Arrows in A represent the flow of secretory fluids and those in B the flow of ingested food. Black triangles indicate contracted and white triangles indicated relaxed muscles. Further explanations in text. See Fig. 25B for description of the muscles involved.

Secretory fluids are then seen to be flushed forward through the ducts at great speed (Wyss 1977b, Wyss et al. 1988, Wyss and Inst. Wiss. Film 1988). Within the food canal these fluids will be forced forward (Fig. 27A) to be injected into the perforated root cell because the radial muscles attached to the pump chamber are then relaxed and the lumen within this chamber is closed (Fig. 24B). As soon as pumping is resumed, fluids of the food cell will be drawn into the widely opened lumen of the pump chamber (Fig. 24C) when the radial muscles attached to it contract (Fig. 27B). The function of the basal bulb in longidorid nematodes for food transport toward the intestine has been discussed at length (Towle and Doncaster 1978 Seymour 1983, Robertson et al. 1987).

Figure 28 summarizes a typical sequence of events in the basal bulb and dorsal gland of a *X. index* female, feeding on a column of meristematic cells of a *F. carica* root-tip.

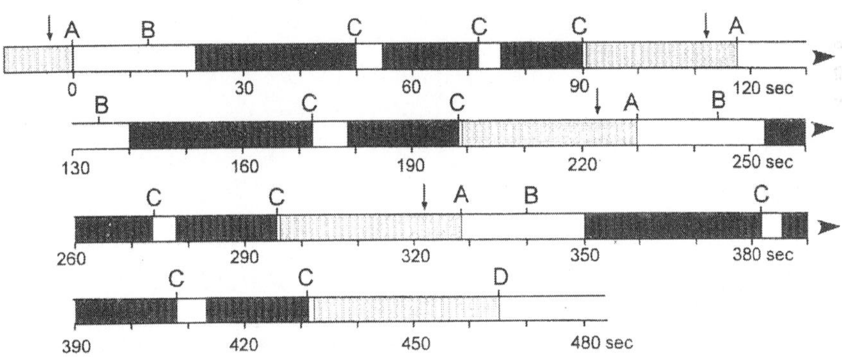

Fig. 28. Characteristic events occurring in the basal bulb and the dorsal gland duct system of a *X. index* female, feeding on a column of four progressively deeper meristematic cells of a *F. carica* root-tip Shaded areas: Bulb pulled forward as odontostyle is pushed deeper. **White areas**: Injection of secretory fluids from dorsal gland. **Dark areas**: Bulb pumping for food ingestion, accompanied by a dilation of the dorsal gland duct system. **Arrows**: Approximate time when the wall of the next deeper cell has been perforated by the odontostyle. **A**: Odontostyle protraction has stopped, followed by bulb stretching and dilation of dorsal gland ducts; **B**: Dorsal gland ducts depleted; **C**: Sudden contraction of dilator muscles that open main duct orifice of the dorsal gland; **D**: Retraction of odontostyle.

The Two Subventral Gland Cells

Very little is known about the structure and function of the duct system of the two subventral glands, which enter the lumen of the pump chamber in the posterior half of the bulb (Fig. 23B). The nucleus of one of these glands and its associated ducts, densely surrounded by secretory granules, is

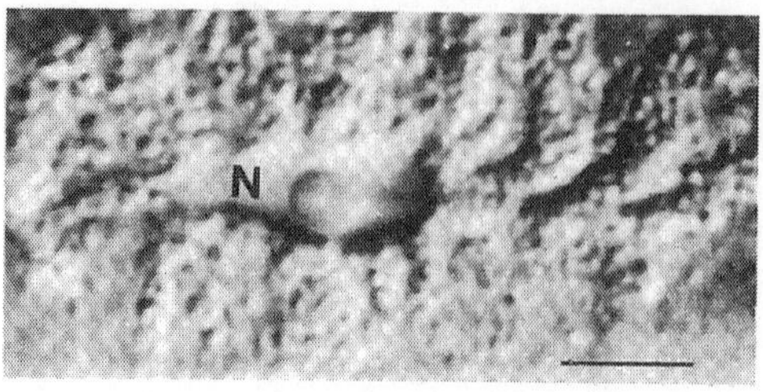

Fig. 29. Part of one of the two subventral gland cell, showing its nucleus (**N**) and associated gland system surrounded by numerous secretory granules. Bar: 5 µm.

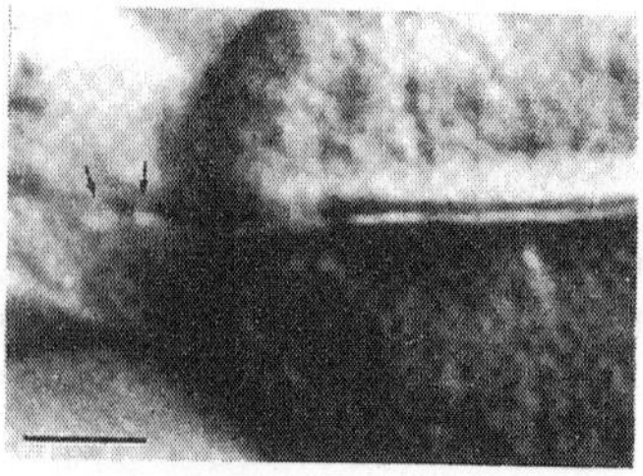

Fig. 30. Posterior end of bulb. Arrows point to secretory fluids flowing through esophago-intestinal valve. Bar: 5 µm.

shown in Fig. 29. It is generally assumed that the secretions of these glands are passed into the intestine for digestive purposes. When the ducts of the dorsal gland dilate for the first time after odontostyle insertion into a root cell, a steady flow of fluids through the esophago-intestinal valve can be seen when the posterior end of the bulb is observed (Fig. 30). The platelets within the closed pump chamber are then slightly opened by the backward flowing fluids, which have been attributed to be products of the two subventral glands (Wyss and Inst. Wiss. Film 1988).

Quiescent Periods

In addition to this general feeding pattern, described in detail, long quiescent periods are commonly observed when *X. index* feeds on unmodified and modified root-tips. These periods of apparent bulb inactivity last on average 15 min (Wyss et al. 1988) and they are also common in other *Xiphinema* species, such as *X. diversicaudatum* (Trudgill 1967, Trudgill et al. 1991), *X. vulgare* (Hunt and Towle 1979) and *X. bakeri* (Sutherland 1969). In *X. index* they appear to occur at random, but are only observed just after stylet insertion into a root cell. The ducts of the dorsal gland then stay slightly dilated and, apart from an occasional twitching of muscles near the gland's nucleus and a slight movement of secretory granules then not seen to break down, the dorsal gland appears to be completely inactive. The ducts of the two subventral glands are, as always, surrounded by numerous secretory granules, and some of them are actually seen to break down. Quiescent periods are always followed by much prolonged periods of basal bulb pumpings, which can last up to 39 min in *X. index* and are still typically intermittent, i.e., periods of ingestion alternate with short pauses during which the ducts of the dorsal gland are depleted as usual (Wyss et al. 1988). In *X. diversicaudatum*, however, such pauses have not been observed (Trudgill et al. 1991).

In vivo Observations of Root Cell Response to Feeding

High resolution video-enhanced contrast light microscopy, which proved to be an excellent tool to study the dynamics of cellular responses to juveniles of a cyst nematode, *Heterodera schachtii*, feeding in the vascular cylinder of the root (Wyss and Zunke 1986b), was used to analyze the impact of the dorsal gland secretions on root cells fed upon by *X. index*. Previous to the application of this technique (Wyss and Zunke 1986a), it was not possible to ascertain with conventional interference light of these secretions are actually injected into food cells and nothing was known about their effect on the cell's protoplast. The video system used is composed of the following elements (Fig. 31):

The observations with this technique confirmed that secretions from the dorsal gland are injected into root cells and that these secretions are highly destructive as they rapidly liquefy the cytoplasm for food removal (Wyss et al. 1988, Wyss and Inst. Wiss. Film 1988). The thin walls of meristematic cells are perforated within seconds by the stylet-tip, which is then pushed just few µm deep into the cytoplasm, where it stays at rest throughout feeding. 12–19 s later secretory fluids are released for the first time through the orifice of the stylet-tip. These secretions cause immediate changes in the cytoplasm around the stylet-tip as evident by a loss of structure and an apparent decrease in viscosity (Fig. 32A). About 10 s later fluids are ingested by basal bulb pumping, recognized by vibrations of the cell wall and oscillations of the nucleus. 15-40 min later secretions are again injected, now however, with much greater volume and force. They further liquefy the cytoplasm and also the nucleoplasm of the nearby nucleus. After 2–3 additional injections and ingestion pumpings the cell in nearly empty and

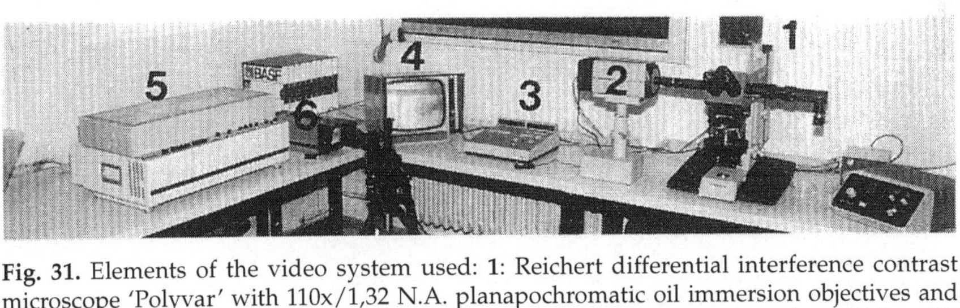

Fig. 31. Elements of the video system used: **1**: Reichert differential interference contrast microscope 'Polyvar' with 110x/1,32 N.A. planapochromatic oil immersion objectives and optovar up to 2x; **2**: Monochrome camera system Grunding FA 76I, converted to 875 lines/50 Hz vertical frequency and equipped with a 1-inch Newvicon pickup tube (light sensitivity 0.1 1x); **3**: Remote control Grunding FB 400 E to control all functions; **4**: Picture monitor Grunding BG 330 (screen diagonal 31 cm) converted to 875 lines/50 Hz vertical frequency; **5**: Professional 1-inch video recorder Grunding BK 411, 10 MHz resolution, converted to 875 lines/50 Hz. For recordings 1-inch chromium dioxide BASF tapes CV 26R are used; **6**: Picture monitor Grunding BG 12 (screen diagonal 12.8 cm), converted to 875 lines/50 Hz vertical frequency. From this monitor photographs or ciné film pictures with a 35 mm camera are taken.

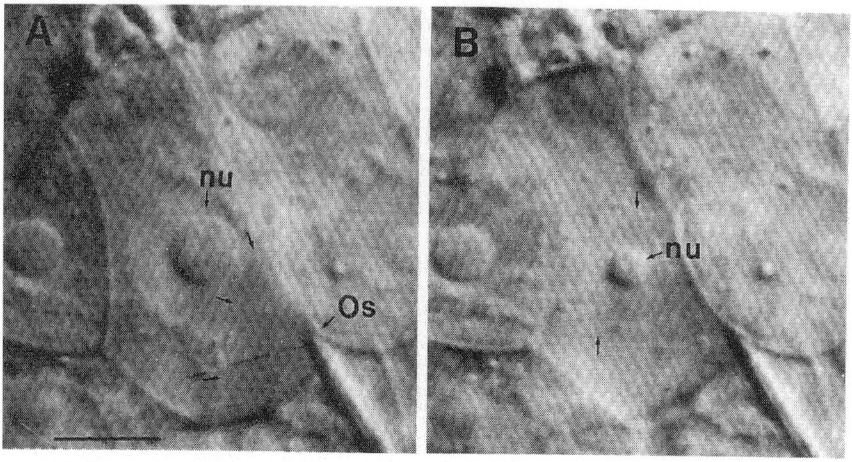

Fig. 32. Responses of a meristematic root-tip cell of *F. carica* to dorsal gland secretions of a *X. index* female. **A**: 25 sec after cell wall perforation by the tip of the odontostyle (**Os**). Zone of cytoplasm liquefied by the first injection of secretory fluids marked by arrows. **B**: 74 sec later; cell now nearly emptied by three further injections and nucleolus (**nu**) reduced to a dense core within the still intact nuclear envelope marked by arrows. Bar: 5 μm.

the nucleolus, still retained in the nuclear envelope, has shrunk to a dense core (Fig. 32B). The odontostyle is then pushed through the dead cell by a series of short jabs at a rate of 2–3 per s until it hits the wall of the next deeper cell, which, when perforated, suffers the same fate. In general the contents of meristematic cells are consumed within two minutes by 2–4 ingestion spells and a total of 3–5 injections of dorsal gland secretions (Wyss et al. 1988).

The pattern of events in individual meristematic cells, fed upon by *X. index*, corresponds precisely to the observed in the dorsal gland within the basal bulb, both of which, however, cannot be recorded simultaneously at high magnification. The first injection of secretory fluids into a root cell, at the time when secretory granules around the ducts break down, is rather gentle. The following injections, especially the second, are much more forceful, and they coincide with the rapid discharge of secretory fluids through the ducts and duct orifice. Figure 33 summarizes the coincidence of the patterns within the bulb and a food cell.

Figure 34 illustrates a typical example of a forceful injection of secretory fluids.

One day after the initial attack, binucleate cells are clearly visible in swollen root-tips. Figure 35 compares binucleate cells in semi-thin sections of roots embedded in resin with those observed *in vivo*, demonstrating the potential of the video system used. In this respect it is interesting to note that the amoeboid-shaped nuclei, commonly seen in semi-thin sections

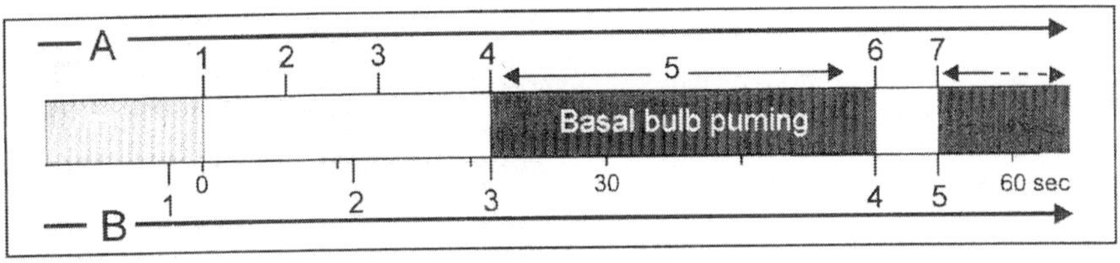

Fig. 33. Combined analysis of **A**: Events occurring in the basal bulb and dorsal gland ducts of a *X. index* female (same first min as in Fig. 28) and **B**: Events occurring simultaneously in a meristematic root cell fed upon by a *X. index* female. **Explanation of numbers in A**: **1**: Bulb no longer pulled forward, ducts of dorsal gland normal size; **2**: Bulb suddenly stretched, ducts dilated, exocytosis of secretory granules; **3**: Ducts depleted, bulb still stretched; **4**: Bulb shortened, followed immediately by bulb pumping; **5**: Ducts become progressively dilated; **6**: Sudden contraction of main duct dilator muscles, secretory fluids discharged through ducts and duct orifice at great speed; **7**: Bulb pumping resumed. **Explanation of numbers in B**: **1**: Cell wall just perforated by odontostyle; **2**: Gentle release of secretory fluids through odontostyle orifice, rapid liquefaction of adjacent cytoplasm; **3**: Commencement of food ingestion; **4**: Forceful injection of secretory fluids; **5**: Ingestion of liquefied cell contents.

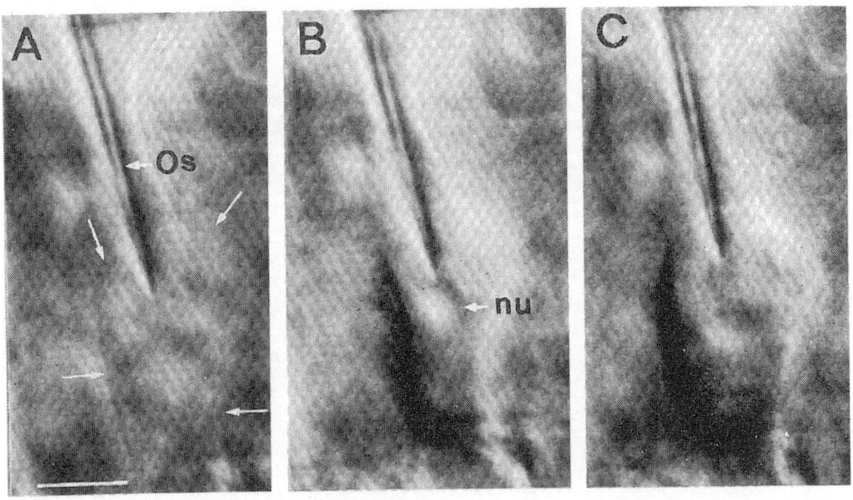

Fig. 34. Responses of a meristematic root-tip cell of *F. carica* to dorsal gland secretions *X. index* female. **A**: Tip of odotostyle (**Os**) just inserted into a meristematic cell marked by four arrows; **B**: 58 sec later, contents of cell partially ingested, nucleolus (**nu**) within nuclear membrane pulled toward tip of the odontostyle, here just before second injection of secretions; **C**: One s after B, indicating force and amount of secretion injected. Bar: 5 µm.

as shown here, and in root-tip swelling of other host plants fed upon by another species, *X. diversicaudatum* (Griffiths and Robertson 1988) are not evident in living cells.

Feeding on binucleate cells is similar to that on uninucleate meristematic cell but now lasts longer and more injections of secretory fluids from the dorsal gland (on average 12) are required until these cells are emptied (Wyss et al. 1988). This indicates an increased density of the cytoplasm in modified cells, which is now more 'resistant' against a rapid liquefaction. As shown by Griffiths and Robertson (1984a), the DNA content of nuclei increases in binucleate cells and this is accompanied by an increase in the proportion of cytoplasm of the root-tip volume (Griffiths and Robertson 1988).

Figure 36 shows a binucleate cell into which secretions have already been injected seven times, but the cytoplasm is not yet fully liquefied and the two nucleoli are not yet degraded to dense residues.

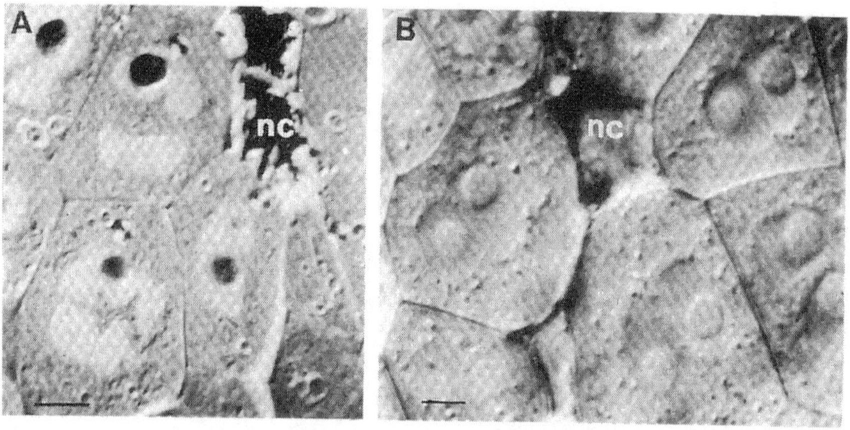

Fig. 35. Binucleate cells in a *F. carica* root-tip swelling surrounding necrotic cells (**nc**), 1 day after the first attack by a *X. index* female. **A**: Semi-thin section through a root embedded in resin; **B**: As seen *in vivo* with video-enhanced light microscopy. Bar: 5 µm.

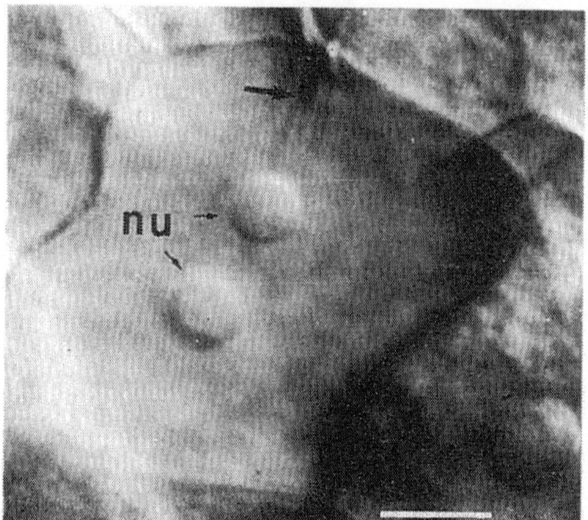

Fig. 36. *X. index* feeding on a binucleate cell which is less rapidly degraded than a uninucleate cell. Arrow points to tip of inserted odontostyle. **nu**: Partially degraded nucleoli. Bar: 5 µm.

Cell responses in multinucleate cells within fig root-tip galls are difficult to observe in detail with the video system used, but feeding on such cells lasts many minutes and requires numerous injections of dorsal gland secretions.

Quiescent Periods

As previously mentioned, these periods of apparent bulb inactivity occur at irregular intervals after odontostyle insertion into cell. Secretions are then seen to emanate through the orifice of the odontostyle, forming a plug-like structure around the protruded stylet-tip (Wyss et al. 1988), but otherwise no obvious changes are visible within the perforated cell (Fig. 37A). When bulb pumping finally commences, the usual forceful injections of secretory fluids from the dorsal gland are attenuated by the presence of the plug, which has never been seen to become dislodged from the stylet-tip. Degradation of the cytoplasm and nucleus within the cell containing the plug is considerably retarded, which explains the much prolonged ingestion duration after a quiescent period. As the forceful injections of dorsal gland secretions are impeded by the plug, some of these secretions obviously escape through the narrow slit of the odontostyle into an unaffected cell through which the odontostyle has been pushed without removing food.

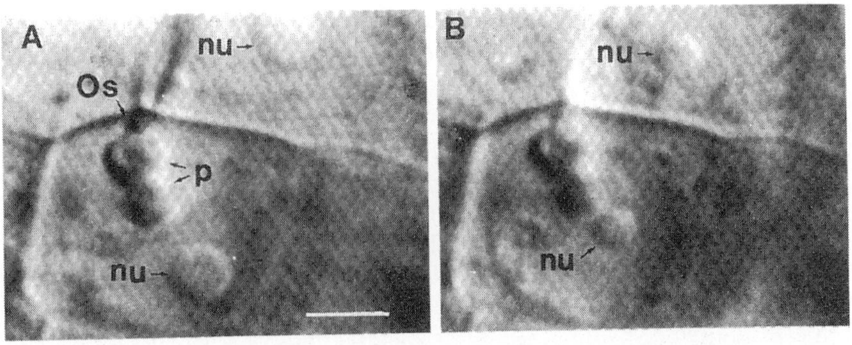

Fig. 37. Response of a meristematic cell just at the end of a quiescent period (**A**) and after subsequent injections of dorsal gland secretions during food withdrawal (**B**). **A**: Just prior to food ingestion after a 20 min quiescent period. Secretions steadily released through the odontostyle orifice during this period have hardened into a plug-like material (**p**) surrounding the tip of the odotostyle (**Os**). Nucleolus (**nu**) of penetrated cell and cell above, through which the odontostyle has been pushed without removing food, still unaffected. **B**: 8 min after commencement of food ingestion and 22 dorsal gland secretions with stylet-tip just retracted. Nucleoli of both cells now degraded. Bar: 5 μm.

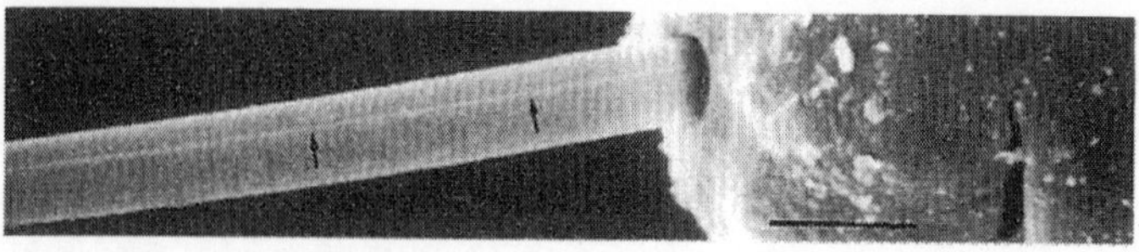

Fig. 38. Scanning electron micrograph of a protruded *X. index* odontostyle. Slit along stylet marked by arrows. Bar: 2 μm.

The protoplast of this cell is then gradually degraded in a similar way as the cell with plug (Fig. 37B).

The narrow slit observed in EM cross sections through the odontostyle of longidorid nematodes runs throughout the length of the odontostyle (Robertson and Taylor 1975) and can also be recognized in scanning electron microsopy (Fig. 38).

Concluding Remarks and Future Prospects

Xiphinema index has developed remarkable strategies to exploit root tips of its principal host plants, grape and fig. It feeds preferably on undifferentiated cells rich in cytoplasm, the contents of which are ingested within minutes by a series of extracorporal digestive secretions produced in a highly active and large dorsal gland cell that occupies about 34% of the muscular basal oesophageal bulb (Robertson 1987). In order to avoid a rapid depletion of food resources by killing these cells in numerous successive feeding episodes, *X. index* induces the formation of large hypertrophied multinucleate cells that predominate in attractive root tip galls, the actual food sources for reproduction. The mechanisms by which these metabolically highly active multinucleate cells are induced are unknown. They resemble in many respects those induced by root-knot nematodes, which also arise by repeated mitosis without cytokinesis (e.g., Sijmons et al. 1994). However, whereas multinucleate cells produced by root-knot nematodes stay alive until completion of the nematode's life cycle, those produced by *X. index* are killed by nematode feeding, although at a much slower rate than uninucleate meristematic cells (Wyss et al. 1988). Root cell responses induced by *X. index* are identical to those induced by *X. diversicaudatum*, which also transforms root-tips of its host plants into terminal galls containing multinucleate nurse cells that reveal increased amounts of DNA, RNA and proteins (Griffiths and Robertson 1984a, 1988). Gheysen et al. (1997) reviewed the basic mechanisms involved in DNA amplification in the specialized feeding structures (nurse cell systems) of root-knot and cyst nematodes. Based on their results and conclusions it may be hypothesized that nurse cell-inducing *Xiphinema* spp. may also activate cell cycle genes in order to transform meristematic cells into multinucleate units with a high transcriptional and translational capacity. Great progress in elucidating the role played by host genes in the induction and maintenance of root-knot and cyst nematode-specific nurse cell systems has recently been achieved by using *Arabidopsis thaliana* as a model host (e.g., Fenoll et al. 1997). In this respect the *X. diversicaudatum* may be the most suitable model nematode among nurse cell-inducing Xiphinematinae to analyse plant-nematode interactions at the molecular level. This species has a much broader host range than *X. index*, and hence it also feeds on *A. thaliana*, where it induces the formation of root-tip galls containing multinucleate cells (Barthels et al. 1997).

Compared to sedentary endoparasites, such as cyst and root-knot nematodes, where in addition to oesophageal gland secretions other nematode secretions may also be involved in triggering nurse cell induction (e.g., Jones and Robertson 1997), there is no doubt that these trigger molecules originate from oesophageal gland secretions in nurse cell-indicating *Xiphinema* spp. These nematodes are true ectoparasites as they are only capable of penetrating root tissues by means of their long stylets. It was originally speculated that only secretions from the dorsal gland are injected into the root and that the trigger molecules of these secretions should have a low molecular weight to enable their symplastic transport via plasmodesmata into adjacent yet unmodified cells (Wyss et al. 1980). Concurrent studies, using micro-electrophoresis of isolated basal oesophageal bulbs of starved and well fed *X. index* females with few and respectively many secretory granules in the dorsal gland showed a significant increase in the amounts of two water soluble acid glycoproteins in the bulbs of well fed nematodes. It was assumed that these glycoproteins are constituents of secretory granules in the dorsal gland cell (Poehling and Wyss 1980). Now however, with the knowledge that the secretions of the dorsal gland are highly destructive in order to facilitate food removal, the assumption that they could play a decisive role in inducing multinucleate cells during feeding may no longer be maintained. It appears more probable that the trigger for nurse cell induction is released during the quiescent periods described in some detail in this chapter and that the plug-like structure formed around the protruded stylet-tip may originate from secretions produced in the two subventral glands. In spite of the fact that these gland cells occupy about 25% of the basal bulb (Robertson 1987) nothing is known about their precise structure and function. Future research should prove if their presumed functions (Wyss et al. 1988) can be supported.

The dorsal gland in *X. index*, currently probably one of the best examined secretory glands in root parasitic nematodes, requires further research to determine the chemical nature of its secretions that quickly liquefy the cyto- and nucleoplasm of fed on cells without disrupting the plasma membrane and nuclear membrane as visualized by *in vivo* observations with video-enhanced light microscopy. An intriguing question to be answered is how the duct system of this gland is so quickly re-filled with secretory fluids during the spells of intermittent ingestion pumping as a breakdown of secretory granules around the ducts is only evident prior to ingestion.

It appears to be a general feature of longidorid nepovirus vectors that they kill the cells from which they derive their nutrients. Feeding sites of *Longidorus* spp. in root-tip galls of their host plants are for instance characterized by a cluster of empty cells (Griffiths and Robertson 1984b) or lysigenous cavities (e.g., Bleve-Zacheo et al. 1982, Andres et al. 1989) that are surrounded by enlarged uninucleate cells with signs of an increased metabolic activity. Therefore the question remains how these vectors are efficient virus vectors if they kill the cells from which they feed and into which inject virus particles together with oesophageal gland secretions that dissociate these particles from their specific retention sites in the feeding apparatus of the nematodes. In *Xiphinema* spp. virus particles ingested from infected root cells are absorbed to retention sites on the surface of the cuticle that lines the lumen of the food canal from the most anterior part of the odontophore to the posterior end of the oesophageal basal bulb (Taylor and Brown 1997). In *X. index* particles of grapevine fanleaf virus are adsorbed in a monolayer to these retention sites (Raski et al. 1973) but sometimes particles of *Xiphinema*-transmitted nepoviruses are also adsorbed in multiple layers (Robertson and Wyss 1983). The mechanisms by which nepoviruses may become attached to their retention sites and may be dissociated from them have been discussed in recent reviews (Brown et al. 1995, Taylor and Brown 1997). Wyss et al. (1988) and Trudgill et al. (1991) speculated that the force with which dorsal gland secretions are expelled during the brief ingestion pauses in *X. index* and *X. diversicaudatum* may be sufficient to dislodge nepoviruses from their retention sites without any specific release mechanisms. However, as the recipient host cell is then rapidly killed and emptied, this kind of virus transmission would be unsuccessful with regard to both, virus replication and cytoplasmic tubule formation by means of which nepoviruses can breach the cell wall barrier for cell-to-cell movement (Gilbertson and Lucas 1996). The rather gentle release of dorsal gland secretions just prior to food ingestion should be equally unsuitable for a successful transmission as it initiates the liquefaction of the host cell cytoplasm. Therefore, as suggested by Wyss et al. (1988) and Trudgill et al.

(1991), successful virus transmission may be associated with the quiescent periods, when host cells are killed after considerable delay by dorsal gland secretions. Future research should prove if these quiescent periods are indeed essential for both, the induction of multinucleate nurse cells and successful virus transmission and if, as assumed, the subventral gland secretions may then play a decisive role in both processes.

References

Andres, M., T. Bleve-Zacheo and M. Arias. 1989. The ultrastructure of cereal and leguminous root tips parasitized by *Longidorus belloi*. Revue Nematol. 12: 365–374.

Aumann, J. 1997. Monoxenic culture of *Xiphinema index* (Nematoda: Longidoridae) on *Ficus carica*. Nematol. Medit. 25: 209–211.

Barthels, N.F., M. van der Lee, F.M.J. Klap, O.J.M. Goddijn, M. Karimi, P. Puzio, F.M. Grundler, S.A. Ohl, K. Lindsey, L. Robertson, W.M. Robertson, M. Van Montagu, G. Gheysen and P.C. Sijmons. 1997. Regulatory sequences of *Arabidopsis* drive reporter gene expression in nematode feeding structures. Plant Cell 9: 2119–2134.

Bird, A.F. 1979. Histopathology and physiology of syncytia. pp. 155–171. *In*: F. Lamberti and C.E. Taylor (eds.). Root-knot Nematodes (*Meloidogyne* species). Systematics, Biology and Control. Academic Press, New York.

Bleve-Zacheo, T., G. Zacheo, M.T. Melilo, F. Lamberti and O. Arrigoni. 1982. Ultrastuctural response of celery root cells to *Longidorus apulus*. Nematol. Medit. 10: 141–155.

Bleve-Zacheo, T. and G. Zacheo. 1983. Early stage of disease in fig roots induced by *Xiphinema index*. Nematol. Medit. 11: 1975–187.

Bleve-Zacheo, T., G. Zacheo and C.E. Taylor. 1987. A comparison of histological changes induced by *Xiphinema basiri* and *X. ifacolum* in the roots of tomato. Nematol. Medit. 15: 235–251.

Brown, D.J.F., W.M. Robertson and D.L. Trodgill. 1995. Transmission of viruses by nematodes. Ann. Rev. Phytopathol. 33: 223–249.

Coiro, M.I. and D.J.F. Brown. 1984. The status of some plants as hosts for four populations of *Xiphinema index*. (Nematoda: Dorylaimida). Revue Nematol. 7: 283–286.

Dalmasso, A. and T. Younes. 1969. Ovogénèse et embryogénèse chez *Xiphinema index* (Nematoda: Dorylaimida). Ann. Zool. Ecol. Anim. 1: 265–272.

Fenoll, C., F.M. Grundler and S.A. Ohl. 1997. Cellular and molecular aspects of plant-nematode interactions. Kluver Academic Publishers, Dordrecht. 266 pp.

Fisher, J.M. and D.J. Raski. 1967. Feeding of *Xiphinema index* and *X. diversiaudatum*. Proc. Helminth. Soc. Wash. 34: 68–72.

Gheysen, G., J. de Almeida Engler and M. Van Montagu. 1997. Cell cycle regulation in nematode feeding sites. pp. 120–132. *In*: C. Fenoll, F.M.W. Grundler and S.A. Ohl (eds.). Cellular and Molecular Aspects of Plant-nematode Inter-actions. Kluver Academic Publishers, Dordrecht.

Gilbertson, R.L. and W.J. Lucas. 1996. How do viruses traffic on the "vascular highway"? Trends Plant Sci. 1: 260–268.

Griffiths, B.S. and W.M. Robertson. 1984a. Nuclear changes induced by the nematode *Xiphinema diversicaudatum* in root-tips of strawberry. Histochem. J. 16: 265–273.

Griffiths, B.S. and W.M. Robertson. 1984b. Morphological and histochemical changes occurring during the life-span of root-tip galls on *Lolium perenne* induced by *Longidorus elongatus*. J. Nematol. 16: 223–229.

Griffiths, B.S. and M. Robertson. 1988. A quantitative study of changes induced by *Xiphinema diversicaudatum* in root-tip galls of strawberry and ryegrass. Nematologica 34: 198–207.

Hewitt, Wm. B., D.J. Raski and A.C. Goheen. 1958. Nematode vector of soil-borne fanleaf virus of grapevines. Phytopathology 48: 586–595.

Hunt, D.J. and A. Towle. 1979. Feeding studies on *Xiphinema vulgare* Tarjab, 1964 (Nematoda: Longidoridae). Revue Nematol. 2: 37–40.

Hoagland, D.R. and D.L. Arnon. 1938. The water-culture method for growing plants without soil. Univ. Calif. Coll. Agri., Agri. Exp. Station, Berkeley, Calif. Circular 347: 39 pp.

Jones, J.T. and W.M. Robertson. 1997. Nematode secretions. pp. 98–106. *In*: C. Fenoll, F.M. Grundler and S.A. Ohl (eds.). Cellular and Molecular Aspects of Plant-nematode Interaction. Kluver Acadmic Publishers, Dordrecht.

Poehling, H.-M. and U. Wyss. 1980. Microanalysis of proteins in the aseptic host- parasite system: *Ficus carica-Xiphinema index*. Nematologica 26: 230–242.

Raskei, D.J., A.R. Manggenti and N.O. Jones. 1973. Location of grapewine fanleaf and yellow mosaic virus particles in *Xiphenema index*. J. Nematol. 5: 208–211.

Raski, D.J. and L.R. Krusberg. 1984. Nematode parasites of grapes and other small fruits. pp. 457–506. *In*: W.R. Nickle (ed.). Plant and Insect Nematodes. Marcel Dekker, New York.

Robertson, W.M. 1976. A possible gustatory organ associated with the odontophore in *Longidorus leptocephalus* and *Xiphinema diversicaudatum*. Nematologica 21: 443–448.

Robertson, W.M. 1987. Serology and carbohydrate histochemistry of the oesophageal bulb of *Xiphinema index* (Nematoda: Xiphinemidae). Nematologica 33: 401–409.

Robertson, W.M. and C.E. Taylor. 1975. The structure and musculature of the feeding apparatus in *Longidorus* and *Xiphinema*. pp. 179–194. *In*: F. Lamberti, C.E. Taylor and J.W. Seinhorst (eds.). Nematode Vectors of Plant Viruses. Plenum Press, London, New York.

Robertson, W.M., P.B. Topham and P. Smith. 1987. Observations on the action of the oesophageal pump in *Longidorus* (Nematoda). Nematologica 33: 43–54.

Robertson, W.M. and U. Wyss. 1979. Observations on the ultrastructure and function of the dorsal oesophageal gland cell in *Xiphinema index*. Nematologica 25: 391–396.

Robertson, W.M. and U. Wyss. 1983. Feeding processes of virus-transmitting nematodes. *In*: K.F. Harris (ed.). Current Topics in Vector Research. Praeger, New York, I: 271–295.

Roggen, D.R., D.J. Raski and N.O. Jones. 1967. Further electron microscopic observations of *Xiphinema index*. Nematologica 13: 1–16.

Rumpenhorst, H.J. and B. Weicher. 1978. Histopathological and histochemical studies on grapevine roots damaged by *Xiphinema index*. Revue Nematol. 1: 217–225.

Seymour, M.K. 1983. Action of the feeding pump in *Longidorus caespiticola* Hooper. Nematologica 29: 65–77.

Sijmons, P.C., H.J. Atkinson and U. Wyss. 1994. Parasitic strategies of root nematodes and associated host cell responses. Annu. Rev. Phytopathol. 32: 235–259.

Staudt, G. and B. Weischer. 1992. Resistance to transmission of grapevine fanleaf virus by *Xiphinema index* in *Vita rotundifolia* and *Vitis mansoniana*. Vitic. Enol. Sci. 47: 56–61.

Sutherland, J.R. 1969. Feeding of *Xiphinema bakeri*. Phytopathology 59: 1963–1965.

Taylor, C.E. and D.J.F. Brown. 1997. Nematode vectors of plant viruses. CAB International, Wallingford. 286.

Thorne, G. and M.W. Allen. 1950. *Paratylenchus hamatus* n. sp. and *Xiphinema index* n. sp., two nematodes associated with fig roots, with a note on *Paratylenchus anceps*. Cobb., Proc. Helminth. Soc. Wash. 17: 27–35.

Towle, A. and C.C. Doncaster. 1978. Feeding of *Longidorus caespiticola* on rye-grass, *Lolium perenne*. Nematologica 24: 277–285.

Trudgill, D.L. 1976. Observations on the feeding of *Xiphinema diversicaudatum*. Nematolgica 22: 417–423.

Trudgill, D.L., W.M. Robertson and U. Wyss. 1991. Analysis of the feeding of *Xiphinema diversicaudatum*. Revue Nematol. 14: 107–112.

Vovlas, N., R.N. Inserra and G.P. Martelli. 1978. Modificazioni anatomiche indotte da *Xiphinema index* e *Meloidogyne incognita* in radici di un ibrido di *Vitis vinifera* x *V. Rotundifolia*. Nematol. Medit. 6: 67–75.

Weischer, B. 1982. Histopathological studies on grape species with different degrees of resistance to *Xiphinema index*. Nematologica 28: 178–179.

Weischer, B. and U. Wyss. 1976. Feeding behaviour and pathogenicity of *Xiphinema index* on grapevine roots. Nematologica 22: 319–325.

Wright, K. 1965. The histology of the oesophageal region of *Xiphinema index* Thorne and Allen, 1950, as seen with the electron microscope. Can. J. Zool. 43: 689–700.

Wright, K.A., R.F. Carter and W.M. Robertson. 1983. The musculature of the anterior feeding apparatus of *Xiphinema* species (Nematoda: Dorylamoidea). Nematologica 29: 49–64.

Wyss, U. 1977a. Feeding mechanisms and feeding behaviour of *Xiphinema index*. Med. Fac. Landbouww. Rijksuniv. Gent. 42: 1513–1519.

Wyss, U. 1977b. Feeding phases of *Xiphinema index* and associated processes in the feeding apparatus. Nematologica 23: 463–470.

Wyss, U. 1978. Root and cell response to feeding by *Xiphinema index*. Nematologica 24: 159–166.

Wyss, U. Ectoparasitic root nematodes: Feeding behavior and plant cell responses. *In*: B.M. Zuckerman and R.A. Rohde (eds.). 1981. Plant Parasitic Nematodes. Academic Press, New York. III: 325–351.

Wyss, U. Video assessment of root cell responses to Dorylamid and Tylenchid nematodes. pp. 211–220. *In*: J.A. Veech and D.W. Dickson (eds.). 1987. Vistas on Nematolgy. Society of Nematologists. Hyattsville, Maryland.

Wyss, U. 1992. Observations on the feeding behaviour of *Heterodera schachtii* throughout development, including events during moulting. Fundam. Appl. Nematol. 15: 75–89.

Wyss, U. and Inst. Wiss. Film. 1977. *Xiphinema index* (Nematoda)—Saugen an Wurzeln von Samlingen (Feige). Film E 2375. IWF Gottingen. 20 pp.

Wyss, U. and Inst. Wiss. Film. 1988. Reaktion von Wurzelzellen (*Ficus carica*) auf die Saugtatigkeit des Nematoden *Xiphinema index*. Film D 1657. IWF Gottingen. 22 pp.

Wyss, U., H. Lehmann and R. Jank-Luwig. 1980. Ultrasrtucture of modified root-tip cells in *Ficus carica*, induced by the ectoparasitic nematode *Xiphinema index*. J. Cell Sci. 41: 193–208.

Wyss, U., W.M. Robertson and D.L. Trudgill. 1988. Oesophageal bulb function of *Xiphinema index* and associated root cell responses, assessed by video-enhanced contrast light microcopy. Revue Nematol. 11: 253–261.

Wyss, U. and U. Zunke. *In vitro* feeding of root parasitic nematodes. pp. 91–96. *In*: B.M. Zuckerman, W.F. Mai and M.B. Harrison (eds.). 1985. Plant Nematology Laboratory Manual. Univ., Massachusets Agric. Exper. Station, Amherst, MA.

Wyss, U. and U. Zunke. 1986a. The potential of high resolution video-enhanced contrast microscopy in nematological research. Revue Nematol. 9: 91–94.

Wyss, U. and U. Zunke. 1986b. Observations of the behaviour of second stage juveniles of *Heterodera schachtii inside* host roots. Revue Nematol. 9: 153–165.

Zacheo, G. and T. Bleve-Zacheo. Plant-nematode interactions: histolopgical, physiological and biochemical interactions. pp. 321–353. *In*: K. Kohmoto, U.S. Singh and R.P. Singh (eds.). Pathogenesis and Host Specificity in Plant Diseases. Histopathological, Biochemical, Genetic and Molecular Bases. Pergamon, Oxford, Vol. II. Eukaryotes.

14

Xiphinema index, Rearing in Greenhouse

Ulrike Ipach

ABSTRACT

The performance of nematode experiments in the greenhouse needs a lot of test animals. A well operating rearing is one of the essential preconditions. *Xiphinema index* is not only one of the important grapevine nematodes worldwide but is the ideal candidate for establishing rearing requirements and conditions necessary to complete its life cycle.

Key Words: Grapewine, fanleaf virus, GFLV, *Xiphinema index* *Ficus carica*

Introduction

Grapevine fanleaf virus (GFLV) transmitted by its specific vector *Xiphinema index* is one of the major viruses affecting grapevine world-wide causing important damage (Bovey et al. 1980). Since 1988 in Germany it is forbidden to combat virus transmitting nematodes by nematicides. Therefore investigations are focussed on alternative combating methods. Beside the breeding programmes for developing resistant rootstocks, research programmes were initiated to find plant substances as a basis for possible new generations of nematicides with different kinds of action. For testing

Phytomedizin–Weinbau, Dienstleistungszentrum Ländlicher Raum–Rheinpfalz, Breitenweg 71, D-67435 Neustadt/Weinstrasse, Germany.

promising substances it is necessary to have enough nematodes for experiments, which requires an efficient rearing. *Xiphinema index* is not well adapted to the temperate zone, the population density is usually low in vineyards of this region (Mohr 2005, Taylor and Brown 1997). But if this nematode has optimal living conditions like in greenhouses, the population density may increase rapidly.

Rearing of *Xiphinema index* in Greenhouse

The rearing of X. *index* can be established on the host plant *Ficus carica* according to Dechet (1991). This plant has two great advantages compared to grapevine: X. *index* shows a higher reproduction rate on *F. carica* and this plant does not host the *Grapevine fanleaf virus*. So you will obtain virus-free nematodes for, e.g., virus-transmitting experiments.

Ficus carica can be easily cultivated from cuttings. Rooted cuttings are planted in 5–10 polyethylene pots. Pure quartz sand (grain size 2) or a mixture of gravel (grain size 6), quartz sand and vineyard soil at the rate of 5 : 4 : 1 is used as substrate. All material used should be sterilized before. Hoagland No. 2 nutrient solution (Hoagland and Arnon 1938) is used to avoid micronutrients deficiency (conductivity 750 µS, pH 6.5). Additionally the leaves are sprayed weekly with a liquid fertilizer. The rearing pots standing at 25°C in a greenhouse chamber are exposed additionally at lower light intensity. *Xiphinema index* life stages are susceptible to very wet conditions due to poor water drainage in rearing pots. So it is very important to use a good drainage in the pots guaranteed here by sand and gravel. Also it is necessary to avoid over watering the pots. It is like a tightrope walk between water deficiency for the plants and water abundance for the nematodes. This method is acceptable to maintain the rearing in the greenhouse over a longer period without to much work. The extraction of the nematodes from soil substrates is carried out according to the decanting and sieving method of Flegg (1967) modified after Dechet (1991).

For gaining a lot of nematodes in a short period of time it is more efficient to use smaller plastic pots with, e.g., 12 cm diameter according to the substrate test after Schaaf (2000). Faced by a plastic bag to prevent the trickling out of the sand, the pots are filled with 675 g pure quartz sand (grain size 2). Rooted cuttings of *Ficus carica* are planted into the pots. The plants are placed in a greenhouse chamber with 25°C temperature and 60 % relative humidity (Fig. 1). For sufficient root development the plants are watered daily to 60% maximum water capacity (WC_{max}). After inoculation with nematodes the plants are watered daily. 40% WC_{max} should not be exceeded because the nematodes are very sensitive to humidity. Especially the daily watering to 40% WC_{max} seems to be an important requirement to reach a constant reproduction rate of X. *index*. For a successful inoculation

Fig. 1. Rearing of *Xiphinema index* on figs in greenhouse.

it is necessary to place the nematodes close to the Ficus-roots. With the described conditions the first adult nematodes of the following generation can be detected not before 60 days. This implies that one generation cycle of *X. index* takes at least eight to nine weeks. Naturally the nematode-rearing pots can be used longer in the greenhouse. Our experiences showed that this is possible up to one year without changing the plant. At this time the figs usually do not have leaves anymore. However, the nematodes are able to develop furthermore if the root system is intact. The use of quartz sand and gravel has a very great advantage: it is very easy to extract the nematodes from the substrate to obtain relatively clean nematodes for further trials.

Nematode experiments can be performed according to the rearing method; the size of the pots can be variable. First deficiency symptoms of the figs may appear approximately after 60 days. Depending on the experiment a soil-fertilization should be avoided due to a possible negative effect on the nematodes. But a reproduction of the nematodes still occurs. To study the influence of substances added to the substrate on the development of nematode eggs or larvae it is useful to disrupt the life cycle of *X. index* 40 days after inoculation. At this time there are many newly developed larvae and the inoculated adults in the substrate. You are able to detect potential effects of, e.g., plant ingredients or newly developed nematicides and obtain results which are reproducible. This is the basis of the so-called substrate test according to Schaaf (2000) used for testing the influence of a lot of plant ingredients and enemy plants on *X. index*.

References

Bovey, R., W. Gärtel, W.B. Hewitt, G.P. Martelli and A. Vuittenez. 1980. Virus and Virus-like Diseases of grapevines. Verlag Eugen Ulmer, 181 pp.

Dechet, F. 1991. Untersuchungen zur Wirkung von Pflanzen und Pflanzeninhaltsstoffen auf *Xiphinema index* Thorne & Allen, 1950 (Nematoda: Dorylaimida). Ph.D. Universität Kaiserslautern. 118 pp.

Flegg, J.J.M. 1967. Extraction of *Xiphinema* and *Longidorus* species from soil by a modification of Cobb's decanting and sieving technique. Ann. Appl. Biol. 60: 429–437.

Hoagland, D.R. and D.L. Arnon.1938. The water-culture method for growing plants without soil. Univ. Calif., Coll. Agri., Agri. Exp. Station, Berkely, Calif., Circular 347: 39 pp.

Mohr, H.D. 2005. Farbatlas Krankheiten, Schädlinge und Nützlinge an der Weinrebe. Verlag Eugen Ulmer. 320 pp.

Schaaf, C. 2000. Untersuchungen zur Wirkung von Pflanzeninhaltsstoffen und Feindpflanzen auf die phytosanitären Nematoden *Xiphinema index* Thorne & Allen, 1950 und *Meloidogyne incognita* (Kofoid White 1919) Chitwood 1949. Ph.D. Universität Hohenheim. 153 pp.

Taylor, C.E. and D.J.F. Brown. 1997. Nematode Vectors of Plant Viruses, CAB International, 286 pp.

15

Maintenance of Fungal Vectors of Plant Viruses

Chuji Hiruki

ABSTRACT

It has been possible to isolate and maintain soil-borne fungi that are obligate parasites of roots of virus-infected plants, and finally identify them as the vectors of certain viruses in question. However, the maintenance of a culture of vector fungus in nutrient broth has not been possible and must, at present, rely on the living plants or tissues under non-sterile conditions using glasshouses or growth chambers. Two main groups of zoosporic fungi have been established as virus vectors, viz., Chytridiomycetes and Plasmodiophoromycetes. To investigate virus-fungus vector relationships, the fungus must be cultured in plant tissues or in seedlings grown in various media including liquid nutrient, silica sand, and soil cultures. In glasshouse experiments, great care must be taken to prevent cross-contamination of cultures by other viruses and fungi. Different types of incubators for use in growth chambers and glasshouses are available for maintaining fungal cultures and studying virus-vector relationships. Environmental factors that influence the proper maintenance of fungal vectors are temperature, humidity, light, pH, supporting medium and soil structure, chemicals and the soil microflora. All vector-fungi from resting spores having a survival function which permits their long-term storage and viability. Thus, viruses that are carried internally in the resting spores probably can persist as long as the resting spores remain viable.

Department of Agricultural, Food and Nuritional Science, 4-10D Agriculture/Forestry Center, University of Alberta, Edmonton, Alberta, Canada T6G 2P5.

Introduction

Intensive research over the last few decades has established that two main groups of root-inhabiting fungi, viz. certain members of Chytridiomycetes and Plasmodiophoromycetes are highly efficient vectors of plant viruses (Barr 1988). They are obligate parasites which require living plants as their hosts to complete simple life cycle. Zoospores, a motile stage in their life cycles, function as virus vectors in the rhizosphere of host plants under appropriate environmental conditions. Two types of virus-fungal vector relationships are known; in the first type, the virus is acquired *in vivo* in the virus-infected plant during replication of the fungus. Newly formed zoospores carry viral genome internally and transmit it to new hosts and/or continue to maintain the virus throughout the dormant stage of resting spores, presumably as long as the resting spores remain viable. In the second type, virus acquisition occurs by virus adsorption to the surface of active zoospores prior to cyst formation. Since the virus does not remain internally, the nature of virus-vector association is short-term and thus the virus does not persist in the resting spores that emerge from virus-infected plants. In this chapter the emphasis will be focused on the methods established for the maintenance of the cultures of known vector-fungi as well as on a brief description of environmental factors that affect the maintenance of the fungal cultures. For detailed discussions of different virus-vector fungus systems, further reading of review paper is recommended (Campbell 1979, 1988, Teakle 1988, Hiruki and Teakle 1987, Brunt 1989).

Maintenance Methods

Taxonomy of the Vector Fungi and Viruses

Known genera of plant vector fungi are confined to the Chytridiomycetes and the Plasmodiophoromycetes. High vector specificity exists between vector genera, *Olpidium* and *Polymyxa*, and plant viruses that they transmit (Table 1). These fungi apparently incite no appreciative damage to host plants and are capable of acquiring viruses efficiently, transmitting them with ease. Some are even capable of preserving acquired viruses internally for long periods of time as they form resting spores.

Chytrid fungi are known for their diversity in vector specificity and transmit a variety of viruses in a range of virus groups including tobacco necrosis necrovirus, lettuce big-vein and tobacco stunt viruses, cucumber necrosis tombusvirus, melon necrotic spot carmovirus, red clover necrotic mosaic dianthovirus. Nine furoviruses are transmitted worldwide by Plasmodiophoromycete vectors. In addition, several filamentous viruses of the bymovirus group are transmitted by *Polymyxa graminis*.

Table 1. Zoosporic fungi as vectors of plant viruses.

Chytridiomycetes vectors		
Olpidium brassicae	TNV	freesia leaf necrosis
	TNV-S	lettuce ring necrosis
	TSV	pepper yellow vein
	LBVV	
Olpidium radicale	CNV	(RCNMV)
	CLVS	
	CLVS-FS	
	CSBV	
	SqNV	
	MNSV	
Plasmodiophoromycetes vectors		
Polymyxa graminis	(BBNV)	BaYMV-NM
	OGSV	BaYMV-M
	PCV	OMV
	IPCV	RNMV
	RSNV	WYMV
	SBWMV	WSSMV
Polymyxa betae	BNYVV	
	BSBV	
Spongospora subterranea	PMTV	
	WYSV	

TNV, tobacco necrosis virus, Teakle 1962, Teakle and Hiruki 1964, Kassanis and MacFarlane 1964, TNV-S, Kassanis and MacFarlane 1968; TSV, tobacco stunt virus, Hidaka and Tagawa 1965, Hiruki 1965; LBVV, lettuce big-vein virus, Campbell 1962, Tomlinson and Garrett 1964; LRNV, freesia leaf necrosis, van Dorst and Peters 1988, lettuce ring necrosis, Bos and Huijberts 1996, Campbell and Lot 1996, pepper yellow vein, Fletcher et al. 1987; CNV, cucumber necrosis virus, Dias 1970; CLSV, cucumber leaf spot virus, Miller et al. 1996; CLSV-FS, fruit streak strain, Gallitelli et al. 1983; CSBV, cucumber soil borne virus, Koenig et al. 1983; SqNV, squash necrosis virus, Campbell 1996; MNSV, melon necrotic spot virus, Hibi and Furuki 1985, Tomlinson and Thomas 1986; RCNMV, red clover necrotic mosaic virus, MacFarlane 1982; BBNV, broad bean necrosis virus, Fujikawa 1955, Lu et al. 1996; OGSV, oat golden stripe virus, Plumb and MacFarlane 1977, Adams et al. 1988; PCV, Thouvenel and Fauguet 1980; IPCV, Indian peanut clump virus, Ratna et al. 1991; RSNV, rice stripe necrosis virus, Fauquet and Thouvenel 1983; SBWMV, soil-borne wheat mosaic virus, Estes and Brakke 1966, Canova 1966; BaYMV-NM, barley yellow mosaic virus-NM, Kusaba and Toyama 1970; BaYMV-M, Huth 1988; OMV, oat mosaic virus, Atkinson 1945; RNMV, rice necrosis mosaic virus, Inouye and Fujii 1977; WYMV, wheat yellow mosaic virus, Sawada 1927; WSSMV, wheat spidle streak mosaic virus, Slykhuis 1970; BNYVV, beet necrotic yellow vein virus, Tamada et al. 1975; BSBV, beet soil-borne virus, Henry et al. 1986; BSBMV, beet soil-borne mosaic virus, Wisler et al. 1996; PMTV, potato mop-top virus, Jones and Harrison 1969; WYSV, watercress yellow spot virus, Walsh et al. 1989.

Life Cycle of Vector Fungi

The biology and life history of selected fungal vectors has been described by Teakle (1969, 1983), Campbell (1979, 1988), Barr (1988), Barr and Asher (1989), and Hiruki (1990, 1993). In spite of the taxonomically diverse orders to which they belong, there are common biological features among these vector fungi. They are obligate parasitic lower fungi, which cannot be maintained in artificial media; they have an actively motile stage (zoospore) that is important in virus acquisition and transmission, and that infects root cells. They transform themselves into membrane-delimited thalli replacing host cytoplasm in the early stages following penetration; at maturity they form zoosporangia and resting spores in the infected cells. Thick-walled resting spores are later released into spoil.

They are capable of withstanding drying and other adverse environmental conditions, and may germinate to produce zoospores, even after prolonged dormancy. Host-parasite specificities may also vary considerably among species of these fungi.

Establishment

Once the soil-borne nature of a virus disease has been established and infested soils have been identified, the next step is to study possible relationships between the virus and the fungus suspected of being its vector.

Preparation of Infested Soils

Soil samples for subsequent tests may be collected from the area immediately surrounding the infected plants. With tobacco stunt, the disease occurrence is the highest when infected soil is located in the top layer of the soil up to 2 cm in depth (Hidaka et al. 1956). In the case of *P. graminis*, soil collected from under diseased plants and tested soon thereafter was a poor inoculum source, whereas its potential was higher if it was sampled after overwintering or if it was dried prior to testing (Brakke and Estes 1967, Slykhuis 1975). Similarly, tobacco stunt-infested soil samples from the previous year, were prepared by air-drying for 2 Wk and pulverized in a mortar with a rubber stopper and sifted through a 3 mm sieve, prior to using as a satisfactory inoculum for tobacco stunt. Steamed soil in a 34 cm pot was covered with 200 g of such a soil inoculum (Hidaka et al. 1956). In another study with tobacco stunt, small amounts of air-dried tobacco roots containing resting spores of *Olpidium brassicae* were used as an inoculum which was spread over an autoclaved soil mixture (one volume sand, two volumes loam, three volumes peat) in plastic trays (36 cm x 36 cm) (Hiruki 1987). *O. brassicae* can be recovered routinely from 5 g of air-dried soil using lettuce seedlings as bait (Campbell 1985). The best bait plant is usually the original host on which the disease occurs in the field. However, in some cases, alternate host species that germinate with ease and grow rapidly may be preferred for tests in glasshouses and growth chambers.

Bulk Culture

A rapid method was developed for the detection of *Polymyxa betae* from infested soils (Beemster and de Heij 1987, Lindsten 1990). The soils to be tested were carefully crumbled and mixed. Fifty grams of the test soil was placed in a 120 ml plastic cup and 50 ml of sterilized water was added. After stirring, the soil-water mixture was heated at 40°C for 30 min. The bait plants (*Beta vulgaris* cv. Hilma) were grown in sterilized sand in 9 cm clay pots. After washing the roots, five one-week-old seedlings in pairs were transplanted into test soils in the cups. The bait plants were incubated at 25°C in a growth room with light intensity of 18,000 μmc/m^2 and a photoperiod of 18 h for one week. The roots were washed free of sand or soil and could be examined in several ways. Plants infected by the fungus may be transplanted into sand culture.

With *O. brassicae*, new zoospores released from sporangia in the roots may be collected in a root washing, and poured onto healthy test seedlings (Alderson and Hiruki 1977). Infected parts of the roots, usually the rapidly growing tissues behind the root cap, can serve as an inoculum when dissected out and placed in pots with healthy seedlings (Hiruki 1967). Bulk cultures established as above should be assumed to be biologically impure, containing undesirable fungi or bacteria and different strains of the target fungal species. If the fungal isolates that were established in a bulk culture show different temperature optima, it is possible to separate them into relatively pure cultures by subjecting the bulk cultures on a common host to different temperatures that affect different components of the fungal population. Thus, virtual unifungal cultures of *O. brassicae* have been produced on lettuce at temperatures below 18°C (Campbell and Grogan 1964, Tomlinson and Garrett 1964). In contrast, *O. radicale* has optimum temperature of 30°C (Teakle and Thomas 1985). *P. betae* predominated on beets at about 25°C (Asher and Blunt 1971), whereas *P. graminis* growth is favored by between 16-19°C (Brakke 1971).

Incubators

Under glasshouse conditions, if zoospores of *O. brassicae* are present together with the plant virus such as tobacco necrosis virus (TNV), the virus is easily acquired from drainage water by the zoospores. The virus then can be transmitted via splashing water in daily routine work in the glasshouse. To avoid such contamination, a polyethylene sheet placed between a test pot and a supporting pot (Hiruki 1967), or a water proof-glazed pot under a test pot (Hiruki 1965) have been used. However, none of these is suitable for routine, large scale maintenance of *O. brassicae*. A simple incubator was therefore developed for use in a growth chamber or glasshouse to maintain a large number of fungal isolates and to use them for virus transmission (Hiruki 1969, Fig. 1).

This incubator has been in use for more than 20 years and was proven very efficient for the above purpose (Alderson 1975). The incubator consists of three major parts: (a) an upper chamber for supporting test seedlings and the fungus; (b) a lower chamber for ½ strength Hoagland's solution (pH 6.0) (Hoagland and Arnon 1950); and (c) a cover tube with an opening covered with a vinyl mesh at the top. The nutrient solution in the lower

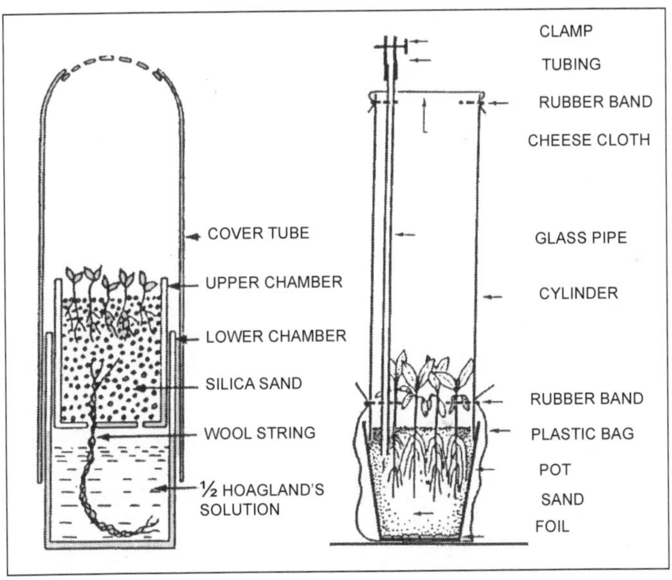

Fig. 1. Incubators useful for maintaining cultures of vector fungi free from chance contamination due to splashing water. In A, an incubator unit consists of an upper chamber for supporting test seedlings and a vector fungus in silica sand and a lower chamber for retaining ½ strength Hoagland's solution (pH 6.0), which is supplied to the upper chamber through a piece of wool string. To prevent chance contamination, the incubator unit is fitted with a cover tube (Hiruki 1969). B, an incubator suitable for glasshouse use (van Dorst and Peters 1988).

chamber is supplied to the test seedlings in the upper chamber by means of a wool string wick. A single incubator can support 10–30 tobacco seedlings up to 40 days after seeding. The seeds of tobacco (*Nicotiana tabacum* cv. Bright Yellow) are disinfected in 70% alcohol. Seedlings infected in the incubators by *O. brassicae* can be transplanted later to larger containers if necessary. Seedlings can be inoculated with virus-carrying *O. brassicae* in the upper container using zoospores or resting spores as follows: (a) An upper container with 7-day-old tobacco seedlings is dipped into a suspension containing 10^3 zoospores/ml for 15–30 min. or the suspension is poured into silica sand. (b) Dry root powder (0.1 mg) containing approximately 3,000 resting spores carrying tobacco stunt virus (TSV) is placed in the center of the surface of the silica sand in the upper container, on which 20 5-day-old tobacco seedlings are growing. The root powder is prepared by pulverizing air-dried tobacco roots containing the viruliferous fungus and filtering the power through a 200-mesh sieve. A wooden rack with 38 holes is used to support the incubator. The holes measure 26 mm in diameter and 30 mm in depth. The rack is 46 mm in height, 252 mm in width, 252 mm in length, and is painted black with waterproof paint. Plants are incubated in a growth period with a 16h photoperiod at an intensity of 20,000 μmc/m^2 and light and dark temperatures of 20°C and 17°C respectively (Fig. 2). Infection with zoospores suspensions can be detected as early as three days after inoculation, whereas the results of inoculation with resting spores are somewhat variable with infections occurring 1 to 3 weeks after inoculation. The incubator also proved useful for the investigation of lettuce big-vein virus (LBVV) and TNV, both of which are transmitted by the same fungus (Campbell 1962, Teakle 1962). To prevent contamination through splash water in a glasshouse, a similar incubator was used to study the transmission of freesia leaf necrosis agent (van Dorst and Peters 1988, Fig. 1B).

This incubator has many advantages: (a) Silica sand makes examination of roots easier, more accurate, and faster. Used sand can be disinfected by heating or by chemicals. (b) Adequate nutrient solution is constantly supplied through a wick and changing of the solution is required only weekly. It also is possible to provide an automated system for nutrient supply. (c) The cover tube serves as a barrier, which prevents chance contamination with viruses and other microbes. (d) The compact size allows an increased number of replications in a limited space. The incubator is very suitable for use in a growth chamber where precise control of environmental conditions and uniform growth of test plants and fungi is possible. Furthermore, the cost of modifying the incubators to suit individual needs is inexpensive.

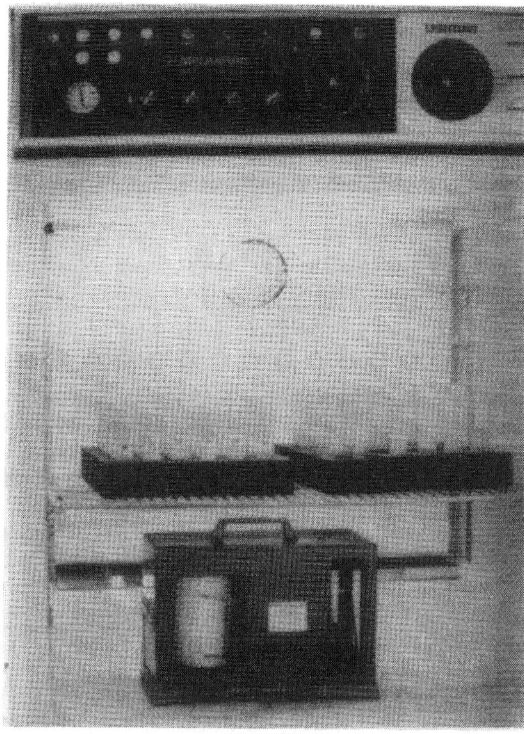

Fig. 2. The growth chamber in which incubators containing isolates of *Olpidium brassicae* are maintained routinely. The incubators are supported by wooden stands.

Unifungal Culture

Unifungal culture of *O. brassicae, O. radicale* and *P. graminis* have been established by using dissected pieces of root tissues infected with a target fungal species as inoculum (Barr 1979). Dilute zoospore suspensions of *O. brassicae* and *O. radicale* may be used to establish isolated infections in root hairs from which single spore cultures may be obtained (Sahtiyanci 1962, Dias 1970). Root tissues infected with *O. brassicae* (Lin et al. 1970) or *Lagena radicola* (Campbell et al. 1975) can be blended in a small volume of cold, 0.5 M glycine-NaOH buffer, pH 7.6, followed by filtration through a 120 or 250 mesh screen to yield a suspension which can be streaked onto 2% agar which results in successful isolation of individual. Virus transmission by means of single-sporangium isolates of *O. brassicae* (Lin et al. 1970, Fig. 3) and *O. radicale* (Campbell and Sim 1994) has been demonstrated.

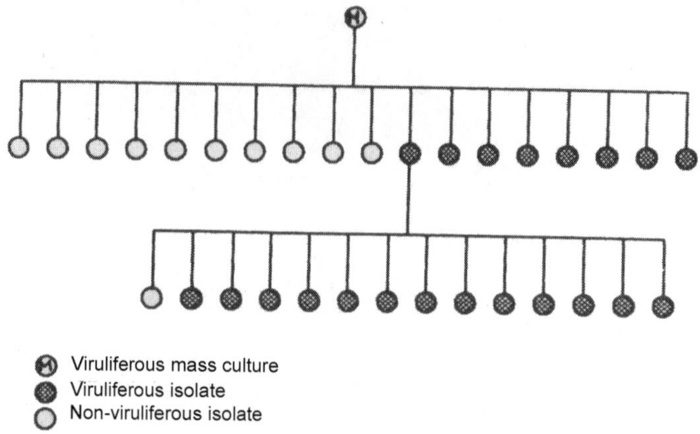

Fig. 3. Transmission of lettuce big-vein virus (LBVV) by single-sporangium isolates derived from an LBVV-transmitting mass culture of *Olpidium brassicae* (Lin et al. 1970).

Molecular Assays

Unifungal cultures contain only one root-infecting fungal species, but there is no guarantee of genetic uniformity. Although developing molecular techniques to study obligate parasites is more difficult than for typical mycelial fungi because they cannot be grown in pure culture, several methods to analyze vector fungi are available: pulsed field gel electrophoresis allows to separate individual fungal chromosomes for detailed genetic studies (Ward et al. 1996). Polymerase chain reaction followed by restriction fragment length polymorphism are routinely used to establish molecular distinction between *P. graminis* and *P. betae* (Ward et al. 1996) and between *O. brassicae* and *O. radicale* (Jiang and Hiruki 1996). The use of single copy DNA probes can yield useful information on genetic purity of established cultures as well as genetic variability at the molecular level among different isolates (Ward et al. 1994, 1996).

Maintenance of Stock Cultures

Once cultures of the vector fungi have been established, stock cultures can be maintained routinely in the laboratory, glasshouse or growth chambers, using sand as a supporting medium. Abundant increase of *O. brassicae* (Campbell et al. 1961, Hiruki 1969) and *P. graminis* (Adams et al. 1986) is possible in sand and a nutrient solution, providing zoospore suspensions as high as 10^6 zoospores/ml. Other supporting media, such as glass beads or Molochite, yield similar results as quarz sand (Jones and Harrison 1969,

Stobbs et al. 1982). Various automatic watering systems have been used to maintain stock cultures (Kassanis and MacFarlane 1964, Smith et al. 1969, Adams et al. 1986, Littlefield 1997). However, the use of circulated nutrient solution from a common reservoir should be avoided to prevent accidental cross contamination.

Harvesting Zoospores

When a batch of zoospores is needed, infected roots are washed completely free of sand and immersed in a container of chlorine-free tap water. Zoospores are released into the water by germination tubes through the root cell walls. Normally such a zoospore suspension is called a 'root washing'. Under appropriate conditions, it is not difficult to obtain root washings of *O. brassicae* with >10^6 zoospores/ml. Although exact quantification of zoospores is difficult, it is possible to count motile zoospores for experimental purposes. At concentrations bellow 10^5 zoospores/ml, a reliable estimate of zoospores can be made in a volume of 10^{-4} ml by counting in a microscope. To establish a quantitative estimate of zoospores in a given specimen, zoospores can be killed or immobilized by iodine (Adams et al. 1986) or mercury (Dias 1970) before counting.

Inoculation with Zoospores

For *in vitro* experiments, inocula are prepared by adding a known amount of virus to a quantified suspension of zoospores. Test plant seedlings may be pre-exposed to the inocula or the suspension is poured onto healthy seedlings in a pot. If a virus is a member of the type that is acquired *in vivo*, seedings are pre-inoculated with the virus alone then inoculated with virus-free zoospores, or the virus is inoculated to host plants which have been exposed to virus-free zoospores. In the case of TSV, tobacco plants are less susceptible to *O. brassicae* zoospores at growth stages older than one month after seed germination (Hidaka et al. 1956, Alderson 1975), therefore the latter method is recommended.

Maintenance of Contaminant-free Fungal Cultures

During the maintenance of vector fungi, some complications may occur due to chance contamination by undesirable viruses or microbes. It should be emphasized that a virus such as TNV can be acquired easily by *O. brassicae* zoospores under normal conditions of glasshouse management. TNV, TSV and LBVV in various combinations are transmitted concomitantly without apparent interference by zoospores of a lettuce strain or a tobacco strain of *O. brassicae* (Hiruki 1994). Therefore, a variety of check procedures are needed to maintain contamination-free stock cultures. Chemical treatment or UV irradiation of resting spores is the fastest, most reliable method to free vector cultures of viruses that are externally borne but it is not effective for those that are internally borne (Campbell and Fry 1966, Hiruki 1987). Single-spore isolates derived from virus-transmitting, bulk or unifungal cultures may yield a significant number of virus-free isolates (Lin et al. 1970). A virus-free fungal culture can be obtained by maintaining the vector for several weeks in a host that is resistant or immune to the virus (Campbell 1962, Adams et al. 1987). Treatments of stock cultures with virus-inhibiting chemicals may also produce fungal isolates that are virus-free (Campbell 1980). Axenic cultures of *P. betae* on transformed roots of beet have been reported, offering the possibility of studying uncontaminated unifungal cultures (Mugnier 1987).

Environmental Conditions

Environmental factors affecting the maintenance of vector fungi are temperature, moisture, pH, light, supporting medium and soil structure, chemicals and soil microflora.

Temperature

Temperature is one of the major factors directly affecting the maintenance of fungus vectors. It also indirectly influences other environmental factors such as soil moisture, soil texture, etc. The direct effect of temperature on the vector fungi can be variously expressed under different temperature regimes, i.e., the minimum, optimum and maximum temperatures for germination, encystment, infection, growth, multiplication, sporulation, survival, etc. (Westerlund et al. 1978a, Claxton et al. 1995). Their temperature requirements will be variable, depending on the fungus involved and the duration of exposure of the fungal cultures to the particular temperature and the interaction with other factors such as moisture. Generally speaking, most vector fungi prevail under relatively cool conditions, i.e., between 15 to 25°C. However, disease incidence is not necessarily regulated by setting constant temperature range. For example, although tobacco stunt does not develop when a constant soil and air temperature of 25°C is maintained, severe disease occurs at constant 17°C. Also, the same severe disease can be produced by providing an alternate cycle of 25°C and 17°C (Hidaka et al. 1956). The optimum temperature range for establishment of *P. graminis* and *P. betae* in their respective hosts also differs. An Indian isolate of *P. graminis* has a narrow optimum temperature range of about 30°C, while that for European and North American isolates of *P. graminis* is approximately 15–20°C (Adams and Swaby 1988, Ledingham 1939, Linford and McKinny 1954, Legreve et al. 1998).

Moisture

Moisture affects the maintenance of vector fungi at different stages of their life cycle. With *O. brassicae*, germination of zoospores can occur only when soil is saturated, while the highest germination of resting spores occurs in a water-saturated soil, but gradually decreases to zero as soil moisture is reduced below –60 millibars (Westerlund et al. 1978b). Maintenance of a suitable moisture content for a minimum of 6 days is required for the germination of *O. brassicae* resting spores (Westerlund et al. 1978b). Fluctuation in soil moisture content can be more favorable to the vector fungus than constant high soil moisture. The resting spores of *S. subterranea* germinate more readily when air-dried for more than a week before remoistening than when kept at a constantly high moisture level (Cooper et al. 1976).

pH

A pH of 6–8 has been found to be optimal for *in vitro* experiments on the transmission of viruses by zoospores of *O. brassicae* and *O. radicale* (Kassanis and Macfarlene 1964, Teakle and Gold 1964, Dias 1970). Germination of zoosporangia of *O. brassicae* was inhibited by a pH of 5.2 (Tomlinson and Garret 1964). Soil reaction is an important factor in the maintenance of the vector fungi. The germination of *S. subterranea* resting spores and infection of seedlings is slightly reduced at pH 5.2–5.8 (Kole 1954), but a pH 5.0 or below, the fungus and potato mop-top virus (PMTV) infection were significantly inhibited (Cooper et al. 1976). When the soil pH was later increased to 6.2 by adding lime, PMTV transmission continued, indicating, that the resting spores had remained viable (Cooper et al. 1976). However, in a water-culture system with a pH range from 5.0 to 8.0, pH of the nutrient solution did not influence resting spore germination of *S. subterranea* (Merz 1989).

Light

Active plant photosynthesis supported by adequate light is essential for the vigorous growth of host plant root. This, in turn, favors infection by the obligate parasitic vector fungi. Light may interfere with the encystment of zoospores of *O. brassicae* and *O. radicale* on roots (Beever and Fry 1970,

Stobbs et al. 1982). Further work is needed to determine the extent of light interference of other vector fungi.

Supporting Medium and Soil Texture

The texture and porosity of a soil may affect the activity of zoospores and the rate of soil drainage. With *P. betae*, infection occurs at soil matric potentials ≥–400 mbars and infection is greater in lighter texture soil at higher matric potentials (Gerik et al. 1990). Sand provide a satisfactory environment for both root development and vector fungi. The ability of zoospores to move is greater in sand than in clay. This may partly explain why Indian peanut clump occurs only in sandy soils and loam soils (Reddy et al. 1983) and was easily adaptable to sand culture in an automatic immersion tank system (Littlefield et al. 1997).

Chemicals

A variety of chemicals that influence the development and activity of vector fungi can be found in water and soil. Fertilizers and fungicides also greatly influence these organisms in every phase of their life cycles. In handling zoospores of vector fungi, the water to which they are released can be tap water but it must be free of undesirable chemicals such as chlorine and other detergents. Certain salts (e.g., potassium phosphate), and amino acids (e.g., glycine), preserve the motility and infectivity of *O. brassicae* zoospores when added to distilled water *in vitro* (Teakle and Gold 1964). The fresh soil water, containing a dilute mixture of inorganic compounds, may be favorable to prolonged zoospore activity *in vitro* (Teakle and Gold 1964) or *in situ* (Westerlund et al. 1978a), when concentrations remain below 0.1M. Some chemicals used as plant nutrients such as potassium chloride and sodium nitrate may interfere with zoospore activity *in vitro* (Teakle 1988). Soil application of chemical or organic fertilizers with high nitrogen contents reduce incidence of wheat spindle streak mosaic, whereas the same fertilizers usually increase incidence at low levels (Slykhuis 1970). Fungicides can be applied to the soil or water to control the vector fungi and prevent virus transmission (Teakle 1988, Hiruki 1989).

Soil Microflora

Except for *S. subterranea*, all vector fungi are weak root parasites and weak competitors if chance contamination occurs. Threrefore, great care must be taken to maintain established culture free of other soil microorganisms that may interfere with the survival of the vector fungi. Parasites of the vector fungi may degrade or destroy their viability, as reported with *P. betae* parasitized by *Trichoderma harzianum* at the cystosori stage (d'Ambra and Mutto 1986). There were no differences regardless of the strains of *O. brassicae* when multiple transmission of plant viruses was tested (Hiruki 1994). More work is required to shed light on the interactions with other microbes including parasites, predators and antagonists.

Discussion

This chapter reflects the significant progress made in the last few decades in identifying vector fungi of a range of plant viruses and in studying their vector-virus relationships. It is now firmly established that certain fungi in the Chytridiomycetes and Plasmodiophoromycetes are the vectors of a range of plant viruses including polyhedral to elongated viruses. Common traits shared among different classes of the lower fungi are that (a) they have a freely motile stage (zoospore) in their life cycles. Prior to penetration though host cell walls, zoospores attached to cell walls undergo profound membrane transformation (Hiruki 1993), a stage presumably involved with virus acquisition *in vitro*. (b) All vector fungi are obligate parasites requiring living host plants for completion of their life cycle and have extremely intimate interactions among host, fungus and virus. Mechanisms involved in *in vivo* virus acquisition and release remain unsolved and warrant further work.

Unifungal culture of certain vector fungi are established (Sahtiyanci 1962, Dias 1970, Lin et al. 1970, Campbell and Sim 1994), and a rapid increase in similar unifungal cultures of other fungi will enable precise analyses of vector fungi-virus interactions. In order to facilitate the maintenance of fungal cultures, both simple (Hiruki 1969, van Dorst and Peters 1988, Merz 1989) and elaborate automatic culture systems (Smith et al. 1969, Adams et al. 1986, Littlefield 1997) have been developed. Further improvement in manipulating several important factors in the culture systems will be possible by constantly progressing computer technology. Axenic cultures of *P. betae* on transformed roots of beet indicate that similar approaches are possible with other vector fungi and should be combined with use of unifungal cultures in the future. The advent of molecular approaches toward the detection, identification and characterization of vector fungi is yielding very valuable information and provides powerful tools in understanding the complex problems associated with fungal transmission of important plant viruses (Ward et al. 1994, Jiang and Hiruki 1996, Mutasa-Gottgens et al. 1996a, 1996b, Ward and Adams 1996, Hiruki 1996, Hiruki 2008, Hiruki 2012).

Acknowledgements

I express my sincere appreciation to Peter Blenis for his critical reading of and useful comments on the manuscript and to Keri Wang for his able assistance in the preparation of illustrations.

References

Adams, M.J. and A.G. Swaby. 1988. Factors affecting the production and motility of zoospores of *Polymyxa graminis* and their transmission of barley yellow mosaic virus (BaYMV). Ann. Appl. Biol. 112: 69–78.

Adams, M.J., P. Jones and A.G. Swaby. 1987. The effect of cultivar used as host for *Polymyxa graminis* on the multiplication and transmission of barley yellow mosaic virus (BaYMV). Ann. Appl. Biol. 110: 321–327.

Adams, M.J., P. Jones and A.G. Swaby. 1988. Purification and some properties of oat golden stripe virus. Ann. Appl. Biol. 112: 285–290.

Adams, M.J., A.G. Swaby and I. MacFarlane. 1986. The susceptibility of barley cultivars to barley yellow mosaic virus (BaYMV) and its fungal vector, *Polymyxa graminis*. Ann. Appl. Biol. 109: 561–572.

Alderson, P.G. 1975. Tobacco stunt: investigations on the complex of disease agent, fungus vector and host plant. Ph.D. Thesis, 234 pages. University of Alberta.

Alderson, P.G. and C. Hiruki. 1977. Scanning electron microscopy of zoospore of *Olpidium brassicae*, free or attached to tobacco roots. Phytopathol. Z. 90: 123–131.

Asher, M.J.C. and S.J. Blunt. 1987. The ecological requirements of *Polymyxa betae*. Proc. 50th Winter Congress of the International Institute of Sugar Beet Research, Brussels. 45–54.

Atkinson, R.E. 1945. A new mosaic chlorosis of oats in the Carolinas . Plant Dis. Rep. 29: 86–89.

Barr, D.J.S. 1979. Morphology and host range of *Polymyxa graminis, Polymyxa betae,* and *Ligniera pilorum* from Ontario and some other areas. Can. J. Plant Pathol. 1: 85–94.

Barr, D.J.S. 1988. Zoosporic parasites as fungal vectors of viruses: taxonomy and life cycles of species involved. *In*: J.L. Cooper and M.J.C. Asher (eds.). Viruses with Fungal Vectors. 123–137.

Barr, K.J. and M.J.C. Asher. 1989. Studies on the life-cycle of *Polymyxa betae* in sugar beet roots. Mycol. Res. 100: 203–208.

Beemster, A.B.R. and A. de Heij. 1987. A method of detecting *Polymyxa betae* and beet necrotic yellow vein virus in soil using sugar-beet as a bait plant. Neth. J. Plant Pathol. 93: 283–288.

Beever, J.E. and P.R. Fry. 1970. The effect of light on the transmission of tobacco necrosis virus by *Olpidium brassicae*. Virology 40: 357–362.

Bos, L. and H. Huijberts. 1996. Lettuce ring necrosis, caused by a chytrid-borne agent distinct from lettuce big-vein "virus". Eur. J. Plant Pathol. 102: 867–873.

Brakke, M.K. 1971. Soil-borne wheat mosaic virus. CMI/AAB Descriptions of Plant Viruses, No. 77, 4 pp.

Brakke, M.K. and A.P. Estes. 1967. Some factors affecting vector transmission of soil-borne wheat mosaic virus from root washings and soil debris. Phytopathology 57: 905–910.

Brunt, A.A. 1989. Viruses and virus-like pathogens transmitted by zoosporic fungi. Bull. OEPP 19: 437–451.

Campbell, R.N. 1962. Relationship between the lettuce big vein virus and its vector, *Olpidium brassicae*. Nature 195: 675–677.

Campbell, R.N. 1979. Fungi as vectors of plant viruses. pp. 8–24. *In*: H.P. Molitoris, M. Hollings and H.A. Wood (eds.). Fungal Viruses. Springer-Verlag, New York.

Campbell, R.N. 1980. Effects of benomyl and ribavirin on the lettuce big vein agent and its transmission. Phytopathology 70: 1190–1192.

Campbell, R.N. 1985. Longevity of *Olpidium brassicae* in air-dry soil and the persistence of the lettuce big-vein agent. Can. J. Bot. 63: 2288–2289.

Campbell, R.N. 1988. Cultural characteristics and manipulative methods. pp. 153–165. *In*: J.J. Cooper and M.J.C. Asher (eds.). Viruses with Fungal Vectors.

Campbell, R.N. and P.R. Fry. 1966. The nature of the associations between *Olpidium brassicae* and lettuce big-vein and tobacco necrosis virus. Virology 29: 222–233.

Campbell, R.N. and R.G. Grogan. 1963. Big-vein virus of lettuce and its transmission by *Olpidium brassicae*. Phytopathology 53: 252–259.

Campbell, R.N. and R.G. Grogan. 1964. Acquisition and transmission of lettuce big-vein virus by *Olpidium brassicae*. Phytopathology 54: 681–690.

Campbell, R.N. and H. Lot. 1996. Lettuce ring necrosis, a virus-like disease of lettuce: evidence for transmission by *Olpidium brassicae*. Plant Dis. 80: 611–615.

Campbell, R.N., R.G. Grogan and D.E. Purcifull. 1961. Graft transmission of big vein of lettuce. Virology 15: 82–85.

Campbell, R.N., O. Lovisolo and V. Lisa. 1975. Soil transmission of petunia asteroid mosaic strain of tomato bushy stunt virus. Phytopathol. Mediter. 14: 82–86.

Campbell, R.N. and S.T. Sim. 1994. Host specificity and nomenclature of *Olpidium bronovanus* (=*Olpidium radicale*) and comparison to *Olpidium brassicae*. Can. J. Bot. 72: 1136–1143.

Canova, A. 1966. Studies on virus disease of Gramineae. III. *Polymyxa graminis* as a vector of wheat soil-borne mosaic virus. Phytopathol. Mediter. 5: 53–58.

Claxton, J.R., D.L. Arnold, D. Blakesly and J.M. Clarkson. 1995. The effect of temperature on zoospores of the crook root fungus *Spongospora subterranean* f. sp. *nasturtii*. Plant Pathol. 44: 765–771.

Cooper, J.I., R.A.C. Jones and B.D. Harrison. 1976. Field and glasshouse experiments on the control of potato mop-top virus. Ann. Appl. Biol. 83: 215–230.

d'Ambra, V. and S. Mutto. 1986. Parasitism of *Trichoderma harzianum* on cystosori of *Polymyxa betae*. J. Phytopathol. 115: 61–72.

Dias, H.F. 1970. Transmission of cucumber mosaic virus by *Olpidium cucurbitacearum* Barr & Dias. Virology 40: 828–839.

Estes, A.P. and M.K. Brakke. 1966. Correlation of *Polymyxa graminis* with transmission of soil-borne wheat mosaic virus. Virology 27: 772–774.

Fauquet, C. and J.C. Thouvenel. 1983. Association d'un nouveau virus en batonnet avec la maladie neécrotique á rayures du riz en Côte-d'Ivoire. Compt. Rend. Séances l'Acad. Sci. 296: 575–580.

Fletcher, J.T., W.A. Walis and F. Davenport. 1987. Pepper yellow vein, a new disease of sweet peppers. Plant Pathol. 36: 180–184.

Fujikawa, T. 1955. Study on broad bean necrosis disease. Agr. Hort. (Nogyo oyobi Elgei) 30: 85–86.

Gallitelli, D., C. Voulas and A. Avgelis. 1983. Some properties of cucumber fruit streak virus. Phytopathol. Z. 106: 149–162.

Gerik, J.S., J.C. Hubbard and J.E. Duffus. 1990. Soil matric potential effects on infection by *Polymyxa betae* and BNYVV. Proc. First Symposium of International Working Group on Plant Viruses with Fungal Vectors. R. Koenig ed., Braunschweig, Germany 75–78.

Henry, C.M., R.A.C. Jones and R.H.A. Coutts. 1986. Occurrence of a soil-borne virus of sugar beet in England. Plant Pathol. 35: 585–591.

Hibi, T. and I. Furuki. 1985. Melon necrotic spot virus. CMI/AAB. Descriptions of plant viruses No. 302. AAB, Wellesbourne, U.K.

Hidaka, Z. and H. Tagawa. 1965. Transmission of tobacco stunt virus by *Olpidium brassicae*. Ann. Phytopathol. Soc. Japan 31: 369–372.

Hidaka, Z., C. Hiruki, K. Nakano, T. Shimizu and T. Uozumi. 1956. Studies on the tobacco stunt disease. Bull. Hatano Tob. Expt. Sta. 40: 1–74.

Hiruki, C. 1965. Transmission of tobacco stunt virus by *Olpidium brassicae*. Virology 25: 541–549.
Hiruki, C. 1967. Host specificity in transmission of tobacco stunt virus by *Olpidium brassicae*. Virology 33: 131–136.
Hiruki, C. 1969. An incubator useful for culturing *Olpidium brassicae* for transmission of plant viruses. Virology 39: 333–335.
Hiruki, C. 1987. Recovery and identification of tobacco stunt virus from air-dried resting spores of *Olpidium brassicae*. Plant Pathol. 36: 224–228.
Hiruki, C. 1989. Tobacco stunt and lettuce big-vein viruses: biology, economic impact and control. Bull. OEPP 19: 471–478.
Hiruki, C. 1990. The ultrastructure of the resting spore formation and germination of *Olpidium brassicae*. Proc. First Symposium of International Working Group on Plant Viruses with Fungal Vectors, R. Koenig (ed.). Braunschweig, Germany. 163–166.
Hiruki, C. 1993. Zoospore encystment of *Olpidium brassicae* is triggered by fusion between the body and flagellum membrane. Proc. Second Symposium of International Working Group on Plant Viruses and Fungal Vectors. C. Hiruki ed., Montreal, Canada pp. 95–98.
Hiruki, C. 1994. Multiple transmission of plant viruses by *Olpidium brassicae*. Can. J. Plant Pathol. 16: 261–265.
Hiruki, C. 1996. Camellia yellow mottle(?) varicose virus. pp. 285–286. *In*: A. Brunt, K. Crabtree, M. Dallwitz, A.W. Gibbs and L. Satson [eds.]. Viruses of Plants, Descriptions and Lists from the VIDE Database. CAB International, Oxon, UK.
Hiruki, C. 2008. Tama-no-ura camellia: the current situation, history, and future possibilities. International Camellia Journal 66–70.
Hiruki, C. 2012. Fungal transmission of a color-breaking virus in Tama-no-ura camellia in Japan. Proc. International Camellia Congress, Chuxiong, China, Z.L. Wang ed. 233–238.
Hiruki, C. and D.S. Teakle. 1987. Soil-borne viruses of plants. Curr. Topics in Vector Res. 3: 177–215.
Hoaglund, D.R. and D.I. Arnon. 1950. The water-culture method for growing plants without soil. Calif. Univ. Agric. Exp. Sta. Bull. No. 347 (revised).
Huth, W. 1988. Barley yellow mosaic virus—a disease in Europe caused by two different viruses. *In*: J.I. Cooper and M.J.C. Asher (eds.). Viruses with Fungal Vectors. AAB Wellsbourne, U.K. 61–70.
Inouye, T. and S. Fujii. 1977. Rice necrosis mosaic virus. CMI/AAB Descriptions of Plant Viruses No. 172: 4.
Jiang, L. and C. Hiruki. 1996. PCR amplification and restriction analysis of the ribosomal DNA of *Olpidium radicale* isolates. J. Microbiol. Methods 26: 87–93.
Jones, R.A.C. and B.D. Harrison. 1969. The behaviour of potato mop-top virus in soil and evidence for its transmission by *Spongospora subterranea*. Ann. Appl. Biol. 63: 1–17.
Kassanis, B. and I. MacFarlane. 1964. Transmission of tobacco necrosis virus by zoospores of *Olpidium brassicae*. J. Gen. Microbiol. 36: 79–93.
Kassanis, B. and I. MacFarlane. 1968. The transmission of satellite viruses of tobacco necrosis virus by *Olpidium brassicae*. J. Gen. Virol. 3: 227–232.
Koenig, R., D.E. Lesemann, W. Huth and M. Makkouk. 1983. Comparison of a new soil-borne virus from cucumber with Tombus-, Diantho-, and other similar viruses. Phytopathology 73: 515–520.
Kole, A.P. 1954. A contribution to the knowledge of *Spongospora subterranea* (Wallr.) Lagerh., the cause of powdery scab of potatoes. Tijd. K. 60: 1–65.
Kusaba, T. and A. Toyama. 1970. Transmission of soil-borne barley yellow mosaic virus. I. Infectivity of diseased root washings. Ann. Phytopathol. Soc. Japan 36: 214–222.
Ledingham, G.A. 1939. Studies on *Polymyxa graminis*, n. gen. n. sp., a plasmadiophoraceous root parasite of wheat. Can. J. Res. Series C 17: 38–51.
Legreve, A., P. Delfosse, B. Vanpee, A. Goffin and H. Maraite. 1998. Differences in temperature requirements between *Polymyxa* sp. of Indian origin and *Polymyxa graminis* and *Polymyxa betae* from temperate areas. Eur. J. Plant Pathol. 104: 195–205.

Lin, M.T., R.N. Campbell, P.R. Smith and J.H.M. Temmink. 1970. Lettuce big vein virus transmission by single-sporangium isolates of *Olpidium brassicae*. Phytopathology 60: 1630–1634.

Lindsten, K. 1990. A rapid method to detect *Polymyxa*-transmitted beet viruses from soil. Proc. First Symposium of International Working Group on Plant Viruses with Fungal Vectors. R. Koenig ed., Braunschweig, Germany pp. 13–16.

Linford, M.B. and H.H. McKinney. 1954. Occurrence of *Polymyxa graminis* in roots of small grains in the United States. Plant Dis. Rep. 38: 711–713.

Littlefield, L.J., P. Delfosse, J.H. Whallon, Z.M. Hassan, J.L. Sherwood and D.V.R. Reddy. 1997. Anatomy of sporosori of *Polymyxa graminis*, the vector of Indian peanut clump virus, in roots of *Sorghum bicolor*. Can. J. Plant Pathol. 19: 281–288.

Lu, X., So. Yamamoto, Y. Tanaka, T. Hibi and S. Namba. 1996. The genome organization of broad bean necrosis virus (BBNV) and heterogeneity of the genus Furovirus. Proc. Third Symposium of International Working Group on Plant Viruses with Fungal Vectors. J.L. Sherwood and C.M. Rush eds., Dundee, U.K. 69–72.

MacFarlane, I. 1982. Virus transmission by fungi. Red clover necrotic mosaic virus. *In*: Rothamsted Experimental Station Report 1981, Harpenden, U.K. 190.

Merz, U. 1989. Infectivity, inoculum density and germination of *Spongospora subterranean* resting spores: a solution–culture test system. Bull. EPPO 19: 585–592.

Miller, J.S., H. Damude, M.A. Robbins, R.D. Reade and D.M. Rochon. 1996. The genome organization and RNA sequence of cucumber leaf spot virus, a tombus-like virus. Proc. Third Symposium of International Working Group on Plant Viruses with Fungal Vectors. J.L. Sherwood and C.M. Rush eds., Dundee, U.K. pp. 21–24.

Mugnier, J. 1987. Infection by *Polymyxa betae* and *Plasmodiophora brassicae* of root containing root-inducing transferred DNA of *Agrobacterium rhizogenes*. Phtopathology 77: 539–542.

Mutasa, E.S., D.M. Chwarszczynska and M.H.C. Asher. 1996. Single-tube nested PCR for the diagnosis of *Polymyxa betae* infection in sugar beet roots and its application in field studies. Phytopathology 86: 493–497.

Mutasa-Gottgens, E.S., D.M. Chwarszczynska, G.E. Williams, E. Ward, M.J. Adams and M.J.C. Asher. 1996. Development and application of molecular methods for the study of *Polymyxa betae*. Proc. Third Symposium of International Working Group on Plant Viruses with Fungal Vectors. J.L. Sherwood and C.M. Rush eds., Dundee, UK pp. 173–176.

Plumb, R.T. and I. MacFarlane. 1977. A "new" virus of oats. Report of the Rothamsted Experimental Station for 1976: 256–257.

Ratna, A.S., A.S. Rao, A.S. Reddy, B.L. Nolt, D.V.R. Reddy, M. Vijayalakshmi and D. McDonald. 1991. Studies on transmission of Indian peanut clump virus disease by *Polymyxa graminis*. Ann. Appl. Biol. 118: 71–78.

Reddy, D.V.R., R. Rajeshwari, N. Iizuka, D.E. Lesemann, B.L. Nolt and T. Goto. 1983. The occurrence of Indian peanut clump, a soil-borne virus disease of groundnuts (*Arachis hypogaea*) in India. Ann. Biol. 102: 305–310.

Sahtiyanci, S. 1962. Stidien uber einige Wurzelparasitaare Opidiaceen. Arch. Microbiol. 41: 187–228.

Sawada, S. 1927. Wheat yellow mosaic prevention. J. Plant Protection (Byochugai Zasshi) 14: 444–449.

Slykhuis, J.T. 1970. Factors determining the development of wheat spindle streak mosaic caused by a soil-borne virus in Ontario. Phytopathology 60: 319–331.

Slykhuis, J.T. 1975. Effects of drying on the transmissibility of wheat spindle streak mosaic virus in soils from wheat fields in Ontario. Can. Plant Dis. Surv. 55: 106–108.

Smith, P.R., R.N. Campbell and P.R. Fry. 1969. Root discharge and soil survival of viruses. Phytopathology 59: 1678–1687.

Stobbs, L.W., G.W. Cross and M.S. Manocha. 1982. Specificity and methods of transmission of cucumber necrosis virus by *Olpidium radical* zoospores. Can. J. Plant Pathol. 4: 134–142.

Tamada, T. 1975. Beet necrotic yellow vein virus. CMI/AAB Descriptions of Plant Viruses 144: 4.

Teakle, D.S. 1962. Transmission of tobacco necrosis virus by a fungus, *Olpidium brasicae*. Virology 18: 224–231.

Teakle, D.S. 1969. Fungi as vectors and hosts of viruses. pp. 23–54. *In:* K. Maramorosch [ed.]. Viruses, Vectors, and Vegetation. Interscience Publ., New York.

Teakle, D.S. 1983. Zoosporic fungi and viruses. Double trouble. pp. 233–248. *In*: S.T. Buczaki [ed.]. Zoosporic Plant Pathogens. Academic Press, London, U.K.

Teakle, D.S. 1988. The effect of environmental factors on fungus-transmitted viruses and their vectors. pp. 167–179. *In*: J.L. Cooper and M.J.C. Asher [eds.]. Viruses with Fungal Vectors.

Teakle, D.S. and A.H. Gold. 1964. Prolonging the motility and virus-transmitting ability of *Olpidium* zoospores with chemicals. Phytopathology 54: 29–32.

Teakle, D.S. and C. Hiruki. 1964. Vector specificity in *Olpidium*. Virology 24: 539–544.

Teakle, D.S. and B.J. Thomas. 1985. Effect of heat on zoospore motility and multiplication of *Olpidium radical* and *O. brassicae*. Ann. Appl. Biol. 107: 11–15.

Thouvenel, J.C. and C. Fauquet. 1980. *Polymyxa graminis* on new *Sorghum* species in Africa. Plant Dis. 64: 957–958.

Tomlinson, J.A. and R.G. Garrett. 1964. Studies on the lettuce big-vein virus and its vector *Olpidium brassicae* (Wor,). Dang. Ann. Appl. Biol. 54: 45–61.

Tomlinson, J.A. and B.J. Thomas. 1986. Studies on melon necrotic spot virus disease of cucumber and on the control of the fungus vector (*Olpidium radicale*). Ann. Appl. Biol. 108: 71–80.

Van Dorst and D. Peters. 1988. Experiences with the freesia leaf necrosis and its presumed vector, *Olpidium brassicae*. pp. 315–322. *In:* J.I. Cooper and M.J.C. Asher [eds.]. Viruses with Fungal Vectors. AAB, Wellesborune, U.K.

Walsh, J.A., C.M. Clay and A.A. Miller. 1989. A new disease of watercress in England. Bull. OEPP 19: 463–470.

Ward, E., M.J. Adams, E.S. Mutasa, C.R. Collier and M.J.C. Asher. 1994. Characterization of *Polymyxa* species by restriction analysis of PCR-amplified ribosomal DNA. Plant Pathol. 43: 872–877.

Ward, E. and M.J. Adams. 1996. Molecular studies of variation amongst isolates of *Polymyxa graminis* and *Polymyxa betae*. Proc. Third Symposium of International Working Group on Plant Viruses with Fungal Vectors. J.L. Sherwood and C.M. Rush eds., Dundee. U.K. 93–96.

Westerlund, F.V., R.N. Campbell and J.M. Duniway. 1978. Soil factor affecting the reproduction and survival of *Olpidium brassicae* and its transmission of big vein agent to lettuce. Phytopathology 68: 927–935.

Westerlund, F.V., R.N. Campbell and R.G. Grogan. 1978. Effect of temperature on transmission, translocation, and persistance of the lettuce big-vein agent and big-vein symptom expression. Phytopathology 68: 921–926.

Wisler, G.D. and J.E. Duffus. 1996. Comparative molecular analyses of several BNYVV and BSBMV-related furoviruses infecting sugar beet. Proc. Third Symposium of International Working Group on Plant Viruses and Fungal Vectors. J.L. Sherwood and C.M. Rush eds., Dundee, UK pp. 53–56.

16

Sterile Techniques for Rearing and Cell Culture of Leafhopper Vectors of Plant Pathogens

Jun Mitsuhashi

ABSTRACT

This article summarizes the method for rearing leafhoppers under sterile conditions and the method for culturing leafhopper cells *in vitro*. The methods for the introduction of leafhopper-borne plant pathogens to these insects and cells are described as well.

Key Words: Rearing on sterile plants, artificial diets, cell cultures, leafhopper, extragerm-free

Introduction

In order to maintain leafhopper -borne plant viruses (hereafter the term "leafhoppers" includes "planthoppers" and "treehoppers" unless otherwise stated) in laboratories, the use of extragerm-free leafhoppers or cultured leafhopper cells is beneficial.

Koishikawa 1-28-13, Bunkyo-ku, Tokyo 112-0002, Japan.

Extragerm-free Rearing of Leafhoppers

Since most of homopteran insects posses various endosymbionts, especially intracellular symbionts, continuous rearing of these insects completely free from other organisms is extremely difficult if not impossible. However, it is possible to rear these insects free from extranous microorganisms. In this sense, axenic or aseptic rearing of these insects is practically impossible.

There are 2 methods of rearing leafhoppers under extranous microorganism-free conditions, the one is rearing leafhoppers on host plants cultivated in sterile conditions, and the other is rearing leafhoppers on sterilized artificial diets.

Rearing Leafhoppers on Sterile Host Plants

In order to supply sterile plants to leafhoppers, the host plants should be grown under sterile conditions. Sterile plant seedlings can be obtained from germination of the surface-sterilized plant seeds. For sterilization of seeds, treatment with 70% ethyl alcohol for 5–10 min is common. However, if seeds are sensitive to ethyl alcohol or 70% alcohol is ineffective, 0.1% aqueous solution of mercuric chloride may be used. In this case, if plant seeds are covered with fine hair, contact of the seed surface to sterilization agent may be incomplete. To make contact perfect, incorporation of sodium chloride as wetting agent to mercuric chloride solution at the concentration of 5% is recommended.

After surface sterilization, the seeds are washed with sterile distilled water, and placed on sterilized agar media. The media will be 1.2% agar containing some minerals such as component of Hogland and Knop's solution (Table 1).

The media are dispersed to flasks, bottles or tubes of appropriate sizes with cotton or plastic plugs and autoclaved at 121°C for 15 minutes. When germinated, the plants are kept at 25°C under long day photoperiod (Fig. 1). Leafhoppers lay their eggs in plant tissues as mass or separately.

Table 1. Modified Hogland and Knop's Nutrient solutions (after Chen et al. 1961).

Chemicals	Amount in 1000 ml	Chemicals	Amount in 1000 ml
$Ca(NO_3)_2 4H_2O$	0.950 g	KNO_3	0.610 g
$MgSO_4 7H_2O$	0.490 g	$NH_4H_2PO_4$	0.120 g
$MnSO_4 4H_2O$	3.000 mg	$ZnSO_4 7H_2O$	0.500 mg
H_3BO_3	0.500 mg	$CuSO_4 5H_2O$	0.025 mg
$Na_2MoO_4 2H_2O$	0.025 mg	H_2SO_4	0.500 µl
$FeC_6H_5O_7 5H_2O$	20.000 mg		

Leafhoppers are allowed to oviposit on their host plants from several hours to several days. The plants are harvested and the leafhopper eggs are taken out from the plant tissues under a dissecting microscope with needles or a pair of forceps.

The eggs taken out can be kept in water. When sufficient number of eggs is obtained, the water is discarded and 70% ethyl alcohol is introduced. The alcohol is agitated for one minute, and then replaced with sterile distilled water. The surface sterilized eggs are placed onto the sterilized seedlings

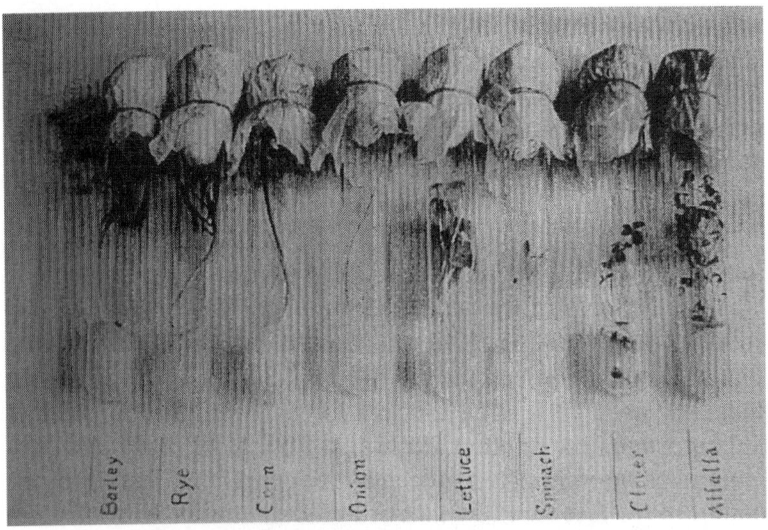

Fig. 1. Plant seedlings grown under aseptic conditions.

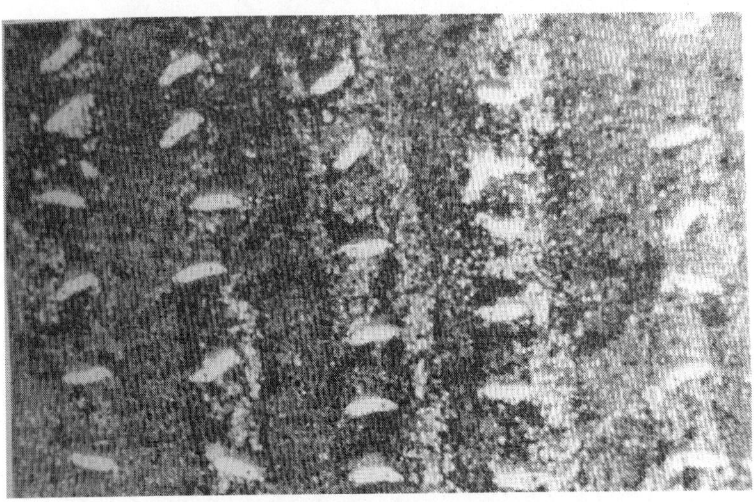

Fig. 2. Leafhopper eggs pasted on a piece of paper.

one by one with a tiny drop of water by means of a pipette. If the eggs fall down from the plants after evaporation of the water, eggs may be glued to a small piece of paper with albumin when the eggs are collected (Fig. 2) and the paper as a whole is sterilized, and then placed onto the seedlings.

Soon nymphs will hatch under sterile conditions, and suck the sap of the seedlings. Care should be taken not to draw the nymphs in drops of condensed water. If too much water condensation wets the inside of the container, a sterile bag of silica gel may be hanged in the container (Fig. 3). When plants wilt or become too old, the leafhoppers should be transferred to another container containing a fresh plant.

Adult leafhoppers may copulate under sterile condition (Fig. 4) and lay eggs into the sterile plants. Successive rearing under sterile conditions is possible in this manner (Mitsuhashi and Maramorosch 1963, Mitsuhashi 1965).

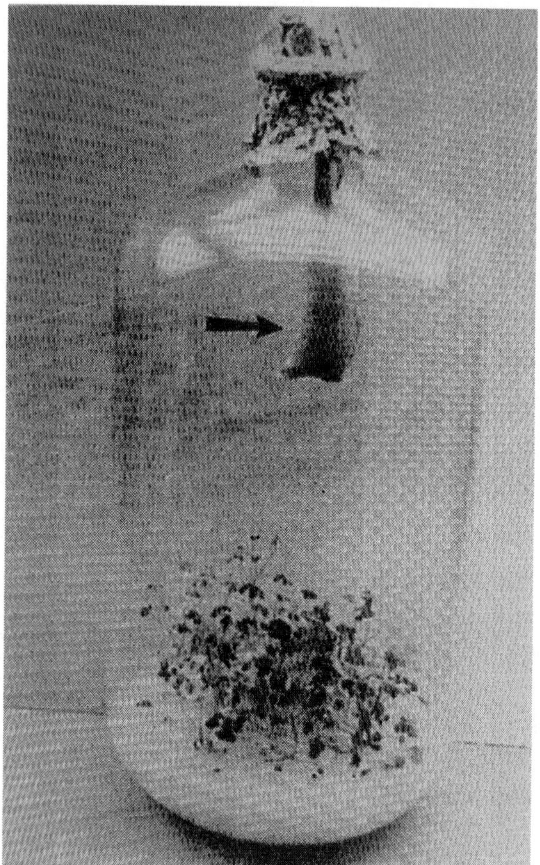

Fig. 3. A cotton bag (arrow) hanged in a sterile bottle.

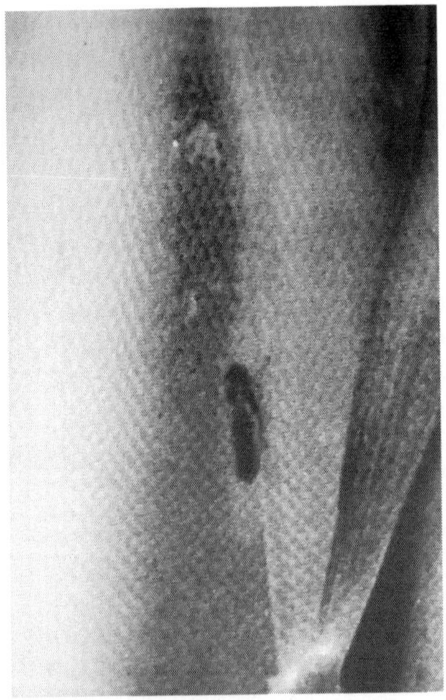

Fig. 4. A leafhopper on a sterile plant.

As a modification of rearing leafhoppers on sterile host plants, the one on plant calli is possible (Fig. 5). In this case, the callus should be obtained from the host plant of the leafhopper. When the callus grows sufficiently large, sterilized leafhopper eggs, or nymphs or adults reared under sterile conditions are introduced onto the callus. Although development of vascular bundles is not certain calli grown on agar media, leafhoppers can suck sufficient nutrient from calli. Successive rearing of leafhoppers on the plant callli is possible, if the calli are suitable as the host of the leafhoppers (Mitsuhashi and Maramorosch 1963).

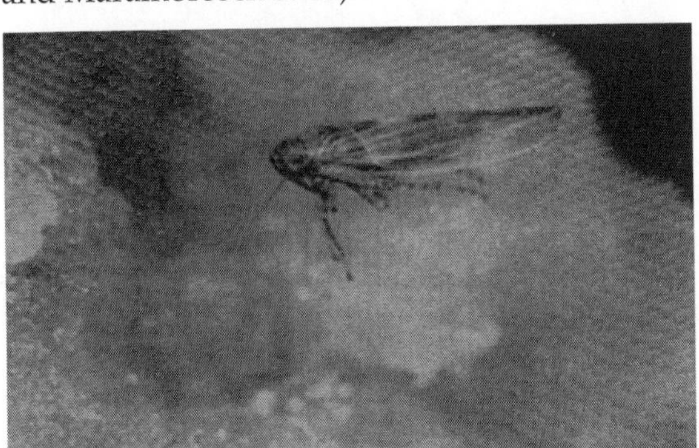

Fig. 5. A leafhopper living on carrot callus under sterile conditions.

Rearing Leafhoppers on Sterile Artificial Diets

Some leafhopper species can be reared on artificial diets (Mitsuhashi and Koyama 1971, Koyama 1973, 1979, Koyama and Mitsuhashi 1980, Koyama et al. 1981, Hou and Brooks 1975, Hou and Lin 1979). The diet is an aqueous solution of nutrients. It consists of minerals, sugars, amino acids and vitamins (Mitsuhashi and Koyama 1971) (Table 2).

From the studies of nutritional requirements, following have been shown to be essential for the growth of the small brown planthopper, *Ladodelphax striatellus*: K, Mg, PO_4 for minerals (Koyama and Mitsuhashi 1991), cysteine, and methionine, for amino acids (Koyama and Mitsuhashi 1975), thiamine, pyridoxine and pantothenate for vitamins (Koyama and

Table 2. Composition of MED-1 Holidic Diet (after Mitsuhashi and Koyama 1971).

Chemicals	Amount (mg in 100 ml)	Chemicals	Amount (mg in 100 ml)
$MgCl_2.6H_2O$	200	L-Tryptophan	100.000
KH_2PO_4	500	L-Tyrosine	20.000
L-Alanine	100	L-Valine	200.000
ɤ-Aminobutryic Acid	20	Thiamine hydrochloride	2.500
L-Arginine hydrochloride	400	Riboflavin	5.000
L-Asparagine	300	Nicotinic acid	10.000
L-Aspiratic acid	100	Pyridoxine hydrochloride	2.500
L-Cysteine	50	Folic acid	1.000
L-Cystine hydrochloride	5	Calcium pantothenate	5.000
L-Glutamic acid	200	Inositol	50.000
L-Glutamine	600	Choline Chloride	50.000
Glycine	20	Biotin	0.100
L-Histidine	200	Sodium L-ascorbate	100.000
DL-Homoserine	800		
L-Isoleucine	200	Sucrose	5,000.000
L-Leucine	200		
L-Lysine hydrochloride	200	$CuCl_2.4H_2O$	0.268
L-Methionine	100	$MnCl_2.2H_2O$	0.793
L-Phenylalanine	100	$ZnCl_2$	0.396
L-Proline	100	$CaCl_2.2H_2O$	3.115
DL-Serine	100	$FeCl_2.2H_2O$	2.228
L-Threonin	200		

pH: 6.5 with KOH

Mitsuhashi 1977). As carbohydrates, several sugars can be utilized; however sucrose was found the best (Mitsuhashi and Koyama 1969). The pH of the nutrient solution is adjusted to 6.5 with KOH, and the solution is sterilized by passing through a membrane filter 0.2 µm pore size. The sterile diet can be stored at 20°C for years. As a representative of synthetic diet, the composition of MED-1 is shown in Table 2.

The extra germ-free leafhoppers are obtained in the same manner as described in the preceding section. The leafhoppers reared on sterile plants or on plant calli can be used too. In case of beginning with newly hatched nymphs, surface-sterilized eggs are kept on a sterilized filter paper wetted with sterilized distilled water in a petri dish until hatching. When the eggs hatch, the hatched leafhoppers are transferred aseptically into a rearing vessel.

There are several methods of supplying liquid diets to leafhoppers. One is wick-feeding method. A rearing tube for wick-feeding under sterile conditions has been devised (Mitsuhashi and Maramorosch 1963). However, in this method leafhoppers make a direct contact with the diet (Fig. 6) and consequently the diet attaches to the bodies of the leafhoppers. When water in the diet evaporates, sucrose in the diet is concentrated resulting in the formation of sticky liquid on the body surface of the leafhoppers. This interferes with the movement of the leafhoppers and often kills leafhoppers by adhering them to the wall of the container and/or the wick.

Feeding of homopteran insects on liquid diets through a membrane has long been attempted. The problem of this method was finding

Fig. 6. A wick feeding apparatus for leafhoppers.

suitable membranes. However, this problem was overcome by the use of Parafilm® (American Can Co., Neenah, Wisconsin) or other similar artificial membranes. Now rearing of aphids on a sachet of Parafilm® is a very common technique (Mittler and Dadd 1962, Auclair 1963). A similar method can be applied for leafhoppers (Koyama and Mitsuhashi 1969) (Fig. 7).

For extragerm-free rearing, glass cups (30 mm, in diameter and 30 mm in height) are sterilized by dry heat at 180°C for 30 minutes, and membranes can be sterilized by dry heat at 180°C for 30 minutes, and membranes can be sterilized by immersion in 70% ethyl alcohol for 10 minutes or longer. For frequent use, the cut membranes can be stored in 70% ethyl alcohol. When a rearing vessel is constructed, a piece of membrane is placed on a sterilized filter paper to remove ethyl alcohol in a laminar flow cabinet. Then the membrane is stretched and placed on to the top of a glass cup into which leafhoppers have already been introduced. A drop of the diet is placed on the center of the stretched membrane and is covered with another stretched membrane.

By this set up, leafhoppers can be reared in an extragerm-free environment. However, it is strongly recommended to replace the sachet at least once a week. Unless contaminated, leafhoppers may survive on the same sachet more than 1 week, but the diet may denature resulting in ultimate death of the insects.

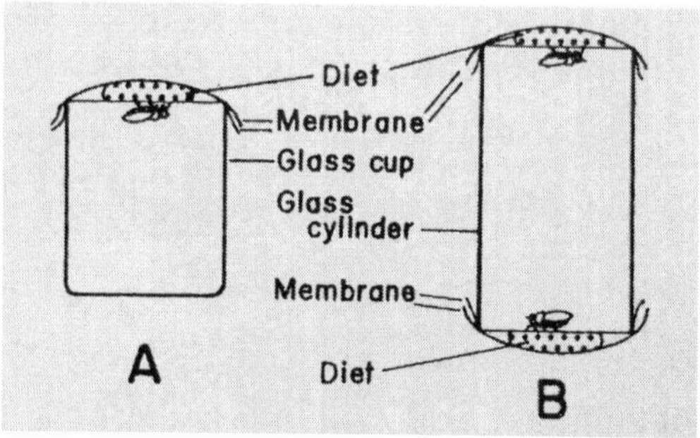

Fig. 7. Membrane feeding apparatus for leafhoppers. **A**: for individual rearing. **B**. for mass rearing.

Acquisition and Transmission of Disease Agents under Sterile Conditions

A sterile viruliferous leafhopper will be obtained by the following method (Fig. 8), First, a host plant is cultivated under sterile conditions (Fig. 8A). When the plant has several leaves, a leaf is pulled out from the culture vessel having leaving remaining parts in the sterile vessel (Fig. 8B). A viruliferous leafhopper is allowed to suck the fluid from the leaf (Fig. 8C). After a certain

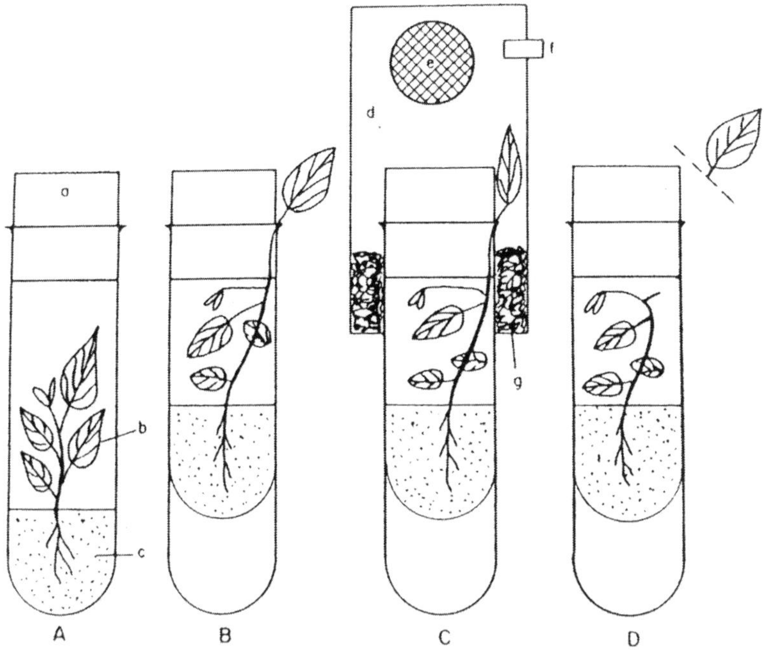

Fig. 8. A method for obtaining sterile viruliferous plant. **A, B, C,** and **D** show a single plant during consecutive steps. Abbreviations: a: plastic stopper; b: aseptically grown seedling; c: agar medium; d: insect cage made of cellulose nitrate; e: window with nylon screen; f: small hole for introducing insects into the cage; g: cotton plug.

feeding periods, the leaf is cut off at the base of the petiole (Fig. 8D). The plant is kept under sterile conditions until a symptom of infection appears. If the symptom appears, sterile health leafhoppers are allowed to feed on the diseased plants. In this way the sterile viruliferous leafhopper can be obtained (Mitsuhashi and Maramorosch 1964).

By transferring these sterile viruliferous leafhoppers to sterile healthy plants, the disease agents can be transmitted to the healthy plants. By this method even a plant callus can be infected, and from it leafhoppers again can be infected with the disease agent.

Leafhopper Cell Cultures

Cell Cultures

Cell cultures of leafhoppers (Cicadellidae) are relatively easy at least at primary culture level, however, those of planthoppers (Delphacidae) are extremely difficult. There are several continuously growing cell lines from cicadellids (Chiu and Black 1967, 1969, Kimura 1984), but no such cell lines have been obtained from Delphacids, except one case, in which the cells from embryos of the smaller brown plant hopper, *Laodelphax striatellus*, are said to be passed more than 100 times (Kimura, personal communication).

In Cidadellids cell cultures, growing cells can be obtained from embryonic, nymphal and imaginal tissues (Mitsuhashi and Maramorosch 1964). Among these developmental stages, embryos may be the most suitable material for cell cultures, because of ease of surface sterilization and induction of cell migration.

The eggs of leafhoppers are surface sterilized as described in the section "Rearing leafhoppers on sterile host plants". The sterilized eggs are washed in sterile distilled water and transferred to physiological saline. In this solution, eggs are cut and embryos are washed with saline and cut into several pieces. These fragments of embryos are transferred into culture media and brought into culture vessels. As the media Mitsuhashi and Maramorosch's medium (Table 3) (Mitsuhaashi and Maramorosch 1964) or MGM-450 medium (Mitsuhashi and Inoue 1988), are recommended. Soon after setting up the cultures, cells will start migration, and small monolayers without continuing of epithelial cell-like cells (Fig. 9), and networks of fibroblastic cells (Fig. 10). These cells all develop by attaching to the substrate. However, sometimes cells, situated in explants which were apart from the substrate come out from the explants without contacting the substrate. Such cells often formed so-called vesicles (Fig. 12), the surface of which consisted of a monolayer of cells.

The inside of the vesicles is filled with a liquid, probably the culture medium itself. The vesicles can enlarge by the increase in number of migrated cells. When the vesicles are broken, some attach to the substrate and form a monolayer of cells.

Table 3. Composition of Mitsuhashi and Maramorosch's medium (Mitsuhashi and Maramorosch 1964).

Chemicals	Amount (mg in 100 ml)	Chemicals	Amount (mg in 100 ml)
NaCl	700.0	Glucose	400.0
$NaH_2PO_4.H_2O$	20.0	Lactalbumin hydrolysat	650.0
$NaHCO_3$	12.0	TC-yeastolate	500.0
KCl	20.0	Fetal Bovine serum	0–20 ml
$MgCl_2.6H_2O$	10.0		
$CaCl_2H_2O$	20.0		

pH: 6.5 with KOH

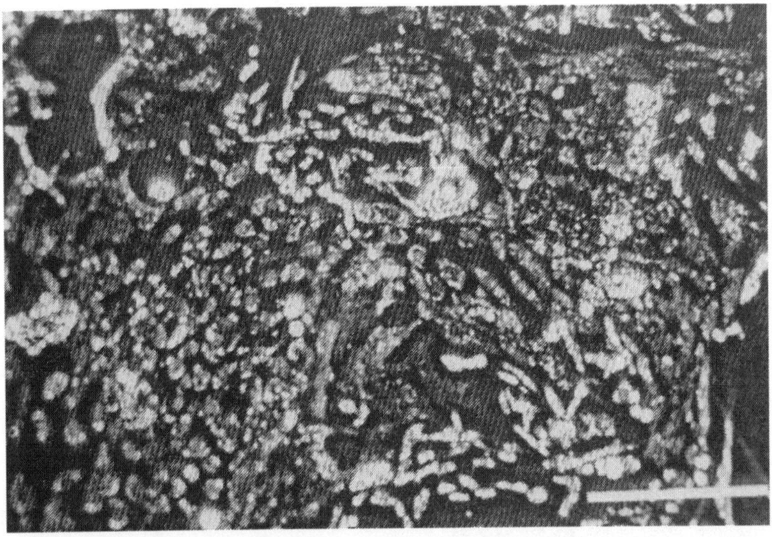

Fig. 9. Epithelial cell-like cells in the primary culture of leafhopper, *Nephotettix cincticeps*, embryonic cells. Bar indicates 100 μm.

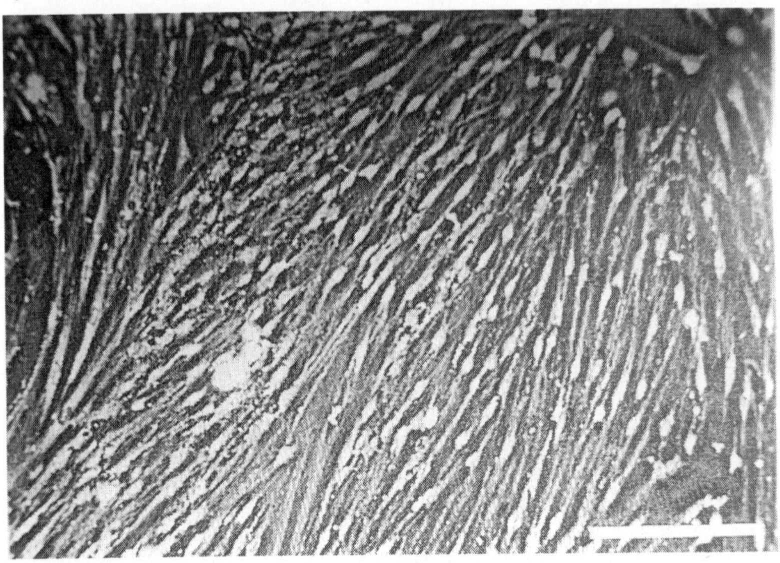

Fig. 10. Fibroblast-like cells in the primary culture of leafhopper, *Recilia dorsalis*, embryonic cells. Bar indicates 100 μm.

Following migration of cells, cells start to proliferate if the culture conditions are favorable. When the cell number increases sufficiently, subcultures will be made. Usually cells are detached from the substrate by the treatment of protein-digesting enzymes, such as trypsin, pancreatin or

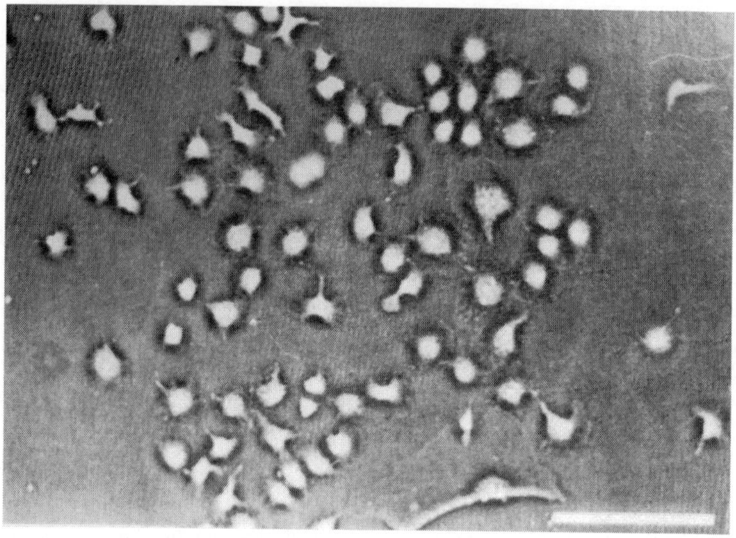

Fig. 11. Hemocyte-like wandering cells in the primary culture of leafhopper, *Recilia dorsalis*, embryonic cells. Bar indicates 100 μm.

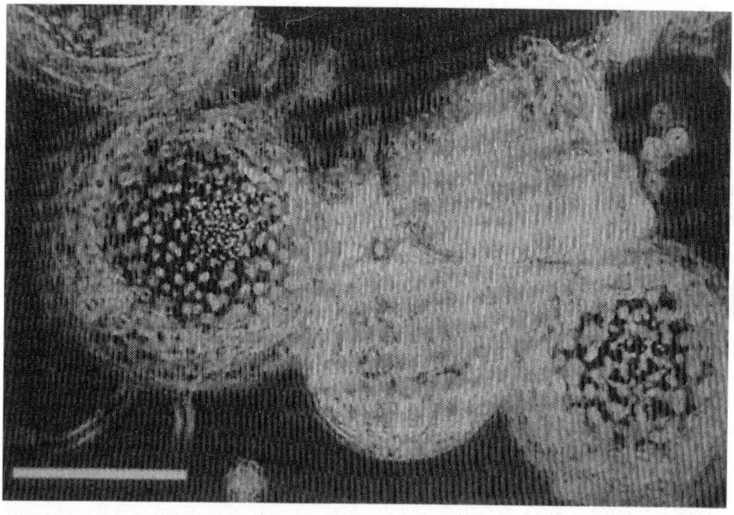

Fig. 12. Hollow spherical vesicles formed in the primary culture of a leafhopper, *Nephotettix apicalis*, embryonic cells. Bar indicates 100 μm.

dispase. When the digestion of cell surface proceeds, the substrate-attached flat cells become three-dimensional. The digestion is stopped by the addition of sera or by the dilution of enzyme solution. The cells with partially digested surfaces can be detached easily from the substrate by pipetting. The detached cells form cell suspensions, and the cells are spun down by centrifugation. The cell pellet is dispersed in fresh medium and dispensed to several vessels. In this manner, cells can be cultured continuously and finally continuous cell lines may be obtained. The continuously growing cells can be stored under -80C with 10% glycerol or DMSO as a cryoprotectant.

Inoculation of Cultured Cells by Plant Disease Agents

For inoculating cultured cells, leafhopper-borne plant disease agents such as viruses and phytoplasmas can be obtained from viruliferous leafhoppers. The viruliferous leafhoppers are surface sterilized by submersion in 70% ethyl alcohol and dissected. The fat bodies, in which most persistent disease agents replicate, are taken out and macerated with the medium for cell cultures under sterile conditions. The homogenate is centrifuged with low speed (150–200Xg) and the supernatant is used as inoculum for cultured cells. Inoculation is made by simply adding a drop of the homogenate to the healthy cell cultures. In most cases where cultured vector insect cells are infected with plant disease agents, they do not show cytopathic effects. They may continue to proliferate (Fig. 13). Therefore, carrier cultures with disease agents will be possible, and by this method disease agents can be passed continuously.

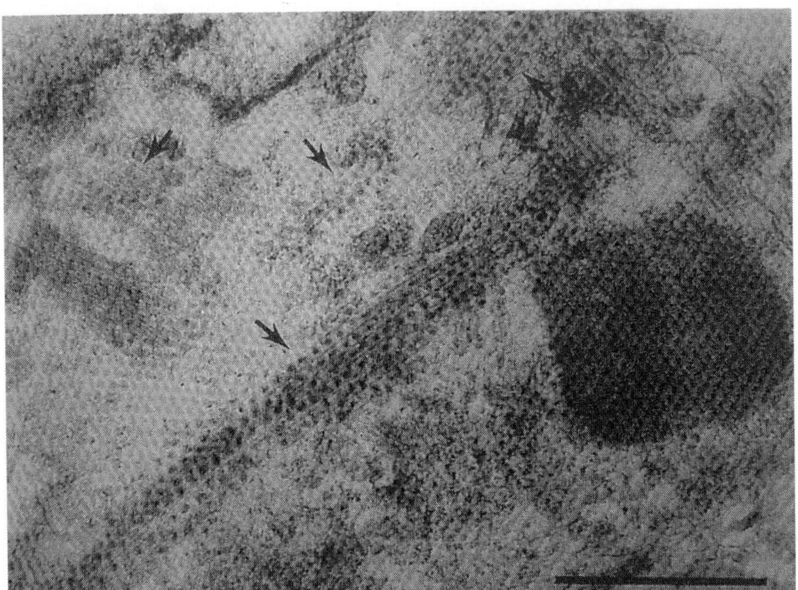

Fig. 13. Rice dwarf virus (arrow) replicated in the inoculated cells of its vector rice green leafhopper, *Nephoteus cincticeps*, cultured *in vitro*. Bar indicates 1.0 μm.

References

Auclair, J.L. 1963. Pea aphid: rearing on a chemically defined diet. Science 22: 1068–1969.

Chen, Tseh-an, R. A. Kilkpatrick and A.A. Rich. 1961. Sterile culture techniques as tools in plant nematology research. Phytopathology 51: 799–800.

Chiu, R. and L.M. Black. 1967. Monolayer cultures of insect cell lines and their inoculation with a plant virus. Nature 215: 1076–1078.

Chiu, R. and L.M. Black. 1969. Assay of wound tumor virus and the fluorescent cell counting technique. Virology 37: 667–677.

Hou, R.F. and A.M. Brooks. 1975. Continuous rearing of the aster leafhopper, *Macrosteles fascifrons*, on a chemically defined diet. J. Physiol. 21: 1481–1483.

Hou, R.F. and L.C. Lin. 1979. Artificial rearing of rice green leafhopper, *Neophotettix cinceciteps*, on a holidic diet. Entomol. Exp. Appl. 25: 158–164.

Kimura, I. 1984. Establishment of new cell lines from leafhopper vector and inoculation of its cell monolayers with rice dwarf virus. Proc. Japan Acad. 60B: 198–201.

Koyama, K. 1973. Rearing of *Inozuma dorsalis* and *Nephotettix cincticeps* on a synthetic diet. Jap. J. Appl. Entomol. Zool. 17: 163–166 (in Japanese).

Koyama, K. 1979. Rearing of the brown planthopper, *Nilaparvata lugens* Såtl (Hemiptera: Delphacidae) on a synthetic diet. Jap. J. Appl. Entomol. 23: 39–49 (in Japanese).

Koyama, K. and J. Mitsuhashi. 1969. Artificial feeding of smaller brown planthopper, *Laodelphax striatellus* Fallen (Hemiptera: Delphacidae). Jap. Appl. Entomol. Zool. 13: 89–90 (in Japanese).

Koyama, K. and J. Mitsuhashi. 1975. Essential amino acids for the growth of the smaller brown planthopper, *Laodelphax striatellus* Fallen (Hemiptera: Delphacidae). Appl. Entomol. Zool. 10: 208–215.

Koyama, K. and J. Mitsuhashi. 1977. Essential vitamins for the nymphal growth of the smaller brown planthopper, *Laodelphax striatellus*. Jap. J. Appl. Entomol. Zool. 21: 23–26 (in Japanese).

Koyama, K. and J. Mitsuhashi. 1980. Rearing of the white-backed planthopper, *Sogatella furcifera* Horvath (Hemiptera: Delphacidae) on synthetic diet. Jap. J. Appl. Entomol. Zool. 24: 117–119 (in Japanese).

Koyama, K. and J. Mitsuhashi. 1991. Essential inorganic salts for nymphal growth of planthopper, *Laodelphax striatellus* Fallen (Homiptera: Delphacidae). Jap. J. Appl. Entomol. Zool. 35: 137–143 (in Japanese).

Koyama, K., J. Mitsuhashi and S. Nasu. 1981. Rearing of the planthopper, *Sogatella longifurcifera* Esaki et Ishihara (Hemiptera: Delphacidae) on a synthetic diet. Jap. J. Appl. Entomol. Zool. 35: 198–200 (in Japanese).

Mitsuhashi, J. 1965. Aseptic rearing of leafhoppers and planthoppers (Hemiptera: Cicadellidae and Delphacidae). Kontyu 33: 271–274.

Mitsuhashi, J. and H. Inoue. 1988. Obtainment of a continuous cell line from the larval fat bodies of the mulberry tiger moth, *Siplosoma imparilis* (Lepidoptera: Arctiidae). Appl. Entomol. Zool. 23: 488–490.

Mitsuhashi, J. and K. Koyama. 1969. Survival of smaller brown planthopper, *Laodelphax striatellus* Fallen, on carbohydrate solutions (Hemiptera: Delphacidae). Appl. Entomol. Zool. 4: 185–193.

Mitsuhashi, J. and K. Koyama. 1971. Rearing of plant hoppers on a holidic diet. Entomol. Exp. Appl. 14: 93–98.

Mitsuhashi, J. and K. Maramorosch. 1963. Aseptic cultivation of four virus transmitting species of leafhoppers (Cicadellidae). Contrib. Boyce Thompson Inst. 22: 165–173.

Mitsuhashi, J. and K. Maramorosch. 1964. Leafhopper tissue culture: embryonic, nymphal, and imaginal tissues from aseptic insects. Contrib. Boyce Thompson Inst. 22: 435–460.

Mitsuhashi, J. and K. Maramorosch. 1964. Inoculation of plant tissue cultures with aster yellows virus. Virology 23: 277–279.

Mittler, T.E. and R.H. Dadd. 1962. Artificial feeding and rearing of the aphid, *Myzus persicae* (Sulzer), on a completely defined synthetic diet. Nature 195: 404.

17

Psyllid Vectors of Plant Pathogens: Collection, Rearing and Use in Laboratory Research Studies

Barbara Jarausch[1,*] and *Phyllis Weintraub*[2]

ABSTRACT

Psyllids are important direct pests and vectors of two genera of plant pathogenic bacteria that cause extensive economic damage: *Candidatus* Phytoplasma species and *Candidatus* Liberibacter species. This chapter deals with collection techniques, rearing and use of psyllids in diverse research studies. Due to their status, it will focus in particular on the psyllid vector species.

Key Words: *Bactericera cockerelli, Cacopsylla* spp., *Diaphorina citri, Trioza apicalis, Trioza erytreae*

[1] AlPlanta Institute for Plant Research, RLP AgroScience, Breitenweg 71, D-67435 Neustadt, Germany.
Email: barbara.jarausch@agroscience.rlp.de.

[2] Agricultural Research Organization, Department of Entomology, Gilat Research Center, D.N. Negev 85280, Israel.

* Corresponding author

Introduction

The Hemiptera are an agriculturally important order because of the large number of direct and indirect pests; particularly, because they transmit a number of plant pathogenic bacteria. Psyllids, or jumping plant-lice, constitute a family within the suborder Sternorrhyncha with more than 3,000 described species worldwide. Psyllids are small plant-feeding insects that have recently garnered much attention as vectors of plant pathogenic bacteria, particularly *Candidatus* Phytoplasma species and *Candidatus Liberibacter* species. As such, it is imperative to be able to rear them and learn not only about the idiosyncrasies of their biology, but the means by which they are such capable vectors.

General Psyllid Biology

Psyllidae occur in all biogeographical habitats from the sea to alpine level. All of them are phloem feeders, mainly monophagous with highly specialized host requirements. Some species are oligophagous on closely related host plants and only a few are polyphagous on different host families (Ossiannilsson 1992). The great majority are associated with perennial dicotyledonous angiosperms (Hodkinson 2009). Closely related psyllid species often develop on related plant taxa indicating coevolution between psyllids and their host plants (Burckhardt 2005). Many tropical and southern temperate species are polyvoltine while northern temperate species tend to be univoltine or bivoltine (Burckhardt 1994). They overwinter as eggs, young nymphal instars or adults. The most important univoltine vector species overwinter as adults with an obligate alternation of host plants for reproduction and overwintering (Jarausch and Jarausch 2010).

The psyllid life cycle typically comprises of an egg stage, five nymphal instars and a bisexually reproducing adult stage (Hodkinson 2009). Facultative parthenogenesis is only known from a few species (Hodkinson 1974, Burckhardt 2005). Females produce several hundred oblong-shaped eggs with a basal pedicel which is inserted in the plant tissue (Hodkinson 1974). Psyllid nymphs are dorso-ventrally flattened insects and mobile in all instars (Ossiannilsson 1992). The different instars can be distinguished by their increasing size and the differentiation of morphological structures. Later instars possess wing buds, and the last one usually has a tarsal segment separated from the tibiotarsus. All immature stages feed intensively in the phloem and thus produce enormous amounts of liquid excreta (honeydew) which favors fungal infestation and may cause damage to the plant (Fig. 1).

Fig. 1. Psyllid nymphs with white curling honeydew exudation.

Bactericera cockerelli was identified just over 100 years ago. It was originally and most frequently found on potatoes in North America but has a host range of about 20 plant families and is especially damaging to solanaceous plants (Liu et al. 2006). Information was conflicting and very little is actually known about its biology. The nymphs are light yellow and eventually develop to pale green adults; all stages feed on the underside of leaves and, with their cryptic coloring are easily unnoticed.

Cacopsylla pyri and *C. pyricola* are polyvoltine and oligophagous on *Pyrus* species (Burckhardt 1994). *Cacopsylla pyri* can produce 4–5 generations in Central Europe and up to 8 generations in Southern France. Two morphologically distinct forms can be distinguished: a darkish winter form (*C. pyri* f. *pyri*) and a light summer form (*C. pyri* f. *pyrarboris*). *Cacopsylla pyricola* has 4–5 generations in France and 3–4 in the USA with the darker winter form (*C. pyricola* f. *simulans*) appearing as one and the lighter summer form (*C. pyricola* f. *pyricola*) as 3–4 generations per year, respectively. The first oviposition of the winter form coincides with raising temperatures in early spring on leave buds and midribs of the leaves (Burckhardt 1994). In contrast, *C. pyrisuga* is univoltine; the adults overwinter on conifers and re-migrate to *Pyrus* by middle March to April. Egg deposition in the middle of May is followed by a 6 week long nymphal development and the emergence of new adults in June.

Cacopsylla picta is strictly monophagous on *Malus* spp., *C. pruni* is oligophagous on *Prunus* spp. and *C. melanoneura* is widely oligophagous on Rosaceae such as *Crataegus*, *Malus* and *Pyrus* spp. All are univoltine with an obligate alternation between their host plants for reproduction and their overwintering sites: After remigration from their overwintering habitats on conifers in February–March they start the reproduction cycle on their host plants within around 2 months. The newly hatched adults (emigrants) leave their host plants within the next 4 weeks and migrate to their overwintering sites.

Diaphorina citri is a non-diapausing polyvoltine species with up to 10 overlapping field generations. Females may lay over 800 eggs. The life cycle takes 15–47 days; adults live up to several months. Population fluctuations are closely related to flushing rhythm of their host because eggs are exclusively laid on young flush points and nymphs develop on young leaves (Burckhardt 1994). When on plants, nymphs are difficult to find because they are flatten on the shoots and are green or orangish in color. Adults can be easily identified because they have a distinctive pattern on their forewings. When they occur in large populations and excrete copious amounts of honeydew these psyllids can also cause direct plant damage.

Similar to *D. citri*, *Trioza erytreae* is a multivoltine tropical/subtropical non-diapausing species with free-running life cycles that may complete between 8 and 16 generations per annum (Hodkinson 2009). One female lays 200–2500 eggs. The nymphs form pit galls on the lower leaf surface; heavy infestation causes severe leaf deformation (Burckhardt 1994). Psyllids need young, actively-growing foliage (flush) for development, and their population peaks at flush periods. The adult female is capable of laying fertile eggs immediately. Eggs of *T. erytreae* are seen mostly along edges of young leaves and nymphs develop on the underside of leaf blades. The *T. erytreae* nymphs are embedded in pits or nests on the underside of the leaves which look like bumps on the upper side.

Trioza apicalis exhibits a life cycle similar to *Cacopsylla* vectors of phytoplasmas in that it overwinters on conifers: primarily, *Picea abies*, but also *Pinus silvestris* and *Juniperus communis* (Kristoffersen and Anderbrant 2007). In spring adults return to the host, mate and females lay eggs on the young leaves. The main developmental host is *Daucus carota sativus* (cultivated carrot) and *D. c. carota* (wild carrot) but it was also reported from parsley (*Petroselinum crispum*) and caraway (*Carum carvi*) (Burckhardt 1994, Laska 2011). The nymphal development is during summer; the new adults emerge in autumn and soon fly to the overwintering plants (Burckhardt 1994). Newly emerged adults have light green heads and emigrate from carrots soon after eclosing; but overwintered adults which infest carrots in late spring to early summer are mostly bluish-green in color.

Psyllids as Vectors of Plant Pathogens

Plant pathogenic bacteria of the group '*Candidatus* Phytoplasma spp.' are responsible for devastating diseases of herbaceous and woody crops throughout the world. These cell wall-less Mollicutes are phloem-limited and are naturally transmitted in a persistent and propagative manner by three groups of phloem-feeding insects: leafhoppers, planthoppers and psyllids. Psyllid-vectored phytoplasmas are major economic problems in European fruit growing areas. All important phytoplasma diseases of *Pyrus*, *Malus* and *Prunus* cultivars are vectored by psyllids belonging to the genus *Cacopsylla*. Among the west Palaearctic *Pyrus*-feeding psyllids, *C. pyricola* was first identified as vector of '*Candidatus* Phytoplasma pyri' in California (Jensen et al. 1964) and the UK (Davies et al. 1992) while recently *C. pyri* has been described as vector of this phytoplasma in other European countries (Carraro et al. 1998, Garcia-Chapa et al. 2005).

Cacopyslla picta and *C. melanoneura* are confirmed vectors of '*Candidatus* Phytoplasma mali' in different geographical regions (Jarausch et al. 2003, Tedeschi et al. 2002) while *C. pruni* is the only identified vector of '*Candidatus* Phytoplasma prunorum' (Carraro et al. 1998, Jarausch et al. 2001, Thébaud et al. 2009).

A second group of unculturable plant pathogenic bacteria which causes major threats to the world citrus industry and solanaceous plants is *Candidatus* Liberibacter. Asian citrus greening (aka Huanglongbin) was known in China (caused by *Candidatus* Liberibacter asiaticum) and Africa (caused by *Candidatus* Liberibacter africanum) for about a century, but it was only with the appearance of symptoms in North and South America (Bove 2006) that interest was focused on determining the pathogen and vectors. The Asian citrus psyllid, *Diaphorina citri*, was discovered in North America (Florida) in 1998; however, citrus greening was first reported in the New World in Brazil in 2004 (Halbert and Manjunath 2004).

The Asian citrus psyllid, *D. citri*, originated in the Far East; first being described in Taiwan in 1907 (Kuwayama 1908) was discovered in North America (Florida) in 1998. At some point it migrated to South America and was noted to cause damage only occasionally (Nava et al. 2007 and references therein). By 2004 citrus greening was observed in Brazil and confirmed to be transmitted by this psyllid (Halbert and Manjunath 2004, Bove 2006).

The second citrus psyllid, *T. erytreae*, was considered a minor direct pest of citrus in Africa since its discovery in 1987 (van den Berg 1990). The distribution was originally limited to the eastern half of Africa, from Ethiopia to South Africa, but has since been found in the Middle East (van den Berg 1990 and references therein). Once it was determined to be the vector of the African form of citrus greening, '*Candidatus* Liberibacter africanus', by McClean and Oberholzer (1965) its status changed to a major pest.

Bactericera cockerelli known as potato or tomato psyllid is a native American species. Soon after its introduction to New Zealand, symptomatic tomato and then pepper plants were observed. Recently '*Candidatus* Liberibacter solanacearum' was identified as causative agent (Liefting et al. 2009) for the disease and the transmission of the agent by *B. cockerelli* has been demonstrated (Munyaneza et al. 2007, 2008, Secor et al. 2009). Zebra chip, first observed in Mexico and the United States, is named for the appearance of the potatoes in cross-section, as potato chips are cut: lines of necrosis radiate out from the center and are very apparent when the potatoes are cooked or fried. Only when the psyllid started appearing in significant outbreaks and was associated as a pathogen vector, were controlled studies initiated. Liu et al. (2006) speculate that this sudden appearance as a pathogen vector is an indication of a new biotype, possibly from imported plant material.

The carrot psyllid *Trioza apicalis*, originally described in 1848 (Foerster), was considered rare in appearance. By the 1920's however, several outbreaks were reported in Scandinavian countries (Laska 2011, and references therein) and nowadays this psyllid is a serious pest of carrots in northern Europe (Nehlin et al. 1994). Feeding damage causes the leaves to curl, even after short feeding periods, and the growth of carrots becomes stunted. Although it was observed that the carrot psyllid caused leaf discoloration (Burckhardt 1986, Nissinen et al. 2007), it was only recently that this was associated with *Candidatus* Liberibacter. With the advent of *Candidatus* Liberibacter in other crops and places world-wide, an analysis of carrots and psyllids was undertaken (Munyaneza et al. 2010). Munyaneza and coworkers noted that the leaf curling and discoloring symptoms and the abundance of psyllids resembled the situation in potatoes and peppers and tested plants for '*Candidatus* Liberibacter solanacearum' using specific PCR primers. Additionally, they tested both field-collected and specimens from the colony and found infective individuals. Not surprisingly, '*Candidatus* Liberibacter solanacearum' was found in field carrots, field collected *T. apicalis*, but also from a psyllid colony which was started 10 years prior.

Psyllid Collection

Depending on the plant type, the collection site and the requirements on the quality of the material, adult psyllids can be collected by different means. Generally, non-destructive sampling can be done using the beating tray method (Müther and Vogt 2003) (Fig. 2A) or by sweep netting (Fig. 2B).

Fig. 2. Psyllid collection methods. Fig. 2A. Beating tray method. **Fig. 2B.** Collecting psyllids from conifers.

The beating tray method is biased towards species that are easily dislodged but which do not fly when disturbed and can be used to assess different leaf dwelling arthropods quantitatively. This method is advantageous for collecting psyllid species on perennial woody host plants with strong branches such as fruit trees. For this sampling a defined number of branches per sample are strongly tapped with a stick (3 times per branch), and the dislodged arthropods are caught in a beating funnel held beneath. The psyllids can be collected in a collection bag (polyethylene bag) or vessel fixed at the bottom of the funnel that can be transferred immediately to a freezer in order to kill them and for storage. Alternatively, the collection vessel can also be filled with 70% alcohol in order to kill and conserve the dislodged psyllids directly in the field. If collected specimens should be maintained alive for further usage they can immediately be removed from the collection bag with a suction apparatus and transferred to another vessel.

For psyllid sampling on herbaceous plants, bushes or conifers a sweep net is more appropriate. The technical version of the net must be adapted to the plant type and collection site. For instance, for psyllid collection on coniferous trees a net on a long pole (3–5 meters) has been proven to be the most appropriate instrument. Live psyllids are immediately removed from the net with an aspirator and handled according to the particular purpose (alive or dead). Munyaneza et al. (2011) collected *T. apicalis* from commercial carrot field by sweep net sampling weekly in late June and early July.

The univoltine species can be collected during the period of their presence on their host plants in orchards using the beating tray method or net sweeping of young trees. Generally, the overwintered adults of these species can be more easily captured just after remigration to their host plants in early spring (March, April) when they are less mobile due to their copulation and egg deposition activities while the newly hatched imagines in summer (June, July) usually whirl around within the orchards before emigration.

Similar to other arthropods, psyllids are attracted to yellow-colored sticky traps. This method can be used for regular monitoring in the orchards or for long-term field surveys in various habitats. However, this is a destructive method and can only serve as a rough quantitative evaluation since psyllids are very small and fragile and by removal from sticky traps (by use of hexanes or other solvent to dissolve the glue) important morphological structures for specific identification are often lost.

Alternative collection methods have been described for *D. citri* which can be collected as the egg stage by taking branches of citrus from the field and putting them in wet cotton in a cage. Newly emerged nymphs can be aspirated as they eclose from eggs. This method was used, e.g., by Xu et al. (1988) to obtain nymphs for transmission trials with *D. citri* or by van den Berg et al. (1990) to provide material for different transmission studies with *T. erytreae*. Theoretically, this method can also be adopted to other psyllids on fruit trees but is less convenient, as direct collecting of adult morphs is usually the basis for establishment of continuous cultures.

Psyllid Rearing Techniques

There are a number of reasons to establish a stable psyllid colony, among which are: acquisition and transmission trials, determination of developmental parameters, determination of physiological parameters, host plant choice tests, and reaction to plant infochemicals. Some techniques and parameters for establishing stable colonies are described below.

Compared with polyvoltine psyllid species producing several generations per year, the austere life cycle and obligate host change of those univoltine species makes collection and rearing difficult and hampers the establishment of a permanent colony. However, despite broad variations within the psyllid life cycle, efforts have been undertaken during recent years to find the most appropriate collection methods and to improve the conditions for mass rearing of all those important vector species in the greenhouse.

The most common way to establish psyllid colonies starts from pairs or groups of field collected adults which are then caged on specific host plants for copulation and oviposition. Standardized healthy material of adequate host plant species should be derived from seeds or produced *in vitro* from tissue cultures to insure that it is pathogen-free. Plantlets of a certain size (e.g., 4–6 leaf stage) are placed under insect proof conditions before infestation with psyllids.

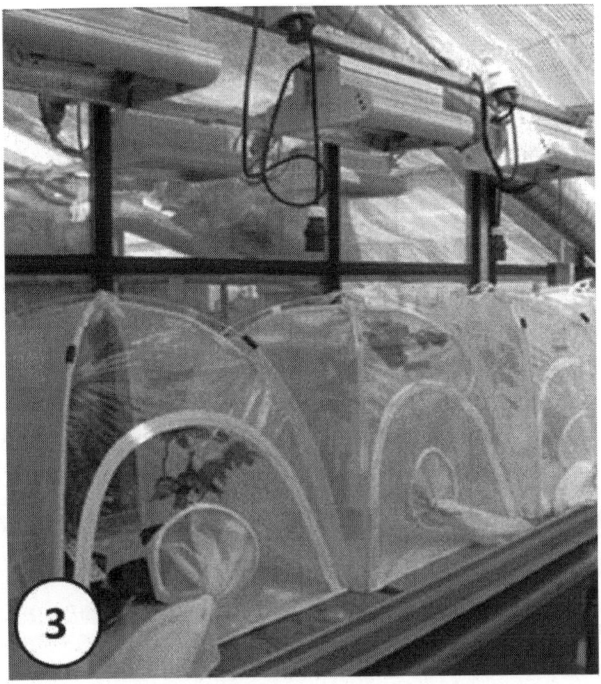

Fig. 3. BugDorms® with psyllid colonies.

Appropriate receptacles are cages, BugDorms® (Taiwan) (Fig. 3), glass vessels, rearing bags or sleeves depending on the particular host plant species and environment. Although all psyllids pass through 5 instars, the period for complete nymphal development can be highly variable depending on the species-specific life history including climatic zone and voltinism status (Hodkinson 2009). This inconvenience can be eliminated by establishment of colonies in the greenhouse under controlled and standardized culture conditions.

Bactericera cockerelli

Abdullah (2008) studied psyllids reared on tomato, *Lycopersicon esculentum*, grown from seeds. Tomatoes were placed in insect proof cages. When plants reached the 4–6 leaf stage (ca. 30 days old), they were infested with 40–50 adults collected from a tomato greenhouse in southern Arizona. Adults from this colony were serially transferred to other seed-grown tomato plants to maintain the colony. The abiotic conditions were 26–27°C, 60–70% relative humidity and a photoperiod of 12:12. It was under these conditions that studies on mating, oviposition and life histories were conducted. To achieve accurate counts and measurements leaf-clip cages were attached to the lower leaf surface to contain individuals in a small area (3 cm diameter, 2 cm high). Paired adults were allowed access to the leaf for 24 hrs before being removed. Females laid an average of 232 eggs. A maximum of 20 eggs were allowed to develop to prevent overcrowding. It was observed that copious amounts of honeydew was produced, but crystallized soon after being excreted.

The potato/tomato psyllid can also be easily reared on potatoes (*Solanum tuberosum*), peppers (*Capsicum annuum*) and eggplant (*Solanum melongena*) in addition to tomatoes (Gao et al. 2009). Plants were grown in insect-exclusion cages at 23–27°C, 50–70% relative humidity and a photoperiod of 14:10 (light:dark). It seems that the psyllids are very adaptable as other researchers established colonies on potatoes under conditions of long day (16:8, light:dark) (Casteel et al. 2006, Munyaneza et al. 2007) as well.

Cacopsylla Species

Colonies of the most important univoltine *Cacopsylla* vector species *C. picta*, *C. melanoneura* and *C. pruni* were installed in cages, BugDorms® or glass vessels infested with mature females and males collected from their natural host plants from February until April (Carraro et al. 1998, Jarausch et al. 2004, Mayer et al. 2008). Cages were covered with an anti-aphid mesh nylon net which allows good ventilation and inhibits excessive water condensation. Recommended culture conditions for these northern temperate species were a temperature between 20–25°C day and around 15°C night, a relative humidity between 50–80% and natural day light conditions of light:dark, 16:8 h. Under these conditions copulation happened efficiently and gravid females soon grew larger due to the maturing eggs (indicated by a swollen

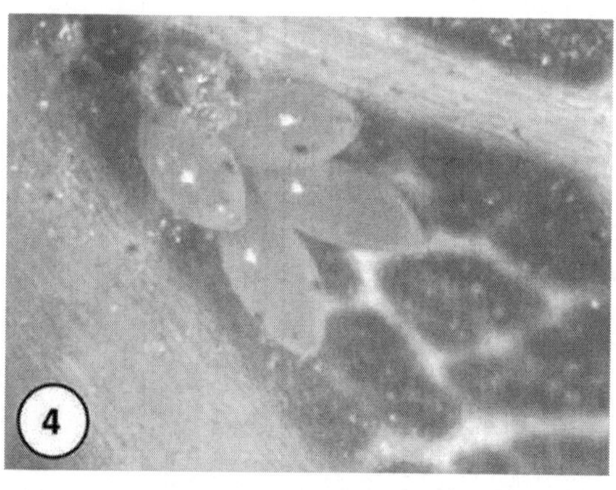

Fig. 4. *Cacopsylla pruni* eggs.

and often deep red colored abdomen). Egg deposition occurred mainly along the veins of the upper or lower surface of the leaves (Fig. 4). The nymphal development under these conditions takes about 5 weeks until emergence of new adults. Thus, continuous cultures in the greenhouse could be maintained for several months, at least until autumn of the same year.

Another rearing approach for *C. pruni* was applied by Thébaud et al. (2008) and by Jarausch (unpublished) using sleeve cages on healthy or infected *Prunus* plants. Overwintered adults of *C. pruni* were inserted into covered branches on healthy and '*Candidatus* Phytoplasma prunorum' infected *P. armeniaca* trees in an experimental field and thus created colonies under natural conditions. Tedeschi et al. (2006) started colonies of *C. pruni* and *C. melanoneura* in plexiglass vessels in the greenhouse from twigs with remainder of eggs which were individually tied to branches of young, potted apple plants. The merged nymphs could thus move to the new plants and complete the nymphal development.

In order to obtain healthy psyllid populations for laboratory use, standardized plant material from tissue culture or from seedlings was used, e.g., from *Malus* cultivars Golden Delicious or Royal Gala for *C. picta* and *C. melanoneura* cultures (Jarausch et al. 2004, Mayer et al. 2008) or from *Prunus* cultivars *P. marianna*, *P. salicina* or *P. cerasifera* for *C. pruni* (Carraro et al. 2004, Thébaud et al. 2009), respectively. For subsequent acquisition and transmission trials, colonies were directly installed on infected *Malus* or *Prunus* cultivars under similar culture conditions.

Breeding of polyvoltine species without host alternation such as *C. pyri* can be started from field collections on *Pyrus* spp. all around the year. However, multivoltine species such as *C. pyricola* or *C. pyri* undergo a reproductive diapause in the autumn generation during winter (Hodkinson 2009).

Diaphorina citri

To determine the effect of different citrus varieties on the growth and development of psyllids at different temperatures, Nava et al. (2007) tested three varieties of citrus: rangpur lime (*Citrus limonia*), orange jasmine (also called orange jessamine) (*Murraya paniculata*) and mandarin orange (*Citrus sunki*) at temperatures ranging from 18 to 32°C at 2 degree intervals, and Liu and Tsai (2000) tested only jessamine orange from 10 to 30°C at 5 degree intervals. The former group produced pathogen-free plantlets or seedlings from seeds and allow them to grow until they were ca. 25 cm long then placed them in acrylic cages (50 x 50 x 70 cm), and the later group used plants grown from seeds at the 2 to 3 leaf stage that were kept in plastic containers 6 x 6 x 8 cm. Paired adults were allowed to oviposit for limited time periods then removed. The environmental conditions were varied from

60–80% RH and a long day of 14:10 light:dark and 13:11, respectively. As expected, both groups found that the duration of all stages was longest at low temperatures and shortest at 30°C, but viability of all stages dropped severely above 30°C, as was also found by Nakata (2006). Liu and Tsai found that psyllids only developed to the 3rd instar at 10°C, and Nakata found that nymphs did not develop to adults at 15°C; therefore, this seems to be a lower limit and a limitation for the geographic distribution of the species. To determine the lower developmental limits, Liu and Tsai (2000) tested the development of groups of 400–500 adults on 12–15 Jessamine orange seedlings at temperatures from 10 to 30°C at 5 degree intervals, 75–80% relative humidity and a photoperiod of 13:11 light:dark. The psyllids only developed to the 3rd instar at 10°C.

While the psyllid can be efficiently reared on whole plants, even small trees, these facilities take considerable space. In an attempt to reduce the area required for rearing for short periods of time, Ammar and Hall (2011) developed a method of containing both nymphs and adults on detached sweet orange (*Citrus sinensis*) leaves put in wet cotton and inserted in the bottom of 50 ml conical centrifuge tubes. Before closing the tube cap was perforated to allow ventilation. They found that partially expanded young leaves would support survival of about 75% of the young adults for up to 4 weeks, with higher survival rates for shorter durations. The use of mature leaves produced significantly poorer results.

Initial attempts at developing an artificial diet for adults consisted of using various concentrations of sucrose (Hall et al. 2010). Psyllids were placed in a tube which was covered with Parafilm over the top. Sucrose was put in a small indentation in the film and a second layer of Parafilm was placed over that to secure the liquid and keep it clean. Interestingly, they found that the adult psyllids would not feed unless coloring was added and ultimately a yellow-green color was used. With this method, these researchers achieved a survival rate of about 70% for 14 days.

Trioza apicalis

The main developmental hosts are wild and cultivated carrots. According to Laska (2011, and references therein), embryonic development in the laboratory at temperatures from 22–26°C lasts for 8–10 days. The optimal relative humidity for hatching was investigated in studies using saturated solutions of several salts. Optimal hatch occurred at a relative humidity of 90%. The authors investigated the duration of the entire nymphal period in natural conditions: in the first trial, mean temperature 17.3°C it took 34–88 days, while in the second trial (mean temperature 16.3°C) the median was also 40 days, but only 11 specimens completed development.

In an effort to learn more about host-plant choice and other behavior, a colony for experimental purposes was needed. Since the carrot psyllid uses two very different host plant species, it took some time to obtain a continuous colony on carrots in 2000 (Nissinen, personal communication). Healthy carrot seedlings are placed in 33 x 33 x 60 cm plexiglass cages on a long day cycle, 20:4, with alternating light:dark temperatures of 20:15°C and a relative humidity of ca. 50% and inoculated with colony adults.

Trioza erytreae

Pioneer work on the bionomics of *T. erytreae* was done by Catling (1970). The citrus psyllid feeds and breeds on citrus and also on indigenous Rutaceae plants (van den Berg et al. 1991, and references therein). Lemon (*Citrus limon*) is supposed to be a good host of *T. erytreae*. At 14°C, incubation requires 14 days; nymphal development requires 40 days. At 24°C, incubation is 7 days and nymphal development lasts about 19. Van den Berg et al. (1991) installed colonies for transmission studies based on branches of field collected sweet orange trees which contained nymphs in their final instars. Adults that emerged thereafter were collected and placed on greened sweet orange trees where they could mate and feed. Rearing conditions were not precisely given.

Laboratory Research Studies

Acquisition and Transmission Trials

Psyllid colonies and psyllid maintenance techniques are the prerequisite for multiple laboratory examinations under controlled conditions where homogeneous insect material is required; e.g., for acquisition and transmission trials, in host plant choice experiments, development experiments or within olfactory studies towards plant infochemicals.

The most important concern in view of pathogen vectors is the proof of the acquisition and transmission capability under laboratory conditions. Although the general processes during pathogen ingestion, multiplication and reinjection by insect vectors is well described, each pathogen-vector system has particular characteristics. Both pathogen genera, *Candidatus* Phytoplasma and *Candidatus* Liberibacter, are acquired passively during feeding; the duration necessary to acquire a sufficient titer is the acquisition access period (AAP). The AAP can be as short as a few minutes but is generally measured in hours, and the longer the AAP, generally the greater the chance of acquisition. The time that elapses from initial acquisition to the ability to transmit the bacteria is known as the latent period (LP) and is sometimes called the incubation period. During the LP the bacteria moves through and replicates in the competent vector's body; but eventually, to be transmitted, it must penetrate specific cells of the salivary glands. This LP can take from 10 days to a few months depending on the vector-bacteria relationship, temperature and other factors. The inoculation access period (IAP) is the length of time that the vector is allowed to feed on healthy plants, and generally the longer the period the greater the chance of inoculating healthy plants.

There is no standard method for conducting transmission trials (Fig. 5). When trials are first started an abundance of psyllids may be used to 'push' the possibility of transmission occurring, especially if the infection rate is not known, and a long feeding period of 2 or more days is used. Once it is established that a particular species can transmit the bacteria, a specified number of vectors and specified feeding periods should be studied to learn more about the epidemiology.

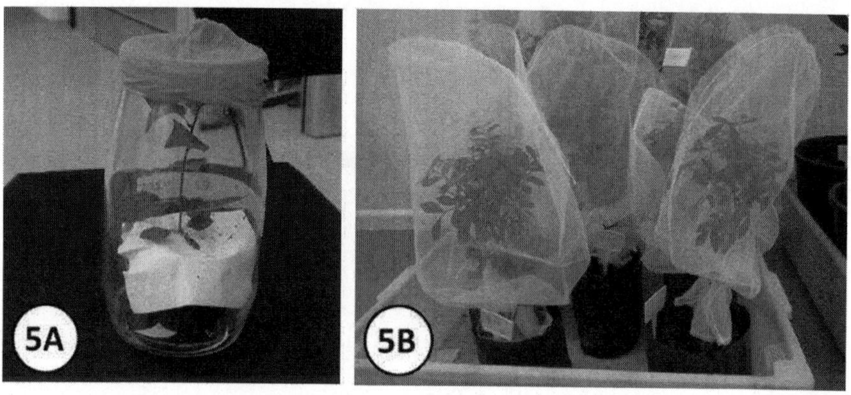

Fig. 5. Transmission and rearing designs. Fig. 5A. Transmission trial with single test plants (*C. pruni* on *Prunus marianna*). **Fig. 5B.** *D. citri* on citrus.

Bactericera cockerelli

Transmission studies were conducted by Munyaneza et al. (2007, 2008) and by Secor et al. (2009) using field collected psyllids from symptomatic potato plants. They isolated nymphs in the third to fifth instar and placed 10–20 of them on pathogen-free potato and tomato plants. Psyllids were allowed to reproduce; then serial transfers were made to healthy plants. About six weeks after transfer, plants showed symptoms; to confirm that the healthy plants became infected with the *Candidatus* Liberibacter, samples were collected and analyzed by PCR for the presence of amplification products.

Because of the way these trials were conducted significant AAP, LP and IAP times could not be determined.

Cacopsylla spp.

Researchers have studied the acquisition and transmission capacity of the fruit tree phytoplasma vector species in various experimental layouts. Detailed reports are described for *C. pruni*, *C. pyri* and *C. picta* by Carraro et al. (1998, 2004), for *C. picta*, *C. melanoneura* and *C. pruni* by Jarausch et al. (2004, 2010) and Mayer et al. (2009), for *C. melanoneura* by Tedeschi et al. (2002) and for *C. pruni* by Thébaud et al. (2008). Depending on the particular system and question, acquisition and transmission events have been studied either in cages with a larger number of specimens and test plants or in small vessels with single plants and individuals, e.g., one-plant/one psyllid transmission trial in glass vessels as conducted by Jarausch et al. (2004). In a common manner, individuals have been recollected from the test system at the end of the study and tested by PCR for pathogen infection in groups or individually by real-time for determining the phytoplasma titer. Details can be seen in the cited references.

Diaphorina citri

Transmission trials have been conducted on *D. citri* adults and nymphs (Xu et al. 1988, Hung et al. 2004, Inoue et al. 2009). Researchers are in agreement that nymphs can acquire and transmit the bacteria effectively and with short LP, but healthy adults feeding on infected plants are poor vectors. Once infective, both nymphs and adults can transmit the pathogen their entire life.

Trioza apicalis

Munyaneza et al. (2011) studied only the natural infection with *Candidatus* Liberibacter of field collected *T. apicalis* from commercial carrot field and thus found infected insects but formal transmission studies have not been reported to date.

Trioza erytreae

Studies by McClean and Oberholzer (1965) and Moll and Martin (1973) indicated that the adult citrus psyllid can transmit the greening disease pathogen but that nymphs failed to do this. However, McClean (1974) found that citrus psyllid nymphs can at times also acquire the pathogen when feeding on diseased tissue, but there may be some delay in the adult stage before such carriers are fully infective and capable of transmitting the disease. Several years later, van den Berg et al. (1991) studied in detail the transovarial transfer of greening disease, the transmission during oviposition and by nymphs of *T. erytreae*.

Trials Using Infochemicals

The roles of olfactory cues and the influence of volatiles on developmental and migratory behavior have been investigated for different vector

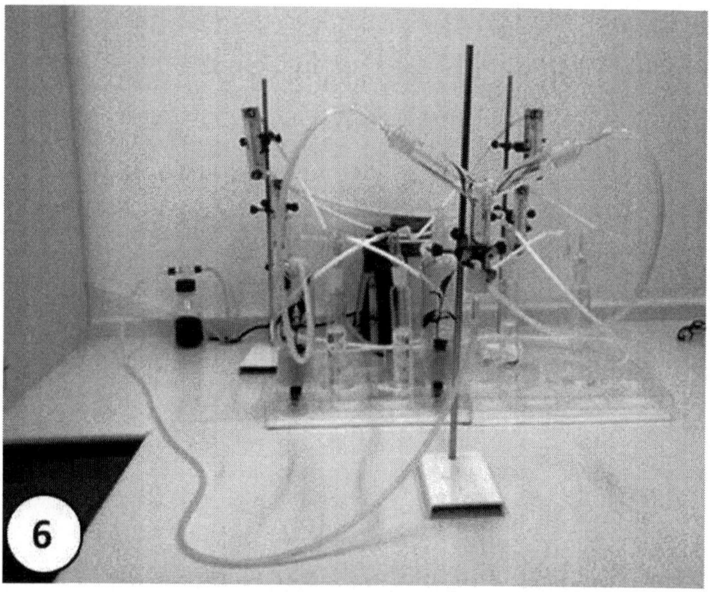

Fig. 6. Y-tube apparatus.

species. These trials rely primarily on the use of Y-tube apparatus (Fig. 6) to determine attractiveness and chemical analysis of volatiles produced by host plants and/or psyllids.

Bactericera cockerelli

Guédot et al. (2010) examined the role of chemical signals in sex attraction, assessing male and female response to male- and female-produced volatile chemicals using Y-tube olfactometry and no host plants. They found that males exposed to volatiles from conspecific males and females were more attracted to females. Alternatively, females avoided volatiles from conspecific females and males and extracts of conspecific females and males that were placed on filter paper. They speculate that the volatiles produced by males may be aggregation pheromones and the fact that female-female repellency serves to limit competition for feeding or breeding sites.

Cacopsylla picta

Little is known about the effects of infected plants on the corresponding insect vector. Working with '*Candidatus* Phytoplasma mali' infected apple trees, Mayer et al. (2008) found that there are changes in the plant volatiles, more ß-caryophyllene is produced, that actually attracts the vector, *C. picta*. This attraction was true for both non-infected and infective psyllids. Most interestingly, psyllids that developed on infected trees but themselves were not infected are more attracted to non-infected apple trees. The authors suggest that this attraction to the volatiles of infected trees is a means of the bacteria manipulating the tree to further spread the pathogen.

Cacopsylla pyricola

It has been observed that the pear psylla has a highly clumped distribution in orchards. Working to understand mate location, Horton and Landolt (2007) and Horton et al. (2007) performed trials with diapausing (collected from pear orchards in late October) or post-diapausing females (collected from pear orchards in late February) on and off of host plants. Only the volatiles from post-diapausing female-infested shoots attracted males. Diapause in males was manipulated in the laboratory by altering the light cycle, but they were still attracted only to post-diapause females.

Diaphorina citri

To elucidate the factors involved in mate location and subsequent host plant location for egg laying, studies were performed. Wenninger et al. (2008) found that female psyllids produced volatiles attractive to males. However, females are not attracted to volatiles from conspecific males or females. To understand more about host plant location Wenninger et al. (2009) examined the role of olfactory and visual cues for the orientation of *D. citri*. Not surprisingly, mated males and females were more attracted to plant volatiles than virgin individuals, and visual cues enhanced the attraction.

Trioza apicalis

Interesting work was done by Nehlin et al. (1994) in which they found that sawdust from spruce and pine, applied along seedling carrot rows could reduce the damage caused by *T. apicalis*. Further investigation showed that specific turpentine and monoterpene hydrocarbons could mimic the sawdust and afford protection by repelling the psyllid. Other work examined the influence of host plant volatiles on *T. apicalis* (Nehlin et al. 1996). Closely related species of carrots, parsley and caraway, were evaluated for plant volatiles and acceptance for oviposition. Each plant produced a unique volatile signature but specific monoterpenes seemed to be the primary cue.

Conclusion

In this chapter we have shown that only a few psyllid species are responsible for transmitting very economically destructive bacterial plant pathogens. For half of these species, the psyllid has two very different hosts, either a fruit tree and a conifer or a vegetable and a conifer. This makes working with the psyllid and learning about the biotic and abiotic factors affecting migration and pathogen transmission difficult. While the *Cacopsylla* spp. are native to Europe, *D. citri* has been artificially transported from the Far East to both North and South America where it has no natural biological control agents and is wreaking havoc on the citrus industry. The more temperature sensitive *T. erytreae* is limited in its distribution and transmits less pathogenic bacteria. It is possible that some of these psyllid-bacteria associations are relatively new and therefore all the more destructive. It was demonstrated that the establishment of colonies and the improvement of culture conditions in the greenhouse is of crucial evidence for providing homogeneous insect material which can be used for diverse biological and molecular studies. Reproducible and reliable laboratory studies are the basis for the development of applied research with psyllid vectors.

References

Abdullah, N.M.M. 2008. Life history of the potato psyllid *Bactericera cockerelli* (Homoptera: Psyllidae) in controlled environment agriculture in Arizona. African J. Agric. Res. 3: 60–67.

Ammar, E.-D. and D.G. Hall. 2011. A new method for short-term rearing of citrus psyllids (Hemiptera: Psyllidae) and for collecting their honeydew excretions. Florida Entomol. 94: 340–342.

Bove, J.M. 2006. Huanglongbing: a destructive, newly emerging century-old disease of citrus. J. Plant Pathol. 88: 7–37.

Burckhardt, D. 1986. Taxonomy and host plant relationship of the *Trioza apicalis* Forster complex (Hemiptera: Triozidae). Entomol. Scandinavica 14: 415–432.
Burckhardt, D. 1994. Psylloid pests of temperate and subtropical crop and ornamental plants (Hemiptera, Psylloidea): A review. Entomol. Trends Agri. Sci. 2: 173–186.
Burckhardt, D. 2005. Biology, ecology, and evolution of gall-inducing Psyllids (Hemiptera: Psylloidea). pp. 143–157. *In*: A. Raman, C.W. Schaefer and T.M. Withers (eds.). Biology, Ecology, and Evolution of Gall-inducing Arthropods. Science Publishers, Inc. Enfield (NH), USA.
Carraro, L., R. Osler, N. Loi, P. Ermacora and E. Refatti. 1998. Transmission of European stone fruit yellows phytoplasma by *Cacopsylla pruni*. J. Plant Pathol. 80: 233–239.
Carraro, L., F. Ferrini, G. Labonne, P. Ermacora and N. Loi. 2004. Seasonal infectivity of *Cacopsylla pruni*, the vector of European stone fruit yellows phytoplasma. Ann. Appl. Biol. 144: 191–195.
Casteel, C.L., L.L. Walling and T.D. Paine. 2006. Behavior and biology of the tomato psyllid, *Bactericera cockerelli*, in response to the *Mi-1.2* gene. Entomol. Exp. Appl. 121: 67–72.
Catling, H.D. 1970. Distribution of the psyllid vectors of citrus greening disease, with notes on the biology and bionomics of *Diaphorina citri*. F.A.O. Pl. Prot. Bull. 18: 8–15.
Davies, D.L., C.M. Guise and A.N. Adams. 1992. Parry's disease is similar to pear decline and is associated with mycoplasma-like organisms transmitted by *Cacopsylla pyricola*. Plant Pathol. 41: 195–203.
Gao, F., J. Jifon, X. Yang and T.-X. Liu. 2009. Zebra chip disease incidence on potato is influenced by timing of potato psyllid infestation, but not by the host plants on which they were reared. Insect Sci. 16: 399–408.
Garcia-Chapa, M., J. Sabaté, A. Lavina and A. Batlle. 2005. Role of *Cacopsylla pyri* in the epidemiology of pear decline in Spain. Eur. J. Plant Pathol. 111: 9–17.
Guédot, C., D.R. Horton and P.J. Landolt. 2010. Sex attraction in *Bactericera cockerelli* (Hemiptera: Triozidae). Environ. Entomol. 39: 1302–1308.
Halbert, S.E. and K.L. Manjunath. 2004. Asian citrus psyllids (Sternorrhyncha: Psyllidae) and greening disease of citrus: a literature review and assessment of risk in Florida. Florida Entomol. 87: 330–353.
Hall, D.G., R.G. Shatters, J.E. Carpenter and J.P. Shapiro. 2010. Research toward an artificial diet for adult Asian Citrus psyllid. Ann. Entomol. Soc. Am. 103: 611–617.
Hodkinson, I.D. 1974. The biology of the Psylloidea (Homoptera): a review. Bull. Entomol. Res. 64: 325–339.
Hodkinson, I.D. 2009. Life cycle variation and adaptation in jumping plant lice (Insecta: Hemiptera: Psylloidea): a global synthesis. J. Nat. Hist. 43: 65–179.
Horton, D.R., C. Guedot and P.J. Landolt. 2007. Diapause status of females affects attraction of male pear psylla, *Cacopsylla pyricola*, to volatiles from female-infested pear shoots. Entomol. Exp. Appl. 123: 185–192.
Horton, D.R. and P.J. Landolt. 2007. Attraction of male pear psylla, *Cacopsylla pyricola*, to female-infested pear shoots. Entomol. Exp. Appl. 123: 177–183.
Hung, T.H., S.C. Hung, C.N. Chen, M.H. Hsu and H.J. Su. 2004. Detection by PCR of *Candidatus* Liberibacter asiaticus, the bacterium causing citrus huanglongbing in vector psyllids: application to the study of vector-pathogen relationships. Plant Pathol. 53: 96–102.
Inoue, H., J. Ohnishi, T. Ito, K. Tomimura, S. Miyata, T. Iwanami and W. Ashihara. 2009. Enhanced proliferation and efficient transmission of *Candidatus* Liberibacter asiaticus by adult *Diaphorina citri* after acquisition feeding in the nymphal stage. Ann. Appl. Biol. 155: 29–36.
Jarausch, W., J.L. Danet, G. Labonne, F. Dosba, J.M. Broquaire, C. Saillard and M. Garnier. 2001. Mapping the spread of apricot chlorotic leaf roll (ACLR) in southern France and implication of *Cacopsylla pruni* as a vector of European stone fruit yellows (ESFY) phytoplasmas. Plant Pathol. 50: 782–790.

Jarausch, B. and W. Jarausch. 2010. Psyllid vectors and their control. pp. 250–271. *In*: P.G. Weintraub and P. Jones [eds.]. Phytoplasmas: Genomes, Plant Hosts and Vectors. CABI, London, UK.

Jarausch, B., N. Schwind, W. Jarausch, G. Krczal, E. Seemüller and E. Dickler. 2003. First report of *Cacopsylla picta* as a vector for apple proliferation phytoplasma in Germany. Plant Dis. 87: 101.

Jarausch, B., N. Schwind, W. Jarausch and G. Krczal. 2004. Overwintering adults and springtime generation of *Cacopsylla picta* (synonym *C. costalis*) can transmit apple proliferation phytoplasmas. Acta Hort. 657: 409–413.

Jensen, D.D., W.H. Griggs, C.Q. Gonzales and H. Schneider. 1964. Pear decline virus transmission by pear psylla. Phytopathology 54: 1346–1351.

Kristoffersen, L. and O. Anderbrant. 2007. Carrot psyllid (*Trioza apicalis*) winter habitats —insights in shelter plant preference and migratory capacity. J. Appl. Entomol. 131: 174–178.

Kuwayama, S. 1908. Die Psylliden Japans. Trans. Sopporo Nat. Hist. Soc. 2: 149–189.

Laska, P. 2011. Biology of *Trioza apicalis*—a review. Plant Protect. Sci. 47: 68–77.

Liefting, L.W., P.W. Sutherland, L.I. Ward, K.L. Paice, B.S. Weir and G.R.G. Clover. 2009. A new '*Candidatus* Liberibacter' species associated with diseases of solanaceous crops. Plant Dis. 93: 208–214.

Liu, D., J.T. Trumble and R. Stouthamer. 2006. Genetic differentiation between eastern populations and recent introduction of potato psyllid (*Bactericera cockerelli*) into western North America. Entomol. Exp. Appli. 118: 177–183.

Liu, Y.H. and J.H. Tsai. 2000. Effects of temperature on biology and life table parameters of the Asian citrus psyllid, *Diaphorina citri* Kuwayama (Homoptera: Psyllidae). Ann. Appl. Biol. 137: 201–206.

Mayer, C.J., A. Vilcinskas and J. Gross. 2008. Phytopathogen lures its insect vector by altering host plant odor. J. Chem. Ecol. 34: 1045–1049.

Mayer, C.J., B. Jarausch, W. Jarausch, W. Jelkmann, A. Vilcinskas and J. Gross. 2009. *Cacopsylla melanoneura* has no relevance as vector of apple proliferation in Germany. Phytopathology 99: 729–738.

McClean, A.P.D. 1974. The efficiency of citrus psylla, *Trioza erytreae* (Del G.) as a vector of greening disease of citrus. Phytophylactica 6: 45–54.

McClean, A.P.D. and P.C.J. Oberholzer. 1965. Greening disease of the sweet orange: evidence that it is caused by a transmissible virus. South African J. Agric. Sci. 8: 253–276.

Moll, J.N. and M.M. Martin. 1973. Electron microscope evidence that citrus psylla *(Triozaerytreae)* is a vector of greening disease in South Africa. Phytophylactica 5: 41–44.

Munyaneza, J.E., J.L. Buchman, J.E. Upton, J.A. Goolsby, J.M. Crosslin, G. Bester, G.P. Miles and V.G. Sengoda. 2008. Impact of different potato psyllid populations on zebra chip disease incidence, severity, and potato yield. Subtrop. Plant Sci. 60: 27–37.

Munyaneza, J.E., J.M. Crosslin and J.E. Upton. 2007. Association of *Bactericera cockerelli* (Homoptera: Psyllidae) with "Zebra Chip," a new potato disease in southwestern United States and Mexico. J. Econ. Entomol. 100: 656–663.

Munyaneza, J.E., T.W. Fisher, V.G. Sengoda, S.F. Garczynski, A. Nissinen and A. Lemmetty. 2010. Association of "*Candidatus* Liberibacter solanacearum": with the psyllid, *Trioza apicalis* (Hemiptera: Triozidae) in Europe. J. Econ. Entomol. 103: 1060–1070.

Munyaneza, J.E., A. Lemmetty, A.I. Nissinen, V.G. Sengoda and T.W. Fisher. 2011. Molecular detection of aster yellows phytoplasma and "*Candidatus* Liberibacter solanacearum" in carrots affected by the psyllid *Trioza apicalis* (Hemiptera: Triozidae) in Finland. J. Plant Pathol. 93: 697–700.

Müther, J. and H. Vogt. 2003. Sampling methods in orchard trials: A comparison between beating and inventory sampling. IOBC WPRS Bull. 26(5): 67–72.

Nakata, T. 2006. Temperature-dependent development of the citrus psyllid, *Diaphorina citri* (Homoptera: Psylloidea), and the predicted limit of its spread based on overwintering in the nymphal stage in temperate regions of Japan. Appl. Entomol. Zool. 41: 383–387.

Nava, D.E., M.L.G. Torres, M.D.L. Rodrigues, J.M.S. Bento and J.R.P. Parra. 2007. Biology of *Diaphorina citri* (Hem., Psyllidae) on different hosts and at different temperatures. J. Appl. Entomol. 131: 709–715.

Nehlin, G., I. Valterov and A.K. Borg-Karlson. 1994. Use of conifer volatiles to reduce injury caused by carrot psyllid, *Trioza apicalis* Forster (Homoptera, Psylloidea). J. Chem. Ecol. 20: 771–783.

Nehlin, G., I. Valterov and A.K. Borg-Karlson. 1996. Monoterpenes released from Apiaceae and the egg-laying preferences of the carrot psyllid, *Trioza apicalis*. Entomol. Exp. Appli. 80: 83–86.

Nissinen, A., P. Vanhala, J.K. Holopainen and K. Tilikkala. 2007. Short feeding period of carrot psyllid (*Trioza apicalis*) females at early growth stages of carrot reduces yield and causes leaf discolouration. Entomol. Exp. Appli. 125: 277–283.

Ossiannilsson, F. 1992. The Psylloidea (Homoptera) of Fennoscandia and Denmark. *Fauna* Entomol. Scandinavica 26: 1–347.

Secor, G.A., V.V. Rivera, J.A. Abah, I.M. Lee, G.R.G. Clover, l.w. Liefting, X. Li and S.H. de Boer. 2009. Association of '*Candidatus* Liberibacter solanacearum' with zebra chip disease of potato established by graft and psyllid transmission, electron microscopy, and PCR. Plant Dis. 93: 574–583.

Tedeschi, R., D. Bosco and A. Alma. 2002. Population dynamics of *Cacopsylla melanoneura* (Homoptera: Psyllidae), a vector of apple proliferation phytoplasma in northwestern Italy. J. Econ. Entomol. 95: 544–551.

Tedeschi, R., V. Ferrato, J. Rossi and A. Alma. 2006. Possible phytoplasma transovarial transmission in the psyllids *Cacopsylla melanoneura* and *Cacopsylla pruni*. Plant Pathol. 55: 18–24.

Thébaud, G., M. Yvon, G. Labonne and R. Alary. 2008. European stone fruit yellows: consequences of the life cycle of the vector and of the multiplication of the phytoplasma in the insect on the epidemiology of the disease. Acta Hort. 781: 423–428.

Thébaud, G., M. Yvon, R. Alary, N. Sauvion and G. Labonne. 2009. Efficient transmission of '*Candidatus* Phytoplasma prunorum' is delayed by eight months due to a long latency in its host-alternating vector. Phytopathology 99: 265–273.

Van den Berg, M.A. 1990. The citrus psylla, *Trioza erytreae* (Del Guercio) (Hemiptera: Triozidae): A review. Agric. Ecosyst. Environ. 30: 171–194.

Van den Berg, M.A., S.P. Van Vuuren and V.E. Deacon. 1991. Studies on the greening disease transmission by the citrus psylla, *Trioza erytreae* (Hemiptera: Triozidae). Israel J. Entomol. 25-26: 51–56.

Wenninger, E.J., L.L. Stelinski and D.G. Hall. 2008. Behavioral evidence for a female-produced sex attractant in *Diaphorina citri* Kuwayama (Hemiptera: Psyllidae). Entomol. Exp. Appl. 128: 450–459.

Wenninger, E.J., L.L. Stelinski and D.G. Hall. 2009. Roles of olfactory cues, visual cues, and mating status in orientation of *Diaphorina citri* Kuwayama (Hemiptera: Psyllidae) to four different host plants. Environ. Entomol. 38: 225–234.

Xu, C.F., Y.H. Xia, K.B. Li and C. Ke. 1988. Further study of the transmission of citrus huanglongbin by a psyllid, *Diaphorina citri* Kuwayama. pp. 243–248. *In*: Proceedings, 10th Conference of International Organization Citrus Virology Riverside, CA USA University of California Riverside.

Chapter 7: Addendum

Acoustic Quality Control in Laboratory-Raised Tsetse Flies

Helmut Kratochvil

Mass releases of sterilized male tsetse flies are used to control tsetse flies, the main vector of sleeping sickness in Africa. Laboratory reared sterilized males when released in nature mate with wild virgin females and prevent their eggs from fertilization. Eggs developed by such mated females do not hatch and decrease the wild population of tsetse flies in areas with endemic sleeping sickness. It is an expensive and laborious ecologically friendly method resulting in absolute control. To make this very complex biological control method effective in nature, quality-control of the sterilized males is conducted by checking their weight, size, and flight performance. During mating tsetse flies are attracted to each other by selecting three acoustic sounds including mating call sound probably produced by males, blood feeding sound produced by both sexes, and larvi position sound of females. Thus it was considered essential to develop an acoustic sound detection criterion for quality control of the above healthy mass reared sterile males. *Glossina pallidipes* and *Glossina morsitans* were used to select the acoustic sound preferred for seeking a mate. Sound produced during blood feeding was the most important acoustic sound used by tsetse flies for finding sexual partners. It took 10 years of research between two cooperating institutions, IAEA—International Atomic Energy Agency, Institut for Schallforschung der Osterreichischen Akademie der Wissenschaften, and Universitat Wien—Department for Evolutions biologie, to attain the above goal.

A special program (STX—Glossina, Austrian Academy of Sciences, Institute for Acoustic Research) was developed with which (in % of time) the occurrence of the males feeding sounds can be measured automatically

Universìtät Wien, Dr.-Karl-Lueger-Ring 1, A-1010 Wien Redaktìon: Alexandra Frey, Michaela Hafner Mìtarbeit: Heídrun Huber, Bernadette Ralser.
Email: helmut.kratochvil@univie.ac.at

under standard conditions (random sample, humidity, temperature, light intensity). The activity value is one of the test methods which should make a quality index which still has to be calculated. The index will be designed to serve as the basis for the pest control management.

References

Kratochvil, H., J. Bolldorf and R. Machan. 2005. Sound analysis, A potential Strategy of Quality Control for Sterilized Insects Pests. FAO/IAEA International Conference on Area Wide Control of Insects Pests.

Noll, A. and H. Kratochvil. 2007. A software for automatic registration of acoustic activity of tsetse flies. FAO/IAEA International Conference on Area-Wide Control of insects Pests. In Press.

Kratochvil, H., A. Noll, J. Bolldorf and A.G. Parker. Acoustic method for testing the quality of sterilized male tsetse flies *(Glossina pallidipes)*. TECDOC publications of the International Atomic Energy Agency.

Kratochvil, H. Oxygen consuption as a quality criterion. TECDOC publications of the International Atomic Energy Agency.

Index

W

X

Y

Z

About the Editors

Dr. Karl Maramorosch, D.h.c, is the Robert L. Starkey Professor of Microbiology and Emeritus Professor of Entomology at Rutgers—The State University of New Jersey, New Brunswick, New Jersey. He obtained his Ph.D. degree from Colombia University. His scientific career began at Rockefeller University, where he worked on plant viruses and insect vectors. Later he became Program Director of Virology and Insect Physiology at the Boyce Thompson Institute and since 1974, he has served as a Distinguished Professor at Rutgers University.

He is a Fellow, former Recording Secretary and Vice-President of the New York Academy of Sciences, a Fellow of the American Association for Advancement of Science, the American Phytopathological Society, Fellow and Honorary Member of the Entomological Society of America, Honorary Fellow of the Indian Virological Society, Foreign Fellow of the Indian National Science Academy, member of the American Virological Society, Leopoldina Academy, Microscopy Society, International Organization for Mycoplasmology, Society of Invertebrate Pathology and Society for *In Vitro* Biology. He won the 1980 Wolf Prize in Agriculture, often called the Agriculture Nobel Prize, and numerous other awards and honors, including the Jurzykowski Award in Biology, AIBS Award of Distinction, the Waksman Award, AAAS-Campbell Award, the 2012 SGGW Award of Distinction, and others. He was nominated by the Entomological Society of America and the Phytopatholgical Society for the National Medal of Science and he held assignments from the Rockefeller Foundation, the Agency for International Development, U.S. Department of Agriculture, Ford Foundation, and U.S.P.H. in Mexico, India, Kenya, Philippines, former Soviet Union, Japan, and Philippines.

Dr. Maramorosch has edited more than 90 volumes on viruses, vectors, plant diseases, invertebrate cell culture, and is the author or co-author of more than 800 research papers. He served twice as visiting Fulbright Professor in former Yugoslavia and as visiting professor in China, U.S.S.R., The Netherlands, Germany, Poland, Romania, Japan, Philippines and India.

His major interests include comparative virology, invertebrate cell culture, parasitology, diseases caused by viruses, viroids, phytoplasmas and spiroplasmas. He is an active participant in biotechnology studies and international scientific cooperation.

Dr. Farida Mahmood is an accomplished Medical and Veterinary Entomologist and has research accomplishments in many vector borne diseases including human and avian malaria, dog heartworm, leishmaniasis, chagas disease, arboviruses (eastern equine encephalomyelitis, western equine encephalomyelitis, Saint Louis encephalomyelitis, and west nile virus), and biology, genetics and physiology of their vectors. At present she is an entomologist at the Department of the Army, Public Health Command Region-South, USA. She obtained her MS and PhD degrees from University of Florida and has research experience at International Center for Medical Research and Training, University of Maryland, School of Medicine, Lahore, Pakistan, Florida Medical Entomology Laboratory, Vero Beach, Florida, Yale University, and University of California Davis. In her career she had worked as interim head of the department of Vector Ecology at the International Center for Medical Research and Training, University of Maryland, School of Medicine, Lahore, Pakistan and conducted research on ecology and genetics of vectors of West Nile virus and Malaria. She completed her postdoctoral research at Yale University and University of California Davis where she investigated various aspects of biology of mosquitoes and sand flies in relation to transmission of arboviruses and leishmania. She worked as a research associate and adjunct faculty at Rutgers the State University of New Jersey where she taught medical entomology, in part insect physiology, and mosquito biology to undergraduate and graduate students and served on graduate committees. She was an assistant director of the Department of Mosquito Surveillance and Control at New York City Department of Health and Mental Hygiene and directed mosquito surveillance to control West Nile virus. She is an author and coauthor of more than 75 publications and coeditor of a book.